Lecture Notes in Computer Science 1336
Edited by G. Goos, J. Hartmanis and J. van Leeuwen

Advisory Board: W. Brauer   D. Gries   J. Stoer

Springer
*Berlin
Heidelberg
New York
Barcelona
Budapest
Hong Kong
London
Milan
Paris
Santa Clara
Singapore
Tokyo*

Constantine Polychronopoulos   Kazuki Joe
Keijiro Araki   Makoto Amamiya (Eds.)

High Performance Computing

International Symposium, ISHPC'97
Fukuoka, Japan, November 4-6, 1997
Proceedings

 Springer

Series Editors

Gerhard Goos, Karlsruhe University, Germany

Juris Hartmanis, Cornell University, NY, USA

Jan van Leeuwen, Utrecht University, The Netherlands

Volume Editors

Constantine Polychronopoulos
University of Illinois at Urbana-Champaign, Center for Supercomputing R&D
1308 West Main Street, Urbana, IL 61801, USA
E-mail: cdp@csrd.uiuc.edu

Kazuki Joe
Wakayama University, Faculty of Systems Engineering
930 Sakaedani, Wakayama city 640, Japan
E-mail: joe@center.wakayama-u.ac.jp

Keijiro Araki
Makoto Amamiya
Kyushu University
Graduate School of Information Science and Electrical Engineering
6-1 Kasugakoen, Kasuga, Fukuoka, 816, Japan
E-mail: araki@dontaku.csce.kyushu-u.ac.jp
      amamiya@is.kyushu-u.ac.jp

Cataloging-in-Publication data applied for

Die Deutsche Bibliothek - CIP-Einheitsaufnahme

High performance computing : international symposium ; proceedings / ISHPC '97, Fukuoka, Japan, November 4 - 6, 1997. Constantine Polychronopoulos ... (ed.).
- Berlin ; Heidelberg ; New York ; Barcelona ; Budapest ; Hong Kong ; London ; Milan ; Paris ; Santa Clara ; Singapore ; Tokyo : Springer, 1997
  (Lecture notes in computer science ; Vol. 1336)
  ISBN 3-540-63766-4

CR Subject Classification (1991): C.1-4, D.1-4, F.1-2, G.1-2, H.2

ISSN 0302-9743
ISBN 3-540-63766-4 Springer-Verlag Berlin Heidelberg New York

This work is subject to copyright. All rights are reserved, whether the whole or part of the material is concerned, specifically the rights of translation, reprinting, re-use of illustrations, recitation, broadcasting, reproduction on microfilms or in any other way, and storage in data banks. Duplication of this publication or parts thereof is permitted only under the provisions of the German Copyright Law of September 9, 1965, in its current version, and permission for use must always be obtained from Springer-Verlag. Violations are liable for prosecution under the German Copyright Law.

© Springer-Verlag Berlin Heidelberg 1997
Printed in Germany

Typesetting: Camera-ready by author
SPIN 10647888   06/3142 – 5 4 3 2 1 0   Printed on acid-free paper

Preface

I wish to welcome all of you to the International Symposium on High Performance Computing (ISHPC) and to the historic city of Fukuoka, Japan. I am pleased to serve as Conference Chair at a time when high performance computing has a significant influence in computer science and engineering. In particular, high performance computing has had a significant impact on advanced technologies that are giving rise to a new era in information processing. The many conferences and symposiums that are held on the subject around the world are an indication of the importance of this area and the interest of the research community.

ISHPC was planned as a focused meeting of top researchers in the field to give them the opportunity to exchange ideas and interact with all the participants in the symposium. One of the goals of this symposium is to provide a forum for the discussion of all aspects of high performance computing (from system architecture to applications) in a more informal and personal fashion. We started planning for the symposium one and half years ago, and today we are delighted to have the symposium, which comprises excellent invited talks, tutorials and workshops, as well as high quality technical papers.

This symposium would not have been possible without the significant help of several people who devoted resources and time. In particular I would like to thank the Organizing Chair, K. Araki from Kyushu University, and all members of the organizing committee, who contributed very significantly to the planning and organization of the ISHPC. I must also thank the Program Chair, C. Polychronopoulos of the University of Illinois at Urbana-Champaign, and the program committee members who assembled an excellent program comprising a very interesting collection of contributed papers from many countries. Finally, I thank all those who have worked diligently to make the ISHPC a success.

I hope you will enjoy the symposium, and that you will find the information and interaction useful.

November 4, 1997   Makoto Amamiya
                                                     General Chair

Foreword

The International Symposium on High Performance Computing (ISHPC'97) held in Fukuoka, Japan, November 4-6, 1997, was thoughtfully planned, organized, and supported by the ISHPC Organizing Committee and Kyushu University.

The ISHPC'97 Program consists of a keynote speech, several invited talks, a workshop on HPC and distributed environments, tutorials on parallelizing compilers and MPI, and several technical sessions covering theoretical and applied research topics on high performance computing which are representative of the current research activities in industry and academia. Participants and contributors to this symposium represent a cross section of our research community and major laboratories in this area, including the Center for Supercomputing Research and Development of UIUC, the Swiss Center for Scientific Computing of ETH, the Maui High Performance Computing Center, and the Institute of Systems & Information Technologies Kyushu.

All of us on the Program Committee wish to thank the authors who submitted papers to ISHPC. We received more than 40 technical contributions from various countries. Each paper received at least three peer reviews and, based on the evaluation process, the program committee selected four papers as distinguished papers to appear as 16-page contributions in the proceedings, and sixteen regular (12-page) papers. Given that several additional papers received favorable reviews, the program committee recommended a poster session comprising shorter papers. Ten contributions were selected as short (8-page) papers for presentation in the poster session and inclusion in the proceedings.

We hope that final program will be of significant interest to the participants and will serve as the launching pad for interaction and debate on technical issues among the attendees.

November 1997   Constantine D. Polychronopoulos
                                                                      Program Chair

ISHPC97 Organization

- General Chair
  - Makoto Amamiya (Kyushu Univ.)
- Organizing Committee
  - Organizing Chair
    * Keijiro Araki (Kyushu Univ.)
  - Organizing Committee Members

Eugene Bal (MHPCC)	Akira Fukuda (NAIST)
Martin Gutknecht (ETH)	Hiroshi Hayashi (Fujitsu)
Kei Hiraki (Univ. of Tokyo)	Yoshitoshi Kunieda (Wakayama Univ.)
Yoshimitsu Ido (NKK)	Masaru Kitsuregawa (Univ. of Tokyo)
Yukio Kaneda (Kobe Univ.)	Mitsunori Miki (Doshisha Univ.)
Yoichi Muraoka (Waseda Univ.)	Hiroaki Nishikawa (Tsukuba Univ.)
Yoshio Oyanagi (Univ. of Tokyo)	Hideyuki Ohtawa (Hitachi)
Masaaki Shimasaki (Kyoto Univ.)	Jun-ichi Shimada (RWCP)
Shinji Tomita (Kyoto Univ.)	Katuyuki Takemura (Sumisho Elect.)
Taiichi Yuasa (Kyoto Univ.)	Tadashi Watanabe (NEC)

- Program Committee
  - Program Chair
    * Constantine Polychronopoulos (UIUC)
  - Program Co-Chair
    * Akira Fukuda (Nara Institute of Sci. and Tech.)
    * Alex Nicolau (UCI)
    * Harry Wijshoff (Leiden Univ.)
  - Program Committee Members

Utpal Banerjee (Intel)	Mohammad R. Haghighat (Intel)
Jose Moreira (IBM Watson)	Dean Tullsen (UCSD)
Alex V. Veidenbaum (UIC)	Tao Yang (UCSB)
Mario M. Furunari (CNR-Italy)	Skevos Evripidou (Univ. of Cyprus)
Jesus Labarta (UPC-Spain)	Stratis Gallopoulos (Univ. of Patras)
Hans P. Lüthi (ETH)	Hiroki Honda (Univ. of Elect-Com.)
Yasuhiro Inagami (Hitachi)	Kazuki Joe (Wakayama Univ.)
Yasunori Kimura (Fujitsu)	Hironori Kasahara (Waseda Univ.)
Toshiyuki Nakata (NEC)	Yoshitoshi Kunieda (Wakayama Univ.)

- Local Arrangement
  - Hiroyuki Sato (Kyushu Univ.)
  - Kazuki Joe (Wakayama Univ.)
- Treasury Chair
  - Kazuki Joe (Wakayama Univ.)

List of Referees

Nikos Bellas
Georgios Dimitriou
Paraskevas Evripidou
Hiroaki Fuji
Stratis Gallopoulos
Kazuki Joe
Yasunori Kimura
Hans Lüthi
Costas Mourlas
Hironori Nakajo
George Samaras
Mitsuru Sato
Kenji Taguchi
Alex Veidenbaum
Tao Yang

Carrie Brownhill
Ioanna Doufexi
Skevos Evripidou
Akira Fukuda
Hiroki Honda
Hironori Kasahara
Yoshitoshi Kunieda
Jose Moreira
Toshiyuki Nakata
George Papadopoulos
Mariko Sasakura
Hiroyuki Seki
Dean Tullsen
Yusaku Yamamoto
Harry Wijshoff

Table of Contents

I Invited Papers

1 The Generation of Optimized Codes Using Nonzero Structure Analysis 1
B.A.Marsolf (Demaco Inc.), A.J.C.Bik (Indiana Univ.),
K.A.Gallivan (FSU), H.A.G.Wijshoff (Leiden Univ.)

2 On the Importance of an End-To-End View of Memory Consistency in Future Computer Sysmtems 30
G.R.Gao (Univ. of Delaware), V. Sarkar (MIT)

3 High Performance Distributed Object Systems 42
D.Gannon (Indiana Univ.)

4 Instruction Cache Prefetching Using Multilevel Branch Prediction 51
A.V.Veidenbaum (UIC)

5 High Performance Wireless Computing 71
G.Cybenko (Dartmouth College)

6 High-Performance Computing and Applications in Image Processing and Computer Vision 72
H.R.Arabnia (Univ. of Georgia)

7 Present and Future of HPC Technologies 73
T.Watanabe (NEC)

II System Architecture

8 Evaluation of Multithreaded Processors and Thread-Switch Policies 75
R.J.Eickemeyer, R.E.Johnson, S.R.Kunkel (IBM AS/400 Division)
B.H.Lim, M.S.Squillante, C.E.Wu (IBM Watson Research Center)

9 A Multithreaded Implementation Concept of Prolog on Datarol-II Machine 91
P.Kacsuk (Hungarian Academy of Sciences), M.Amamiya (Kyushu Univ.)

10 Thread Synchronization Unit (TSU): A Building Block for High Performance Computers 107
S.Evripidou (Univ. of Cyprus)

11 Data Dependence Path Reduction with Tunneling Load Instructions 119
T.Sato (Toshiba Microelectronics Engineering Lab.)

12 Performance Estimation of Embedded Software with Pipeline and Cache Hazard Modeling 131
N.Imlig, A.Tsutui (NTT Optical Network Systems Lab.)

III Network

13 An Implementation and Evaluation of a Distributed Shared-Memory System on Workstation Clusters Using Fast Serial Links 143
H.Nakajo, A.Ichikawa, Y.Kaneda (Kobe Univ.)

14 Designing and Optimizing 3-connectivity Communication Networks Using a Distributed Genetic Algorithm 159
J.Ma, R.Huang, E.Tsuboi (Univ. of Aizu)

15 Adaptive Routing on the Recursive Diagonal Torus 171
A.Funahashi, T.Hanawa, H.Amano (Keio Univ.), T.Kudoh (Real World Computing Partnership)

IV Compilers

16 Achieving Multi-level Parallelization 183
C.J.Brownhill, A.Nicolau (UCI), S.Novack, C.D.Polychronopoulos (UIUC)

17 A Technique to Eliminate Redundant Inter-Processor Communication on Parallelizing Compiler TINPAR 195
A.Kubota, S.Tatsumi, T.Tanaka, M.Goshima, S.Mori, S.Tomita (Kyoto Univ.), H.Nakashima (Toyohashi Univ. of Tech.)

18 An Automatic Vectorizing/Parallelizing Pascal Compiler V-Pascal V.3 205
T.Uehara, Y.Kunieda (Wakayama Univ.), T.Tsuda (Hiroshima City Univ.)

19 An Algorithm for Automatic Detection of Loop Indices for Communication Overlapping 217
K.Ishizaki, H.Komatsu, T.Nakatani (IBM Tokyo Research Lab.)

V System Software

20 NaraView: An Interactive 3D Visualization System for Parallelization of Programs 231
M.Sasakura (Okayama Univ.), K.Joe (Wakayama Univ.), K.Araki (Kyushu Univ.)

21 Hybrid Approach for Non-strict Dataflow Program on Commodity Machine 243
K.Inenaga, S.Kusakabe, T.Morimoto, M.Amamiya (Kyushu Univ.)

22 Resource Management Methods for General Purpose
   Massively Parallel OS SSS-Core 255
   Y.Nobukuni, T.Matsumoto, K.Hiraki (Univ. of Tokyo)

23 Scenario-Based Hypersequential Programming: Formulation
   of Parallelization 267
   N.Uchihira, H.Kawata, F.Tamura (Toshiba Systems & Software
   Research Lab.)

VI Application

24 Parallelization of Space Plasma Particle Simulation 281
   Y.Akiyama, M.Saito, T.Noguchi, K.Onizuka, M.Ando (Real World
   Computing Partnership), Y.Omura, H.Matsumoto (Kyoto Univ.)
   Y.Misoo (Inf. & Math. Science Lab.)

25 Implementing Iterative Solvers for Irregular Sparse Matrix
   Problems in High Performance Fortran 293
   E.de Sturler, D.Loher (ETH)

26 Parallel Navigation in an A-NETL Based Parallel OODBMS 305
   L.Mutenda, M.Hiyama, T.Yoshinaga, T.Baba (Utsunomiya Univ.)

27 High Performance Parallel FFT on Distributed Memory
   Parallel Computers 317
   N.Shimizu (Tokai Univ.), Tk.Watanabe (NTT)

VII Poster Session Papers

28 Parallel Computation Model LogPQ 327
   T.Tooyama, S.Horiguchi (JAIST)

29 Cost Estimation of Coherence Protocols of Software Managed
   Cache on Distributed Shared Memory System 335
   T.Nanri, H.Sato (Kyushu Univ.), M.Shimasaki (Kyoto Univ.)

30 A Portable Distributed Shared Memory System on the
   Cluster Environment: Design and Implementation Fully in
   Software 343
   H.Sato, T.Nanri (Kyushu Univ.), M.Shimasaki (Kyoto Univ.)

31 An Object-Oriented Framework for Loop Parallelization ... 351
   Y.Omori, A.Fukuda (NAIST), K.Joe (Wakayama Univ.)

32 A Method for Runtime Recognition of Collective
   Communication on Distributed-Memory Multiprocessors .. 361
   T.Ogasawara, H.Komatsu (IBM Tokyo Research Lab.)

33 **Improving the Performance of Automated Forward Deduction System EnCal** 371
K.Nishi, J.Cheng, K.Ushijima (Kyushu Univ.)

34 **Efficiency of Parallel Machine for Large-Scale Simulation in Computational Physics** 381
H.Mizuseki, K.Esfarjani, Z.-Q.Li, K.Ohno, Y.Akiyama, K.Ichinoseki, Y.Kawazoe (Tohoku Univ.)

35 **Parallel PDB Data Retriever "PDB Diving Booster"** 389
K.Onizuka, T.Noguchi, M.Saito, Y.Akiyama (Real World Computing Partnership)

36 **A Parallelization Method for Neural Networks with Weak Connection Design** 397
A.I.Cristea, T.Okamoto (Univ. of Electro-Communication)

37 **Exploiting Parallel Computers to Reduce Neural Network Training Time of Real Applications** 405
J.Torresen, O.Landsverk (Norwegian University), S.Mori, H.Nakashima, S.Tomita (Kyoto Univ.)

Author Index 415

The Generation of Optimized Codes Using Nonzero Structure Analysis

Bret A. Marsolf[1]
Aart J.C. Bik *[2] Kyle A. Gallivan **[3] Harry A.G. Wijshoff[4]

[1] Demaco Inc., Champaign, Illinois, USA
[2] Computer Science Department
Indiana University
Bloomington, IN, USA
[3] Florida State University
Tallahassee, FL, USA
[4] High Performance Computing Division
Department of Computer Science
Leiden University
Leiden, the Netherlands

Abstract. In this paper we consider techniques for improving the performance of codes for general sparse problems by extracting both local and global structure information from a sparse matrix instance. This information can be used to improve the performance of the primitives through the utilization of specialized methods for the component parts which result from the matrix decomposition. A calculus is defined for controlling the decompositions and algorithms are presented for implementing the techniques within a code development environment.

1 Introduction

The development of libraries for high-performance computers and their effective use in application codes is an iterative process involving the refinement of the algorithms and implementations, and the tuning of parameters to match the machine architecture and the application context. This process includes the complex algorithm prototyping stage and continues during the lifetime of the library, as the algorithms are updated and ported between machines, requiring the modification of the algorithms, the selection of the correct primitives, and tuning to optimize performance. The development of a good library in an efficient manner, therefore, requires the combination of knowledge and expertise from the areas of numerical analysis, computer software, compilers, machine architecture and

---

* Support was provided by the Foundation for Computer Science (SION) of the Netherlands Organization for the Advancement of Pure Research (NWO) and the EC Esprit Agency DG XIII under Grant No. APPARC 6634 BRA III.
** Supported by the National Science Foundation under Grant No. US NSF CCR-9120105 and by ARPA under a subcontract from the University of Minnesota of Grant No. ARPA/NIST 60NANB2D1272.

applications. The design and implementation of an environment which can help the algorithm designer utilize this knowledge can both speed the development process for libraries and improve their performance within application codes. One ongoing attempt at such an environment is described in [4]. The environment is attempting to use novel compilation techniques adapted for languages specifically targeted for rapid prototyping (such as that in MATLAB [5]), high-level algebraic transformations that exploit detailed mathematical information about the operations performed, implementation details of the libraries available on the target machines and new compiler strategies for detecting and exploiting problem structure such as matrix sparsity.

In the context of the design of computational primitives for the multiplication of a sparse matrix and a vector, Agawal, Gustavson and Zubair have advocated an analysis procedure that decomposes a sparse matrix into a sum of several sparse matrices, each of which possess a sparsity structure that can be exploited to enhance performance [1]. In that work, the extraction and exploitation of sparsity structure was limited to the scope of the matrix-vector product primitive. When placed in the context of, say, one iteration of a sparse linear system solver, it is clear that the structural information can be propagated to the other computation primitives that surround the matrix-vector product routine, to enhance code efficiency.

Feature extraction from a sparse matrix and its use to select appropriate control and data structures, and generate efficient Fortran code, was discussed in [3]. However, no algebraic information was used to aid in the code generation.

In this paper, we investigate the use of feature extraction for sparse matrices algebraic information on computational primitives, and the high-level programming constructs and transformations techniques under development for the MATLAB-based rapid prototyping environment mentioned earlier to generate code for sparse numerical linear algebra computations.

In Section 2 the ability to decompose a sparse matrix into the sum of several sparse matrices is examined and a decomposed matrix form is developed which allows the decomposition of the operations. Next, in Section 3 operations are decomposed using the matrix decomposition to determine how the operation properties control the decomposition. These properties are then used to develop a calculus which defines how the decompositions interact with the operations in Section 4. Algorithms to implement code transformation utilizing the calculus system are presented in Section 5, which also includes discussion of other code generation issues in addition to the calculus. Next, in Section 6 the algorithms are applied to two examples to show how the calculus can guide the code transformations and in Section 7 results are provided for performance experiments with the two code examples. Finally, in Section 8 our conclusions are presented.

2 Matrix Decomposition Using Problem Specific Information

In order to utilize the matrix decompositions it is first necessary to define the algebraic relationship between the matrix and its decomposition. For our purposes, two main relationships are possible.

1. If no matrix element is in more than one component and every element is in a component, then the original matrix is equal to the sum of the component matrices.

$$A = \sum A_k \qquad (1)$$

2. If matrix element may be in more than one component and every element is in at least one component, then the original matrix is equal to the merge of the component matrices.

$$A = \amalg A_k \qquad (2)$$

Where:
$\amalg_{k=1}^{m} A_k = A_1 \vee A_2 \vee \cdots \vee A_m$
and $\vee$ is an arithmetic OR defined as follows:
Let Υ be the undefined value, then
$A = A_1 \vee A_2$ is defined as
$a_{i,j} = a_{1\,i,j} \vee a_{2\,i,j} \forall (i,j)$ such that
  If $a_{1\,(i,j)} == a_{2\,(i,j)}$ then $a_{(i,j)} = a_{1\,(i,j)}$
  else if $a_{1\,(i,j)} == \Upsilon$ then $a_{(i,j)} = a_{2\,(i,j)}$
  else if $a_{2\,(i,j)} == \Upsilon$ then $a_{(i,j)} = a_{1\,(i,j)}$
  else an invalid state.

Using these composition rules, several matrix decompositions can be defined. For example, a non-overlapping dense block decomposition could be defined as:

$$A = \sum A_k, \text{ where}$$
$$A_k(il_k : iu_k, jl_k : ju_k) = A(il_k : iu_k, jl_k : ju_k)$$
$$A_k = 0 \text{ elsewhere.}$$

Since this is non-overlapping, no element can appear in more than one component. If multiple component parts could contain the same element, then the composition operator would be a merge rather than a sum and the value of the remaining elements would be Υ instead of 0.

The block description just defined is sufficient to represent all other sparse matrix structures through the selection of suitable blocks. This is guaranteed to be true because each of the nonzero elements in any sparse structure could be represented with a block of size 1×1. For example, in Figure 1 a lower triangular matrix is shown and in Figure 2 a tri-diagonal matrix is shown.

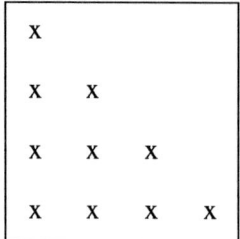 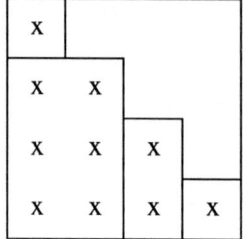

Fig. 1. A lower triangular matrix and its non-overlapping block decomposition.

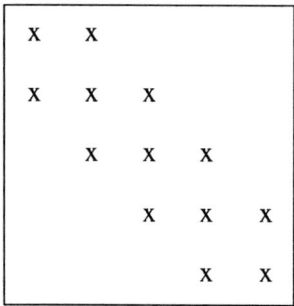 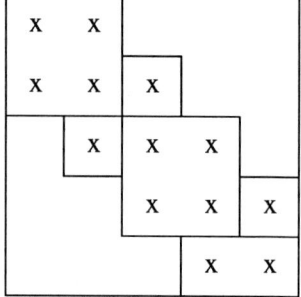

Fig. 2. A tridiagonal matrix and its non-overlapping block decomposition.

Instead of just considering dense blocks, this decomposition can also support *nearly dense* blocks, where some of the elements within the blocks are allowed to be zero. By allowing the constraint of dense blocks to be loosened, the number of blocks required to define the matrix may be reduced and the block sizes may be increased. However, zero elements contained with the blocks will generated additional operations and require additional storage space. The number of extra zeros to allow, therefore, is dependent upon the matrix structure, the operations being performed, and the machine architecture. In Figure 3 the *nearly dense* block decomposition of the lower triangular matrix is shown. In this example at most one zero elements occurs within each dense block and only three evenly-sized blocks are required to represent the matrix, as compared to the four unevenly-sized blocks in Figure 1.

In contrast to the non-overlapping block decompositions, some structures are more easily represented using overlapping blocks. Consider in Figure 4, where the tridiagonal matrix is represented with overlapping blocks. In this case only four blocks are needed, while six non-overlapping blocks were required. Though it may be simpler to represent matrices with overlapping blocks, the subsequent use of the decomposed blocks may be more complicated, requiring extra synchronization, and redundant arithmetic operations will occur with such use. For such reasons, we will only consider non-overlapping decompositions for the remainder of the paper.

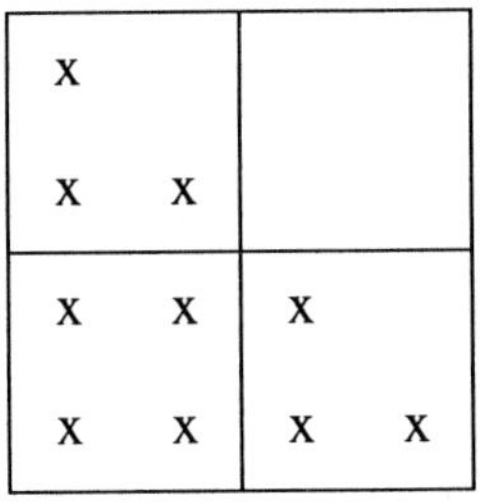

```
  X
  X  X
  X  X  X
  X  X  X  X
```

Fig. 3. A lower triangular matrix and its *nearly dense* block decomposition.

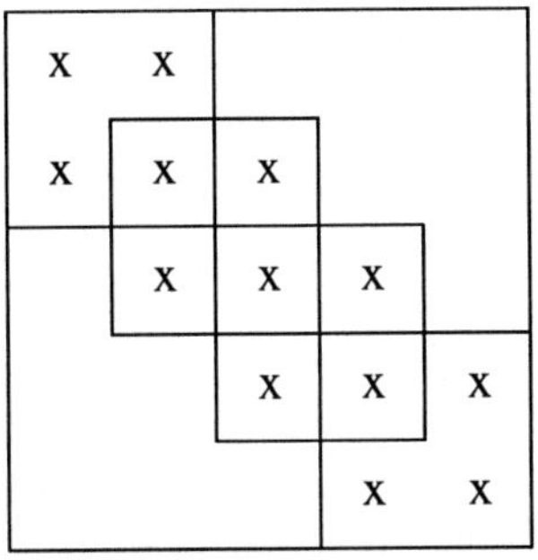

```
  X  X
  X  X  X
     X  X  X
        X  X  X
           X  X
```

Fig. 4. A tridiagonal matrix and its overlapping block decomposition.

Given that such decompositions are possible, there must be a means by which the decomposition can be determined. One possibility is to determine the decomposition by inspection. This may be a reasonable approach when there are only a few component structures in the matrix or when the matrix has an overall global structure. In the case of a general sparse matrix, however, this approach will become very tedious.

However, an analysis engine can be used to collect characteristics of the input data and supply this information to the code generator system automatically. The tool of Bik and Wijshoff, previously mentioned, is able to detect several well-known forms in $O(n+nnz)$ time. In addition, some statistical information about the nonzero structure is computed, comparable with the information that is gathered by the tool of Saad (SPARSKIT) [6]. Although originally developed for a compiler that transforms dense programs into semantically equivalent sparse code at programming level, the tool of Bik and Wijshoff can also be used to control code generation at this higher level.

3   Operator Decomposition

Once the matrix has been decomposed, the next step is the modification of the operations with the matrix to use the component parts. To better understand the issues involved with decomposing the operations, let us consider one operation, a matrix-vector multiplication, $p = A * q$.

1. Replace A with its block decomposition.

$$p = (\sum A_k) * q$$

2. Use the distributive properties of multiplication and addition to distribute the matrix-vector multiplications.

$$p = \sum (A_k * q)$$

3. Use the definition of the block decomposition to eliminate multiplication operations where the component matrix elements are to defined to be zero.

$$p_k(il_k : iu_k) = A_k(il_k : iu_k, jl_k : ju_k) * q(jl_k : ju_k)$$
$$p_k = 0 \text{ elsewhere}$$
$$p = \sum p_k$$

It is important to realize that even though the matrix decomposition was defined to have non-overlapping values, the resulting vectors, p_k do not maintain this property. However, in contrast to the overlapping matrix definition which uses the merge operator for the composition, in this case the overlapping values are combined with the sum operator.

In order to achieve the best results from these decompositions, however, is important to deal with more than a single operation. When a sequence of operations can be decomposed together, it allows the overhead of the decomposition to be amortized over more operations, which can be especially important for parallel machines.

For example, let us consider the following sequence which contains three operations. Here upper case letters represent matrices, lower case letters represent vectors, and greek letters represent scalars.

$$p = r + \beta * p$$
$$q = A * p$$
$$\alpha = p^T * q$$

As in the previous example, the matrix-vector multiplication can be decomposed by the block matrix decomposition. This changes the code sequence to the following.

$$p = r + \beta * p$$
For $k = 1 : m$
$\quad q_k(il_k : iu_k) = A_k(il_k : iu_k, jl_k : ju_k) * p(jl_k : ju_k)$
End For
$$q = \sum q_k$$
$$\alpha = p^T * q$$

This version of the algorithm has now defined a decomposition for the p vector which can be used to only calculate the portion of p that is necessary for the loop iteration.

For $k = 1 : m$
$\quad p_k(jl_k : ju_k) = r(jl_k : ju_k) + \beta * p(jl_k : ju_k)$
$\quad q_k(il_k : iu_k) = A_k(il_k : iu_k, jl_k : ju_k) * p_k(jl_k : ju_k)$
End For
$$p = \amalg p_k$$
$$q = \sum q_k$$
$$\alpha = p^T * q$$

The final step, therefore, is to decompose the inner product, $p^T * q$. However, the index sets which were used to compute the two operands do not align for the inner product. The operand p was calculated for indices $jl_k : ju_k$ and operand q was calculated for indices $il_k : iu_k$. Given that the cost of computing a new set of values for q is much greater than that for p, the index set for p can be modified to compute the values for both the matrix-vector operation and the inner-product operation.

For $k = 1 : m$
$\quad [ijl, iju] = il_k : iu_k \cup jl_k : ju_k$
$\quad p_k(ijl_k : iju_k) = r(ijl_k : iju_k) + \beta * p(ijl_k : iju_k)$
$\quad q_k(il_k : iu_k) = A_k(il_k : iu_k, jl_k : ju_k) * p_k(jl_k : ju_k)$
$\quad \alpha_k = p_k(il_k : iu_k)^T * q(il_k : iu_k)$
End For
$$p = \amalg p_k$$
$$q = \sum q_k$$
$$\alpha = \sum \alpha_k$$

As can be seen from this example, the decomposition of multiple operations becomes complicated as more of the operands are decomposed. The code to perform the operations becomes burdensome and difficult to write by hand. Furthermore, the code transformations must be repeated for each use, making the process even more tedious. To make the process simpler, it would be convenient to have a code generation system which could take the algorithm, decompose the matrices, and generate the subsequent code for the operations.

In order to generate parallel code, the dependencies within the decomposed operations must also be defined. For instance, when a matrix-vector operation is decomposed, each of the component multiplications can be done in parallel. However, when a solve is decomposed, certain operations must be synchronized.

As an example, here is the algorithm for doing a lower triangular solve, $Ly = b$, in parallel using the block decomposition of the matrix. (This example assumes there are k blocks which contain diagonal elements, and that all off-diagonal blocks only contain rows within a single diagonal block. If the matrix decomposition does not obey the condition, it is simple to decompose the off-diagonal block into multiple blocks such that the condition is met.)

$y_1 = b$
Parallel For $i = 1 : k$
　For $j = 1 : i - 1$
　　If $A_{i,j}$ is nonzero then
　　　Wait for y_j
　　　$y_i = y_i - A_{i,j} * y_j$
　　End If
　End For
　$y_i = A_{i,i} \backslash y_i$
　Broadcast y_i
End Parallel For

For a code generation system to perform the automatic code transformations, however, the rules for the composition and decomposition of matrix and vector operations must be well defined. These definitions must also include information about how operations may be combined and the order in which they can be executed, as well as the synchronization required between operations or sequences of operations.

4 The Underlying Calculus

The calculus to support an automatic transformation system must achieve three goals: define the decomposition of operations, define how decomposed operations can be organized, and define how decomposed results can be combined. This is accomplished by defining two levels of rules for the transformation system. At the lowest level, the rules define how the operators perform when using decomposed operands. The second level of rules define how the decomposed results are combined in order to form the final results. With these two levels of rules, as well as the standard algebraic rules, the calculus is defined.

In order to understand the rules, we will first present examples illustrating how a few of the rules were derived. First, let us define the operands and operations that will be used within the calculus.

- Operand types: scalars, vectors, transposed vectors, and matrices.
- Arithmetic Operations: $+$, $-$, $*$, and $/$ (or the equivalent solves).
- Composition Operations: sum and merge.

For a single operation, $p\ op\ q$, there are four possible combinations which use a single composition operation:

1. $p \; op \; \sum(q_k)$,
2. $\sum(p_k) \; op \; q$,
3. $p \; op \; \amalg(q_k)$, and
4. $\amalg(p_k) \; op \; q$.

Now, let us consider the multiplication operator and how it performs for some of the possible operand combinations, using the sum composition. Each operation will be transformed into two statements, the first statement to compute a *partial* value using the decomposed component and a second statement to compute a *final* value by combining the *partial* values. (During this illustration, greek letters will represent scalar values, lower case letters will represent vectors, and upper case letters will represent matrices.)

1. $\alpha = \beta * \sum \gamma_k \Rightarrow$
   $\alpha_k = \beta * \gamma_k \; \forall \; k$
   $\alpha = \sum \alpha_k$
2. $p = q^T * \sum r_k \Rightarrow$
   $p_k = q^T * r_k \; \forall \; k$
   $p = \sum p_k$
3. $p = A * \sum q_k \Rightarrow$
   $p_k = A * q_k \; \forall \; k$
   $p = \sum p_k$

In these examples, notice how the distributive properties of multiplication and addition allow the composition operator to be a sum.

Performing similar operations with the addition operator, however, leads to different results.

1. $\alpha = \beta + \sum \gamma_k \Rightarrow$
   $\alpha_1 = \beta + \gamma_1$
   $\alpha_k = \gamma_k \; \forall \; k > 1$
   $\alpha = \sum \alpha_k$
2. $p = q + \sum r_k \Rightarrow$
   $p_1 = q + r_1$
   $p_k = r_k \; \forall \; k > 1$
   $p = \sum p_k$
3. $A = B + \sum C_k \Rightarrow$
   $A_1 = B + C_1$
   $A_k = C_k \; \forall \; k > 1$
   $A = \sum A_k$

Here, the addition operator can not be distributed over the decomposition, therefore requiring different operations with the decomposed components. Yet, given the right operations, the composition operator is still a sum.

From these examples, it should be apparent that the decomposed operations can be selected, based on the distributive property of the operation, such that the composition will remain a sum. In fact, these rules can be abstracted as follows.

1. *Op* is distributive and commutative.
   (a) $p \; op \; \sum q \to \sum (p \; op \; q)$
   (b) $\sum p \; op \; q \to \sum (p \; op \; q)$
2. *Op* is distributive and not commutative.
   (a) If *op* is right distributive:
   $p \; op \; \sum q \to \sum (p \; op \; q)$
   (b) If *op* is left distributive:
   $\sum p \; op \; q \to \sum (p \; op \; q)$
3. *Op* is associative and not commutative and not distributive.
   (a) $p \; op \; \sum q \to (p \; op \; \sum_{i \le k} q_i) + \sum_{i > k} q_i$
   (b) $\sum p \; op \; q \to \sum_{i < k} p_i + (\sum_{i \ge k} (p_i) \; op \; q)$
4. *Op* is associative and commutative and not distributive.

   Let:
   S = set of all possible instances
   $S_1 \subseteq S$
   $S_2 = S - S_1$

   (a) $p \; op \; \sum q \to (p \; op \; \sum_{i \in S_1} q_i) + \sum_{i \in S_2} q_i$
   (b) $\sum p \; op \; q \to \sum_{i \in S_2} p_i + (\sum_{i \in S_1} (p_i) \; op \; q)$

Therefore, given that such rules can be defined based on the algebraic properties of the operations, the properties must be enumerated for each operation. This can be accomplished using a simple table format as follows for the basic operation form $p + \sum q$. As the '+' operation is commutative, the table is also valid for the basic form $\sum(p) + q$.

Let
   A = associativity, and
   C = commutativity.

$+ \sum$	scalar	vector	vector^T	matrix
scalar	A,C	A,C	A,C	A,C
vector	A,C	A,C		
vector^T	A,C		A,C	
matrix	A,C			A,C

A empty slot in the table indicates the operation is not valid.

The tables of the algebraic properties for all of the operations in combination with the basic operation forms are presented in Appendix A. The list of the compositions rules for all four of the basic forms are presented in Appendix B.

Given that the code transformations are based on the algebraic properties of associativity, distributivity, and commutativity, the transformed code will generate the correct results in true arithmetic. However, when floating point operations are reorganized within computer programs, variations in numerical results can occur.

Having defined the calculus for the compositions, it is important to note that the calculus only defines the high-level code transformations. In order to actually generate code several other issues, which were mentioned in the previous section, must also be resolved. First, strategies must be developed to determine if the redundant computations which can arise when coordinating multiple operation decompositions should be performed. For sequential machines the redundant computations are unnecessary, but for parallel machines the performance trade-off between communicating the values versus recomputing the values must be examined. The second issue is that in order to generate parallel code additional information, such as dependency analysis, is necessary. As was shown in the triangular solve example, it is necessary to sometimes generate synchronization and data communication primitives.

5 Implementation of the Decompositions

Using the calculus just defined, it is now possible to incorporate the matrix decomposition into a code development environment. As was stated earlier, the level at which the transformations will performed is at the algebraic, or algorithmic, level where operations are in terms of matrix and vector primitives. One such environment which operates on this level is the system for the compilation and interactive restructuring of numerical algorithms using MATLAB being developed by DeRose, Gallivan, Gallopoulos, Marsolf, and Padua [4].

The input language to this system is the array language MATLAB. The system parses the MATLAB code and then allows transformations to be performed on the internal format. From the internal format, the system can then generate code in the target language, either Fortran 90, C++, or MATLAB. While the code is in the internal format, however, various transformations can be applied, such as algebraic transformations or primitive selection. It is also at this level that the operations can be decomposed to operate on the matrix components.

The following high-level algorithm illustrates how the decomposition of the sparse matrices could be implemented.

1. Find an operation involving a matrix operand.
2. Decompose the operation using the matrix decomposition.
3. Backtrack to the definition of the other operand.
4. Propagate the decomposed result.

This algorithm requires two other procedures, a backtrack algorithm and a propagate algorithm. These are defined as follows:

– Backtrack Algorithm

   This algorithm starts with a required result index set and a target result variable defined as *target = left op right*.

   If (decomposition(*left*) *op* decomposition(*right*) == *target* index set) then
   Decompose operation using *target* index set

>   Backtrack *left*
>   Backtrack *right*
>   Propagate *target*
> End If

- Propagate Algorithm

   This algorithm starts with a source variable which has been decomposed for a given index set and a list of all the uses of the source variable definition.

   > For each use of *source*
   >   If ((*result = source op other_operand*) can be decomposed) then
   >   Decompose (*result = source op other_operand*)
   >   Backtrack *other_operand*
   >   Propagate *other_operand*
   >   Propagate *result*
   >   End If
   > End For

Within the algorithms, certain tests must be performed to determine if an operation can be decomposed. These tests are based not just on the calculus which defines the decompositions, but also on the comparison of the index sets which are utilized in the operations. As was shown in the previous example, when combining multiple operations it is necessary to resolve any differences in index sets which may arise. If the differences can not be resolved, then the operation can not be decomposed.

6 Examples

To illustrate the effects of these transformations on actual codes, let us take two examples from the solution of linear systems. The first example will be the Conjugate Gradient (CG) iterative solver and the second example is Gaussian elimination without pivoting.

The original code for the CG algorithm, for solving $A * x = b$, was taken from [2]. This code was modified to eliminate the conditional in the loop and a diagonal preconditioner has been incorporated.

$$M = diag(A)$$
$$x = 0$$
$$r = b - A * x$$
$$\rho = 1$$
$$p = 0$$
Loop
$$\quad \rho_1 = \rho$$
$$\quad \text{Solve } Mz = r$$
$$\quad \rho = r^T * z$$

$$\beta = \rho/\rho_1$$
$$p = z + \beta * p$$
$$q = A * p$$
$$\alpha = \rho/p^T * q$$
$$x = x + \alpha * p$$
$$r = r - \alpha * q$$
Test for convergence
End Loop

Within one loop iteration of this algorithm there are two matrix operations. The first operation performs the solve with the preconditioner and the second operation is the matrix-vector multiplication. Let us look at decomposing these operations given that the matrix A is a block tridiagonal matrix.

Given that the preconditioner is the diagonal of the A, the preconditioner can be decomposed using the same block structure. Therefore, the preconditioner solve can be performed by blocks, using a divide operation for the solve. Assuming the matrix has m diagonal blocks, the operation can be decomposed as follows.

For $k = 1 : m$
$\quad z_k(il_k : iu_k) = r_k(il_k : iu_k)./M_k(il_k : iu_k)$
End For
$z = \sum z_k$

Having decomposed z, it is also possible to decompose the inner product which is calculating ρ.

For $k = 1 : m$
$\quad z_k(il_k : iu_k) = r(il_k : iu_k)./M(il_k : iu_k)$
$\quad \rho_k = r(il_k : iu_k)^T * z_k(il_k : iu_k)$
End For
$z = \sum z_k$
$\rho = \sum \rho_k$

Given the block tridiagonal decomposition of the matrix A, the matrix-vector multiplication can be decomposed as follows. (Again, assuming there are m diagonal blocks.)

For $k = 1 : m$
$\quad q_k(il_k : iu_k) = A_{k,k-1}(il_k : iu_k, jl_{k-1} : ju_{k-1}) * p(jl_{k-1} : ju_{k-1})$
$\quad q_k(il_k : iu_k) = q_k(il_k : iu_k) + A_{k,k}(il_k : iu_k, jl_k : ju_k) * p(jl_k : ju_k)$
$\quad q_k(il_k : iu_k) = q_k(il_k : iu_k) + A_{k,k+1}(il_k : iu_k, jl_{k+1} : ju_{k+1}) * p(jl_{k+1} : ju_{k+1})$
End For
$q = \sum q_k$

Using the decomposition of p in the matrix-vector multiplication, the calculation of p can also be decomposed.

For $k = 1 : m$
$\quad p_k(jl_{k-1} : ju_{k+1}) = z(jl_{k-1} : ju_{k+1}) + \beta * p(jl_{k-1} : ju_{k+1})$
$\quad q_k(il_k : iu_k) = A_{k,k-1}(il_k : iu_k, jl_{k-1} : ju_{k-1}) * p_k(jl_{k-1} : ju_{k-1})$
$\quad q_k(il_k : iu_k) = q_k(il_k : iu_k) + A_{k,k}(il_k : iu_k, jl_k : ju_k) * p_k(jl_k : ju_k)$
$\quad q_k(il_k : iu_k) = q_k(il_k : iu_k) + A_{k,k+1}(il_k : iu_k, jl_{k+1} : ju_{k+1}) * p_k(jl_{k+1} : ju_{k+1})$
End For
$p = \amalg p_k$
$q = \sum q_k$

With both p and q decomposed, the calculation of the inner product can also be decomposed.

For $k = 1 : m$
$\quad p_k(jl_{k-1} : ju_{k+1}) = z(jl_{k-1} : ju_{k+1}) + \beta * p(jl_{k-1} : ju_{k+1})$
$\quad q_k(il_k : iu_k) = A_{k,k-1}(il_k : iu_k, jl_{k-1} : ju_{k-1}) * p_k(jl_{k-1} : ju_{k-1})$
$\quad q_k(il_k : iu_k) = q_k(il_k : iu_k) + A_{k,k}(il_k : iu_k, jl_k : ju_k) * p_k(jl_k : ju_k)$
$\quad q_k(il_k : iu_k) = q_k(il_k : iu_k) + A_{k,k+1}(il_k : iu_k, jl_{k+1} : ju_{k+1}) * p_k(jl_{k+1} : ju_{k+1})$
$\quad \alpha_k = p_k(il_k : iu_k)^T * q_k(il_k : iu_k)$
End For
$p = \amalg p_k$
$q = \sum q_k$
$\alpha = \rho / \sum \alpha_k$

Placing the decomposed operations back into the CG algorithm, the following code results.

$r = b - A * x$
$\rho = 1$
$p = 0$
Loop
$\quad \rho_1 = \rho$
$\quad$ For $k = 1 : m$
$\quad\quad z_k(il_k : iu_k) = r(il_k : iu_k)./M(il_k : iu_k)$
$\quad\quad \rho_k = r(il_k : iu_k)^T * z_k(il_k : iu_k)$
$\quad$ End For
$\quad z = \sum z_k$
$\quad \rho = \sum \rho_k$
$\quad \beta = \rho / \rho_1$
$\quad$ For $k = 1 : m$
$\quad\quad p_k(jl_{k-1} : ju_{k+1}) = z(jl_{k-1} : ju_{k+1}) + \beta * p(jl_{k-1} : ju_{k+1})$
$\quad\quad q_k(il_k : iu_k) = A_{k,k-1}(il_k : iu_k, jl_{k-1} : ju_{k-1}) * p_k(jl_{k-1} : ju_{k-1})$
$\quad\quad q_k(il_k : iu_k) = q_k(il_k : iu_k) + A_{k,k}(il_k : iu_k, jl_k : ju_k) * p_k(jl_k : ju_k)$
$\quad\quad q_k(il_k : iu_k) = q_k(il_k : iu_k) + A_{k,k+1}(il_k : iu_k, jl_{k+1} : ju_{k+1}) * p_k(jl_{k+1} : ju_{k+1})$
$\quad\quad \alpha_k = p_k(il_k : iu_k)^T * q_k(il_k : iu_k)$
$\quad$ End For
$\quad p = \amalg p_k$
$\quad q = \sum q_k$

$$\alpha = \rho / \sum \alpha_k$$
$$x = x + \alpha * p$$
$$r = r - \alpha * q$$
Test for convergence
End Loop

As a second example, let us examine Gaussian elimination without pivoting. The original version of the algorithm for dense matrices is as follows.

For $i = 1 : n - 1$
  $A(i+1:n,i) = A(i+1:n,i)/A(i,i)$
  $A(i+1:n,i+1:n) = A(i+1:n,i+1:n) - A(i+1:n,i) * A(i,i+1:n)$
End For

This algorithm is very simple, requiring only two operations, the scaling of a vector and a rank-1 update. In the case of a block tridiagonal matrix, the operations will only involve a limited number of blocks per iteration. As is shown in Figure 5, the scaling operation will update two blocks and the rank-1 update will update four blocks. The original Gaussian elimination code can be rewritten to perform the operations on a block level. First, the scaling operation can be

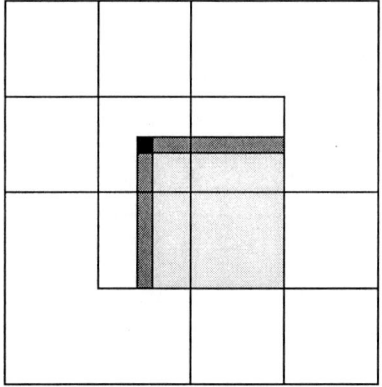

Fig. 5. The values modified during one iteration of Gaussian elimination on a block tridiagonal matrix.

decomposed into two block updates.

$A_{k,k}(i+1:iu_k,i) = A_{k,k}(i+1:iu_k,i)/A_{k,k}(i,i)$
If $il_{k+1} < n$ then
  $A_{k+1,k}(il_{k+1}:iu_{k+1},i) = A_{k+1,k}(il_{k+1}:iu_{k+1},i)/A_{k,k}(i,i)$
End If

Next, the rank-1 update can be decomposed into four rank-1 updates.

$A_{k,k}(i+1:iu_k, i+1:ju_k) = A_{k,k}(i+1:iu_k, i+1:ju_k) - A_{k,k}(i+1:iu_k, i)*$
$A_{k,k}(i, i+1:ju_k)$
If $il_{k+1} < n$ then
$A_{k,k+1}(i+1:iu_k, jl_{k+1}:ju_{k+1}) = A_{k,k+1}(i+1:iu_k, jl_{k+1}:ju_{k+1}) -$
$A_{k,k}(i+1:iu_k, i) * A_{k,k+1}(i, jl_{k+1}:ju_{k+1})$
$A_{k+1,k}(il_{k+1}:iu_{k+1}, i+1:ju_k) = A_{k,k+1}(il_{k+1}:iu_{k+1}, i+1:ju_k) -$
$A_{k+1,k}(il_{k+1}:iu_{k+1}, i) * A_{k,k}(i, i+1:ju_k)$
$A_{k+1,k+1}(il_{k+1}:iu_{k+1}, jl_{k+1}:ju_{k+1}) = A_{k,k+1}(il_{k+1}:iu_{k+1}, jl_{k+1}:ju_{k+1}) -$
$A_{k+1,k}(il_{k+1}:iu_{k+1}, i) * A_{k,k+1}(i, jl_{k+1}:ju_{k+1})$
End If

The decomposed operations can now be placed back into the algorithm.

For $i = 1 : n - 1$
$k = (i \text{ modulo } m) + 1$
$A_{k,k}(i+1:iu_k, i) = A_{k,k}(i+1:iu_k, i)/A_{k,k}(i,i)$
If $il_{k+1} < n$ then
$A_{k+1,k}(il_{k+1}:iu_{k+1}, i) = A_{k+1,k}(il_{k+1}:iu_{k+1}, i)/A_{k,k}(i,i)$
End If
$A_{k,k}(i+1:iu_k, i+1:ju_k) = A_{k,k}(i+1:iu_k, i+1:ju_k) -$
$A_{k,k}(i+1:iu_k, i) * A_{k,k}(i, i+1:ju_k)$
If $il_{k+1} < n$ then
$A_{k,k+1}(i+1:iu_k, jl_{k+1}:ju_{k+1}) = A_{k,k+1}(i+1:iu_k, jl_{k+1}:ju_{k+1}) -$
$A_{k,k}(i+1:iu_k, i) * A_{k,k+1}(i, jl_{k+1}:ju_{k+1})$
$A_{k+1,k}(il_{k+1}:iu_{k+1}, i+1:ju_k) = A_{k,k+1}(il_{k+1}:iu_{k+1}, i+1:ju_k) -$
$A_{k+1,k}(il_{k+1}:iu_{k+1}, i) * A_{k,k}(i, i+1:ju_k)$
$A_{k+1,k+1}(il_{k+1}:iu_{k+1}, jl_{k+1}:ju_{k+1}) = A_{k,k+1}(il_{k+1}:iu_{k+1}, jl_{k+1}:ju_{k+1}) -$
$A_{k+1,k}(il_{k+1}:iu_{k+1}, i) * A_{k,k+1}(i, jl_{k+1}:ju_{k+1})$
End If
End For

Using standard restructuring techniques with dependency analysis, it is possible to reorganize the operations to develop a blocked form of the algorithm.

For $i = 1 : n$ step m
$k = (i \text{ modulo } m) + 1$
For $j = il_k : iu_k - 1$
$A_{k,k}(j+1:iu_k, j) = A_{k,k}(j+1:iu_k, j)/A_{k,k}(j,j)$
$A_{k,k}(j+1:iu_k, j+1:ju_k) = A_{k,k}(j+1:iu_k, j+1:ju_k) -$
$A_{k,k}(j+1:iu_k, j) * A_{k,k}(j, j+1:ju_k)$
End For
If $il_{k+1} < n$ then
For $j = il_k : iu_k - 1$
$A_{k,k+1}(j+1:iu_k, jl_{k+1}:ju_{k+1}) = A_{k,k+1}(j+1:iu_k, jl_{k+1}:ju_{k+1}) -$
$A_{k,k}(j+1:iu_k, j) * A_{k,k+1}(j, jl_{k+1}:ju_{k+1})$
End For
For $j = jl_k : ju_k - 1$

$A_{k+1,k}(il_{k+1}:iu_{k+1},j) = A_{k+1,k}(il_{k+1}:iu_{k+1},j)/A_{k,k}(i,i)$
$A_{k+1,k}(il_{k+1}:iu_{k+1},j+1:ju_k) = A_{k,k+1}(il_{k+1}:iu_{k+1},j+1:ju_k) -$
$A_{k+1,k}(il_{k+1}:iu_{k+1},j) * A_{k,k}(j,j+1:ju_k)$
End For
$A_{k+1,k+1}(il_{k+1}:iu_{k+1},jl_{k+1}:ju_{k+1}) = A_{k,k+1}(il_{k+1}:iu_{k+1},jl_{k+1}:ju_{k+1}) -$
$A_{k+1,k}(il_{k+1}:iu_{k+1},jl_k:ju_k) * A_{k,k+1}(il_k:iu_k,jl_{k+1}:ju_{k+1})$
End If
End For

7 Experimental Results

To test the usefulness of automatically decomposing operations, the algorithms from the previous section were encoded in Fortran. In order to evaluate the performance of the decomposed code, three versions of each code were run. The first version is the code which exactly matches the original algorithm using dense matrices. The second version of the code uses the same algorithm as the first version, but uses sparse data structures and sparse primitives to perform the operations. The code with the operations decomposed for a block tridiagonal matrix is the third version. The tests were performed using block tridiagonal matrices of varying size, where the density within the blocks is 1. An example of such a matrix is shown in figure 6.

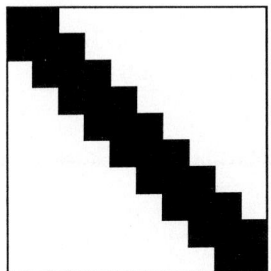

Fig. 6. Block Tridiagonal Form

First, we present timings of CG on one CPU of a Cray C98/4256 and an HP-UX 9000/720. In the following table, Table 1, we show the execution time in seconds of dense CG, an general row-wise implementation, and the block tridiagonal implemention. All codes are compiled with vectorization and default optimizations enabled respectively:

Unfortunately, only a very slight improvement has been obtained an the Cray, while on the HP the specialized algorithm is outperformed by a general sparse row-wise implementation.

In addition, we present timings on the Cray, in Table 2, for three versions of Gaussian elimination: a dense matrix implementation, the block tridiagonal

		block50 (nnz=1300)	block100 (nnz=2800)	block500 (nnz=14800)	block1000 (nnz=29800)	block1500 (nnz=44800)
Cray	dense	$4.8 \cdot 10^{-5}$	$1.6 \cdot 10^{-4}$	$2.0 \cdot 10^{-3}$	$9.7 \cdot 10^{-3}$	$1.6 \cdot 10^{-2}$
	sparse	$6.6 \cdot 10^{-5}$	$1.3 \cdot 10^{-4}$	$6.6 \cdot 10^{-4}$	$1.3 \cdot 10^{-3}$	$2.0 \cdot 10^{-3}$
	btf	$7.1 \cdot 10^{-5}$	$1.5 \cdot 10^{-4}$	$6.0 \cdot 10^{-4}$	$1.2 \cdot 10^{-3}$	$1.9 \cdot 10^{-3}$
HP	dense	$6.0 \cdot 10^{-4}$	$1.2 \cdot 10^{-3}$	$1.6 \cdot 10^{-1}$	$1.2 \cdot 10^{+0}$	-
	sparse	$4.0 \cdot 10^{-4}$	$6.0 \cdot 10^{-4}$	$2.6 \cdot 10^{-3}$	$7.2 \cdot 10^{-3}$	-
	btf	$4.0 \cdot 10^{-4}$	$8.0 \cdot 10^{-4}$	$4.4 \cdot 10^{-3}$	$1.0 \cdot 10^{-2}$	-

Table 1. Performance of the Conjugate Gradient algorithm.

implementation from the previous section, and a straightforward general row-wise implementation (although this algorithm does not have an optimal running time). Obviously, the use of an implementation that exploits the form of the matrix not only yields a substantial gain over the dense implementation, but also over a general sparse row-wise implementation exploiting all zero elements:

	block500 (nnz=14800)	block1000 (nnz=29800)	block1500 (nnz=44800)
dense	$5.3 \cdot 10^{-1}$	$6.0 \cdot 10^{+0}$	$1.3 \cdot 10^{+1}$
sparse	$1.4 \cdot 10^{-1}$	$3.4 \cdot 10^{-1}$	$5.9 \cdot 10^{-1}$
btf	$1.3 \cdot 10^{-2}$	$2.6 \cdot 10^{-2}$	$4.0 \cdot 10^{-2}$

Table 2. Performance of Gaussian elimination.

8 Conclusions

In this paper we have examined techniques for decomposing sparse matrices and how the matrix operations are affected when such decompositions are used. By examining the operations, a calculus was developed to define how the operations interact with the decompositions. This allowed the development of algorithms to support the automatic generation of code for the decomposed matrices within a MATLAB-based rapid-prototyping environment. The ability of the transformations to generate structure specific code was shown for two examples, with the performance results verifying the usefulness of the transformations.

In the future, additional work will be done to implement the transformations within the environment, allowing additional experimentation to be performed. Also, additional research into the matrix decompositions will be performed to allow the recursive use of structure information, i.e. the analysis of the structure within the decomposed blocks, to allow primitives to be selected on a block level or additional decomposition of individual blocks.

References

1. R. C. Agawal, F. G. Gustavson, and M. Zubair. A High Performance Algorithm Using Pre-Processing for the Sparse Matrix-Vector Multiplication. In *Proceedings of the International Conference on Supercomputing*, pages 32–41, 1992.
2. Richard Barrett, Michael Berry, Tony Chan, James Demmel, June Donato, Jack Dongarra, Victor Eijkhout, Roldan Pozo, Charles Romine, and Henk van der Vorst. *Templates for the Solution of Linear Systems: Building Blocks for Iterative Methods.* SIAM Publications, 1993.
3. Aart J. C. Bik and Harry A. G. Wijshoff. Nonzero Structure Analysis. In *Proceedings of the International Conference on Supercomputing*, pages 226–235, 1994.
4. L. DeRose, K. Gallivan, E. Gallopoulos, B. Marsolf, and D. Padua. A MATLAB Compiler and Restructurer for the Development of Scientific Libraries and Applications. To appear in *Proceedings for the 8th International Workshop on Languages and Compilers for Parallel Computing*, Columbus, Ohio, August 1995.
5. The Math Works, Inc. *MATLAB, High-Performance Numeric Computation and Visualization Software. User's Guide*, 1992.
6. Youcef Saad. SPARSKIT: A basic tool for sparse matrix computations. CSRD/RIACS, 1990.

Appendix A. Operator Descriptions

For each combination of operator and composition operator, a table is provided describing the algebraic properties for each variable combination. Within the tables, the following definitions will be used to describe the properties.

A = Associative
A_L = Associative to the left
A_R = Associative to the right
D = Distributive
D_L = Distributive to the left
D_R = Distributive to the right
C = Commutative
C^T = Commutative, but only when the position of the transpose operator is maintained within the operation, i.e. $p^T * q == q^T * p$.
N = None of the above properties are defined.
A blank slot indicates the operation is not valid.

1. One operand is a sum.

   (a) $p + \sum(q) == \sum(p) + q$

$p \backslash q$	scalar	vector	$vector^T$	matrix
scalar	A,C	A,C	A,C	A,C
vector	A,C	A,C		
$vector^T$	A,C		A,C	
matrix	A,C			A,C

   (b) $p - \sum(q)$

$p \backslash q$	scalar	vector	$vector^T$	matrix
scalar	N	N	N	N
vector	N	N		
$vector^T$	N		N	
matrix	N			N

   (c) $\sum(p) - q$

$p \backslash q$	scalar	vector	$vector^T$	matrix
scalar	A	A	A	A
vector	A	A		
$vector^T$	A		A	
matrix	A			A

   (d) $p * \sum(q) == \sum(p) * q$

$p \backslash q$	scalar	vector	$vector^T$	matrix
scalar	D,C	D,C	D,C	D,C
vector	D,C		D	
$vector^T$	D,C	D,C^T		D
matrix	D,C	D		D

(e) $p/\sum(q)$

$p \setminus q$	scalar	vector	$vector^T$	matrix
scalar	N	N		
vector	N	N		
$vector^T$	N		N	N
matrix	N		N	N

(f) $\sum(p)/q$

$p \setminus q$	scalar	vector	$vector^T$	matrix
scalar	D	D		
vector	D	D		
$vector^T$	D		D	D
matrix	D		D	D

(g) $p\backslash\sum(q)$

$p \setminus q$	scalar	vector	$vector^T$	matrix
scalar	D	D	D	D
vector		D		D
$vector^T$	D		D	
matrix		D		D

(h) $\sum(p)\backslash q$

$p \setminus q$	scalar	vector	$vector^T$	matrix
scalar	N	N	N	N
vector		N		N
$vector^T$	N		N	
matrix		N		N

2. Both operands are a sum.

(a) $\sum(p) + \sum(q)$

$p \setminus q$	scalar	vector	$vector^T$	matrix
scalar	A,C	A,C	A,C	A,C
vector	A,C	A,C		
$vector^T$	A,C		A,C	
matrix	A,C			A,C

(b) $\sum(p) - \sum(q)$

$p \setminus q$	scalar	vector	$vector^T$	matrix
scalar	N	N	N	N
vector	N	N		
$vector^T$	N		N	
matrix	N			N

(c) $\sum(p) * \sum(q)$

p \ q	scalar	vector	$vector^T$	matrix
scalar	D,C	D,C	D,C	D,C
vector	D,C		D	
$vector^T$	D,C	D,C^T		D
matrix	D,C	D		D

(d) $\sum(p) / \sum(q)$

p \ q	scalar	vector	$vector^T$	matrix
scalar	N	N		
vector	N	N		
$vector^T$	N		N	N
matrix	N		N	N

(e) $\sum(p) \backslash \sum(q)$

p \ q	scalar	vector	$vector^T$	matrix
scalar	N	N	N	N
vector		N		N
$vector^T$	N		N	
matrix		N		N

3. One operand is a merge.

   (a) $p + \amalg(q) == \amalg(p) + q$

p \ q	scalar	vector	$vector^T$	matrix
scalar	D,C	D,C	D,C	D,C
vector	D,C	D,C		
$vector^T$	D,C		D,C	
matrix	D,C			D,C

   (b) $p - \amalg(q)$

p \ q	scalar	vector	$vector^T$	matrix
scalar	D	D	D	D
vector	D	D		
$vector^T$	D		D	
matrix	D			D

   (c) $\amalg(p) - q$

p \ q	scalar	vector	$vector^T$	matrix
scalar	D	D	D	D
vector	D	D		
$vector^T$	D		D	
matrix	D			D

(d) $p * \text{II}(q) == \text{II}(p) * q$

$p \setminus q$	scalar	vector	$vector^T$	matrix
scalar	D,C	D,C	D,C	D,C
vector	D,C		N	
$vector^T$	D,C	C^T		N
matrix	D,C	N		N

(e) $p / \text{II}(q)$

$p \setminus q$	scalar	vector	$vector^T$	matrix
scalar	D	N		
vector	D	N		
$vector^T$	D		N	N
matrix	D		N	N

(f) $\text{II}(p) / q$

$p \setminus q$	scalar	vector	$vector^T$	matrix
scalar	D	D		
vector	N	N		
$vector^T$	N		N	N
matrix	N		N	N

(g) $p \backslash \text{II}(q)$

$p \setminus q$	scalar	vector	$vector^T$	matrix
scalar	D	D	D	D
vector		N		N
$vector^T$	N		N	
matrix		N		N

(h) $\text{II}(p) \backslash q$

$p \setminus q$	scalar	vector	$vector^T$	matrix
scalar	D	D	D	D
vector		N		N
$vector^T$	N		N	
matrix		N		N

4. Both operands are a merge.

(a) $\text{II}(p) + \text{II}(q)$

$p \setminus q$	scalar	vector	$vector^T$	matrix
scalar	D,C	D,C	D,C	D,C
vector	D,C	D,C		
$vector^T$	D,C		D,C	
matrix	D,C			D,C

(b) $\amalg(p) - \amalg(q)$

$p \setminus q$	scalar	vector	$vector^T$	matrix
scalar	D	D	D	D
vector	D	D		
$vector^T$	D		D	
matrix	D			D

(c) $\amalg(p) * \amalg(q)$

$p \setminus q$	scalar	vector	$vector^T$	matrix
scalar	D,C	D,C	D,C	D,C
vector	D,C		N	
$vector^T$	D,C	C^T		N
matrix	D,C	N		N

(d) $\amalg(p) / \amalg(q)$

$p \setminus q$	scalar	vector	$vector^T$	matrix
scalar	D	N		
vector	D	N		
$vector^T$	D		N	N
matrix	D		N	N

(e) $\amalg(p) \backslash \amalg(q)$

$p \setminus q$	scalar	vector	$vector^T$	matrix
scalar	D	D	D	D
vector		N		N
$vector^T$	N		N	
matrix		N		N

5. First operand is a sum and the second operand is a merge.

(a) $\sum(p) + \amalg(q)$

$p \setminus q$	scalar	vector	$vector^T$	matrix
scalar	A_L, D_R, C	A_L, D_R, C	A_L, D_R, C	A_L, D_R, C
vector	A_L, D_R, C	A_L, D_R, C		
$vector^T$	A_L, D_R, C		A_L, D_R, C	
matrix	A_L, D_R, C			A_L, D_R, C

(b) $\sum(p) - \amalg(q)$

$p \setminus q$	scalar	vector	$vector^T$	matrix
scalar	A_L, D_R	A_L, D_R	A_L, D_R	A_L, D_R
vector	A_L, D_R	A_L, D_R		
$vector^T$	A_L, D_R		A_L, D_R	
matrix	A_L, D_R			A_L, D_R

(c) $\sum(p) * \amalg(q)$

p \ q	scalar	vector	$vector^T$	matrix
scalar	D,C	D,C	D,C	D,C
vector	D,C		N	
$vector^T$	D,C	C^T		N
matrix	D,C	N		N

(d) $\sum(p) / \amalg(q)$

p \ q	scalar	vector	$vector^T$	matrix
scalar	D	D_L		
vector	D	D_L		
$vector^T$	D		D_L	D_L
matrix	D		D_L	D_L

(e) $\sum(p) \backslash \amalg(q)$

p \ q	scalar	vector	$vector^T$	matrix
scalar	D_R	N	N	N
vector		N		N
$vector^T$	D_R		N	
matrix		N		N

6. First operand is a merge and the second operand is a sum.

(a) $\amalg(p) + \sum(q)$

p \ q	scalar	vector	$vector^T$	matrix
scalar	A_R, D_L, C	A_R, D_L, C	A_R, D_L, C	A_R, D_L, C
vector	A_R, D_L, C	A_R, D_L, C		
$vector^T$	A_R, D_L, C		A_R, D_L, C	
matrix	A_R, D_L, C			A_R, D_L, C

(b) $\amalg(p) - \sum(q)$

p \ q	scalar	vector	$vector^T$	matrix
scalar	D_L	D_L	D_L	D_L
vector	D_L	D_L		
$vector^T$	D_L		D_L	
matrix	D_L			D_L

(c) $\amalg(p) * \sum(q)$

p \ q	scalar	vector	$vector^T$	matrix
scalar	D,C	D,C	D,C	D,C
vector	D,C		N	
$vector^T$	D,C	C^T		N
matrix	D,C	N		N

(d) $\text{II}(p)/\sum(q)$

$p \setminus q$	scalar	vector	$vector^T$	matrix
scalar	D_L	D_L		
vector	N	N		
$vector^T$	N		N	N
matrix	N		N	N

(e) $\text{II}(p)\backslash \sum(q)$

$p \setminus q$	scalar	vector	$vector^T$	matrix
scalar	D	D	D	D
vector		D_R		D_R
$vector^T$	D_R		D_R	
matrix		D_R		D_R

Appendix B. Axioms

Defining the rules for combining of operations for $p \; op \; q$ when one or both of the operands is a summation or a merge.

1. One operand is a sum.
   (a) *Op* is distributive and commutative.
      i. $p \; op \; \sum(q) \rightarrow \sum(p \; op \; q)$
      ii. $\sum(p) \; op \; q \rightarrow \sum(p \; op \; q)$
   (b) *Op* is distributive and not commutative.
      i. If *op* is right distributive:
         $p \; op \; \sum(q) \rightarrow \sum(p \; op \; q)$
      ii. If *op* is left distributive:
         $\sum(p) \; op \; q \rightarrow \sum(p \; op \; q)$
   (c) *Op* is associative and not commutative and not distributive.
      i. $p \; op \; \sum(q) \rightarrow (p \; op \; \sum_{i<k}(q_i)) + \sum_{i>k} q_i$
      ii. $\sum(p) \; op \; q \rightarrow \sum_{i<k}(p_i) + (\sum_{i\geq k}(p_i) \; op \; q)$
   (d) *Op* is associative and commutative and not distributive.
      Let:
      S = set of all possible instances
      $S_1 \subseteq S$
      $S_2 = S - S_1$
      i. $p \; op \; \sum(q) \rightarrow (p \; op \; \sum_{i \in S_1}(q_i)) + \sum_{i \in S_2}(q_i)$
      ii. $\sum(p) \; op \; q \rightarrow \sum_{i \in S_2}(p_i) + (\sum_{i \in S_1}(p_i) \; op \; q)$

2. Both operands are sums.
   (a) *Op* is distributive and commutative.
      i. $\sum(p) \; op \; \sum(q) \rightarrow \sum(p \; op \; \sum(q))$, or
      ii. $\sum(p) \; op \; \sum(q) \rightarrow \sum(\sum(p) \; op \; q)$
   (b) *Op* is distributive and not commutative.
      i. If *op* is right distributive:
         $\sum(p) \; op \; \sum(q) \rightarrow \sum(\sum(p) \; op \; q)$
      ii. If *op* is left distributive:
         $\sum(p) \; op \; \sum(q) \rightarrow \sum(p \; op \; \sum(q))$
   (c) *Op* is associative and not commutative and not distributive.
      $\sum(p) \; op \; \sum(q) \rightarrow \sum_{i<j}(p_i) + (\sum_{i\geq j}(p_i) \; op \; \sum_{i<k}(q_i)) + \sum_{i>k}(q_i)$
   (d) *Op* is associative and commutative and not distributive.
      Let:
      S = set of all possible instances
      $S_1 \subseteq S$
      $S_2 = S - S_1$
      $S_3 \subseteq S$
      $S_4 = S - S_2$
      $\sum(p) \; op \; \sum(q) \rightarrow \sum_{i \in S_1}(p_i) + (\sum_{i \in S_2}(p_i) \; op \; \sum_{i \in S_3}(q_i)) + \sum_{i \in S_4}(q_i)$

3. One operand is a merge.
   (a) *Op* is distributive and commutative.
      i. $p \text{ op } \amalg(q) \to \amalg(p \text{ op } q)$
      ii. $\amalg(p) \text{ op } q \to \amalg(p \text{ op } q)$
   (b) *Op* is distributive and not commutative.
      i. If *op* is right distributive:
         $p \text{ op } \amalg(q) \to \amalg(p \text{ op } q)$
      ii. If *op* is left distributive:
         $\amalg(p) \text{ op } q \to \amalg(p \text{ op } q)$
   (c) *Op* is associative and not commutative and not distributive.
      i. $p \text{ op } \amalg(q) \to (p \text{ op } \amalg_{i \leq k}(q_i)) \vee \amalg_{i > k}(q_i)$
      ii. $\amalg(p) \text{ op } q \to \amalg_{i < k}(p_i) \vee (\amalg_{i \geq k}(p_i) \text{ op } q)$
   (d) *Op* is associative and commutative and not distributive.
      Let:
      S = set of all possible instances
      $S_1 \subseteq S$
      $S_2 = S - S_1$
      i. $p \text{ op } \amalg(q) \to (p \text{ op } \amalg_{i \in S_1}(q_i)) \vee \amalg_{i \in S_2}(q_i)$
      ii. $\amalg(p) \text{ op } q \to \amalg_{i \in S_2}(p_i) \vee (\amalg_{i \in S_1}(p_i) \text{ op } q)$
4. Both operands are merges.
   (a) *Op* is distributive and commutative.
      i. $\amalg(p) \text{ op } \amalg(q) \to \amalg(p \text{ op } \amalg(q))$, or
      ii. $\amalg(p) \text{ op } \amalg(q) \to \amalg(\amalg(p) \text{ op } q)$
   (b) *Op* is distributive and not commutative.
      i. If *op* is right distributive:
         $\amalg(p) \text{ op } \amalg(q) \to \amalg(\amalg(p) \text{ op } q)$
      ii. If *op* is left distributive:
         $\amalg(p) \text{ op } \amalg(q) \to \amalg(p \text{ op } \amalg(q))$
   (c) *Op* is associative and not commutative and not distributive.
      $\amalg(p) \text{ op } \amalg(q) \to \amalg_{i < j}(p_i) \vee (\amalg_{i \geq j}(p_i) \text{ op } \amalg_{i \leq k}(q_i)) \vee \amalg_{i > k}(q_i)$
   (d) *Op* is associative and commutative and not distributive.
      Let:
      S = set of all possible instances
      $S_1 \subseteq S$
      $S_2 = S - S_1$
      $S_3 \subseteq S$
      $S_4 = S - S_2$
      $\amalg(p) \text{ op } \amalg(q) \to \amalg_{i \in S_1}(p_i) \vee (\amalg_{i \in S_2}(p_i) \text{ op } \amalg_{i \in S_3}(q_i)) \vee \amalg_{i \in S_4}(q_i)$
5. The first operand is a sum and the second is a merge.
   (a) *Op* is distributive and commutative.
      i. $\sum(p) \text{ op } \amalg(q) \to \sum(p \text{ op } \amalg(q))$, or
      ii. $\sum(p) \text{ op } \amalg(q) \to \amalg(\sum(p) \text{ op } q)$
   (b) *Op* is distributive and not commutative.

      i. If *op* is right distributive:
$$\sum(p) \ op \ \amalg(q) \to \amalg(\sum(p) \ op \ q)$$
      ii. If *op* is left distributive:
$$\sum(p) \ op \ \amalg(q) \to \sum(p \ op \ \amalg(q))$$
  (c) *Op* is associative and not commutative and not distributive.
$$\sum(p) \ op \ \amalg(q) \to \sum_{i<j}(p_i) + (\sum_{i\geq j}(p_i) \ op \ \amalg_{i\leq k}(q_i)) \vee \amalg_{i>k}(q_i)$$
  (d) *Op* is associative and commutative and not distributive.
     Let:
$$S = \text{set of all possible instances}$$
$$S_1 \subseteq S$$
$$S_2 = S - S_1$$
$$S_3 \subseteq S$$
$$S_4 = S - S_2$$
$$\sum(p) \ op \ \amalg(q) \to \sum_{i\in S_1}(p_i) + (\sum_{i\in S_2}(p_i) \ op \ \amalg_{i\in S_3}(q_i)) \vee \amalg_{i\in S_4}(q_i)$$
6. The first operand is a merge and the second is a sum.
  (a) *Op* is distributive and commutative.
      i. $\amalg(p) \ op \ \sum(q) \to \amalg(p \ op \ \sum(q))$, or
      ii. $\amalg(p) \ op \ \sum(q) \to \sum(\amalg(p) \ op \ q)$
  (b) *Op* is distributive and not commutative.
      i. If *op* is right distributive:
$$\amalg(p) \ op \ \sum(q) \to \sum(\amalg(p) \ op \ q)$$
      ii. If *op* is left distributive:
$$\amalg(p) \ op \ \sum(q) \to \amalg(p \ op \ \sum(q))$$
  (c) *Op* is associative and not commutative and not distributive.
$$\amalg(p) \ op \ \sum(q) \to \amalg_{i<j}(p_i) \vee (\amalg_{i\geq j}(p_i) \ op \ \sum_{i\leq k}(q_i)) + \sum_{i>k}(q_i)$$
  (d) *Op* is associative and commutative and not distributive.
     Let:
$$S = \text{set of all possible instances}$$
$$S_1 \subseteq S$$
$$S_2 = S - S_1$$
$$S_3 \subseteq S$$
$$S_4 = S - S_2$$
$$\amalg(p) \ op \ \sum(q) \to \amalg_{i\in S_1}(p_i) \vee (\amalg_{i\in S_2}(p_i) \ op \ \sum_{i\in S_3}(q_i)) + \sum_{i\in S_4}(q_i)$$

On the Importance of an End-To-End View of Memory Consistency in Future Computer Systems

Guang R. Gao[1] and Vivek Sarkar[2]

[1] University of Delaware, Newark DE 19716, USA
[2] MIT Laboratory for Computer Science, Cambridge MA 02139, USA

Abstract. The main purpose of a memory consistency model is to serve as an agreement between hardware system designers and software developers on the semantics of memory operations so as to ensure correct execution of user programs. However, the bulk of past work on memory consistency models has been pursued from the hardware viewpoint. In this viewpoint, a memory consistency model is used to specify certain behavioral properties that are guaranteed by uniprocessor/multiprocessor hardware *e.g.*, the *memory coherence* property. In this paper, we argue that it is essential to adopt an end-to-end view of memory consistency that can be understood at all levels of software and hardware. We believe that this is possible with a memory consistency model based on partial order execution semantics — such as the Location Consistency (LC) model — rather than on the memory coherence assumption.

1 Background

A memory consistency model specifies the semantics of concurrent memory operations (load/store or synchronization operations) in a uniprocessor or multiprocessor system. The bulk of past research on memory consistency models has been pursued from the viewpoint of uniprocessor and multiprocessor hardware in which a memory consistency model is used to specify certain behavioral properties that are guaranteed by the hardware. This has led to hardware-based definitions of memory consistency in which a program is viewed as instructions executing on processors, and in which a total order ("program order") is assumed for instructions that execute on the same processor. Memory consistency is then defined with respect to the relative order in which memory operations are "performed" or "globally performed" in this low-level view of program execution (*e.g.*, see [10]).

While this low-level view of memory consistency has been helpful in the design of more efficient and scalable cache consistency protocols for shared-memory multiprocessors, it is too narrow to be useful in an end-to-end view of computer systems. For example, these memory consistency models cannot be used to specify the semantics of a parallel (multithreaded) program written in a high-level programming language because the notions of total orderings of instructions on processors and of memory operations being "performed" by processors do not make sense in a high-level language.

The hardware memory consistency model that has been most commonly used as a basis for past work is *sequential consistency* (SC) [13]. It has been observed that sequential consistency limits performance by preventing the use of common uniprocessor hardware optimizations such as store buffers and out-of-order memory operations [10, 2]. The main approach taken in recent work on memory consistency models is to allow performance optimizations to be applied, while ensuring that sequential consistency is retained for a restricted class of programs — mainly programs that do not exhibit data races [3]. Therefore, we refer to these weaker memory consistency models as *SC-derived* models. Recently proposed SC-derived models include *weak ordering* (WO) [7], *release consistency* (RC) [10], *data-race-free-0* (DRF0) [2], and *data-race-free-1* (DRF1) [3].

A central assumption in the definitions of all SC-derived memory consistency models is the *memory coherence* assumption, which can be stated as follows [10]: "all writes to the same location are serialized in some order and are performed in that order with respect to any processor". Memory coherence is a less restrictive form of serializability — it enforces a limited serializability on memory operations performed on the same location, rather than serializability on all memory operations.

The general case for an end-to-end view in system design has been eloquently stated by Salzer et al in [17]. However, the memory coherence assumption in past hardware-based SC and SC-derived memory consistency models poses fundamental obstacles to obtaining an end-to-end view of memory consistency in computer systems. First, memory coherence is defined on a hardware-specified granularity, namely, a single memory location/word. Thus, hardware-based memory consistency is not applicable to larger-sized data types (*e.g.*, a double-word complex number) and an artificial distinction is created among different data types based on their representation sizes. Second, there may not be a well-defined total order on all memory operations at the source program level, and there also may be memory operations in the compiled machine code that are just not visible in the source program. This makes it impractical to work with the memory coherence assumption at the source program level. Third, the memory coherence assumption imposes restrictions on the ordering of memory operations that go beyond the partial order defined by synchronization operations in a parallel program. These restrictions couple the ordering of memory operations in different subprograms by requiring that they all observe writes to a memory location in the same order. The coupling due to the memory coherence assumption makes it difficult for the programmer to work with a modular view of the source program. Finally, multithreaded programs are compiled by sequential compilers that are oblivious of the memory coherence assumption. Programmers currently address this mismatch by a trial-and-error process of inserting "volatile" variable declarations and turning off compiler optimizations till the multithreaded program appears to run correctly. These modifications are tedious and also detrimental to performance. Even after making the modifications, the correctness of the modified parallel program will depend on how high-level operations are mapped to low-level memory instructions in the versions of the system software used (*e.g.*, compiler, operating system, etc.) An end-to-end memory consistency model would eliminate the need for making program modifications (such as the declaration of volatile variables) for correctness, and also avoid the inefficiencies that accompany these modifications.

In summary, current memory consistency models are defined at a very primitive level of program execution and include a memory coherence assumption that enforces serializability of memory operations performed on the same memory location. Since the primary purpose of a memory consistency model is to serve as a contract between hardware and software, there is a need for end-to-end definitions of memory consistency that can be understood uniformly at different levels of software and hardware in a computer system.

2 Our Position

The thesis of this position paper is that it is essential to provide an end-to-end view of memory consistency that can be understood at all levels of software and hardware. We believe that this is possible with a memory consistency model based on partial order execution semantics rather than on the memory coherence assumption. The Location Consistency (LC) model [8, 9] is one example of a memory consistency model based on partial order semantics; the Dag Consistency model [5] used in the MIT Cilk project is another such example.

A partial order execution semantics is fundamental to many layers of software and hardware in a computer system. For example:

- **Evaluation order**
  The semantics of many high-level programming languages declares the order of operand evaluation in an expression as being unspecified, and allows the implementation to choose any ordering that it desires. The programmer can easily understand this underspecification as a partial order that represents a set of possible execution sequences. But it would be very difficult for the programmer to work with a memory consistency model that assumes a fixed ordering for memory operations, while the exact ordering of memory operations is left unspecified in the semantics of the programming language. The fact that there may be memory operations in the machine code that are not visible in the source program further exacerbates this problem.
- **Compiler optimizations**
  Many compiler optimizations work with program representations that reflect a partial order. For example, the Sethi-Ullman instruction reordering algorithm for improved register allocation assumes that the partial order is represented by an instruction tree. Similarly, the instruction scheduling phase of an optimizing back-end assumes that the partial order is represented by an instruction dag. It is very difficult for such compiler optimizations to work with a memory consistency model that assumes a fixed instruction ordering, since they would either need to conservatively disable all instruction

reordering or go to great lengths to prove that a change in instruction order (*e.g.,* an interchange of two independent load instructions) does not violate the memory consistency semantics of the program.
- **Virtual machine interpretation**
 Most definitions of memory consistency view hardware as the lowest (primitive) level of instruction execution. However, there may be multiple layers of interpretation even in hardware execution of instructions. Just as in software, one level of an instruction set architecture (or virtual machine) may leave unspecified the execution ordering of the memory operands of a (compound) instruction. As before, it would be difficult to work with a memory consistency model that assumes a fixed ordering of memory operations when defining the semantics of such an instruction set architecture.
- **On-the-fly translation**
 The same phenomenon as in virtual machine interpretation can be observed in the on-the-fly binary translation techniques that have been proposed for some future computer systems. On-the-fly translation is frequently performed as a pattern matching of instruction trees in a source instruction stream. Once again, this raises the possibility of reordering operations and memory accesses during the translation which would make it difficult to work with a memory consistency model that assumes a fixed ordering.

All these examples reinforce the notion that program execution is fundamentally a partial order, and thus motivate the need for a memory consistency semantics that is based on viewing program execution as a partial order.

3 Location Consistency: An Alternative to Memory Coherence

In this section, we give a brief summary of the Location Consistency (LC) model [8, 9], as an example of a memory consistency model that is based on partial order execution semantics and does not use the memory coherence assumption. We expect that other memory consistency models based on partial order semantics will be defined in the future, such as the Dag Consistency model [5] used in the MIT Cilk system.

Location Consistency models the state of a memory location as a partially ordered multiset (pomset). Each element of the pomset corresponds to either a write operation to the location or a synchronization operation. The partial orders in the pomsets are determined entirely by the ordering constraints imposed by the concurrency and synchronization constructs in the program being executed — no extra program modifications (such as the insertion of volatile declarations) need to be made. Thus, the LC model provides a uniform view of memory consistency at all levels of software and hardware in a computer system. Only the partial order defined by the program needs to be obeyed; there is no additional coherence requirement that all writes to the same location be observed in the same order by all read operations.

An ordering in the pomset is created between two write operations to the same location if they originate from the same sequential thread. A synchronization operation inserts an element in the pomsets of all locations affected by the synchronization. The state of a location is observed by read operations: a read operation of location L returns one of a set of possible "most recent write" values from the pomset corresponding to L. Informally, we say that a software/hardware level of a computer system is Location Consistent if each read operation that is executed at that level returns a value that belongs to the most recent write set for the target location. This definition allows for the possibility of different partial orders at different levels of system software and hardware *i.e.*, for the possibility of introducing extra ordering constraints when mapping from a higher level to a lower level. (Note that the pomset model for the state of a location is solely an abstraction used in defining the LC model, and need not be implemented in any software/hardware layer of the computer system.)

There are several benefits that follow from the Location Consistency model due to the fact that it is based on a partial order execution semantics *e.g.*,

- The same memory consistency model can be applied in an end-to-end view of all levels of software and hardware in a computer system.
- All compiler optimizations and transformations (*e.g.*, instruction scheduling) can be performed safely by obeying the partial order constraints as usual, and without requiring all shared data be to treated as volatile variables.
- Interpretation, emulation, or on-the-fly translation of an instruction set architecture or virtual machine can also be performed safely by only obeying the partial order constraints.
- The LC model lends itself to using "multiple-owner" cache consistency protocols that can be more efficient and scalable than "single-owner" protocols. Lazy Release Consistency [12] and Data Merging [11] are examples of two multiple-owner protocols from past work. As outlined in section 4, we have recently gained some initial experience with a more relaxed multiple-owner cache consistency protocol that is also location consistent.

4 Cache Management under the LC model

In cache-based shared-memory multiprocessor systems, the management of the cache is a vital issue that has an impact on system performance. The presence of copies of the same location in multiple caches requires that these copies be managed in a way that does not violate the requirements of the underlying memory consistency model. This is known as the problem of *cache coherence*. Aggressive cache management schemes can exploit loose constraints on memory access ordering by reducing consistency-related cache-coherence traffic.

An interesting challenge is to exploit the semantics of the LC model and design a cache consistency protocol that is more efficient than existing cache consistency protocols. As discussed earlier, a primary difference between LC and other memory consistency models is the elimination of the memory coherence assumption. In most conventional cache-coherence systems, a single-owner protocol is employed: at any time there is a unique owner processor of a cached location. Such a cache protocol serializes memory operations to maintain unique ownership, often generating unnecessary invalidations and other unproductive cache-coherence traffic as a result.

At the time of completing this paper, we have been investigating a cache management mechanism which employs a *multiple-owner cache protocol*. We have already gained some initial experience from a pilot study conducted by Shamir Merali and Hisham Petry in their recently completed MS theses at McGill University [15,16]. Preliminary simulation results show the potential for 1.5x - 10x reductions in # cache misses using this more relaxed LC protocol may be possible for certain programs. The full development of a LC cache management protocol is the main initiative to be undertaken in the near future. The criteria which we will follow when designing and experimenting with different cache consistency protocols in this research are:

- The protocol should correctly implement the LC semantics.
- The protocol should be more efficient than existing cache consistency protocols for SC-derived consistency models.
- The protocol should have a potentially lower hardware cost in its realization.
- The protocol should facilitate software (compiler) optimization.

With a single-ownership cache protocol under most existing memory consistency models, there is unproductive cache traffic to main memory coherence, even when running a program that is free of data races. We believe (or conjecture) that, for a given memory system and a program, an efficient multiple-owner cache management protocol based on LC can be constructed which is "superior" in the sense that no other single-owner cache protocol can have "less" cache traffic to support the correct execution of the program.

5 Related Work

In this section, we briefly comment on a few other memory consistency models. Our discussion will focus on *relaxed* memory models which attempt to relax the restrictions on memory access ordering imposed on the original sequential consistency (SC) model to enhance performance.

In order to alleviate the SC performance limitations, designers have proposed models that guarantee the SC interface for a restricted set of programs, but allow optimizations to be applied safely. The looser models share the common intuition that if a program has enough synchronization then the program appears to execute sequentially consistently. In other words, the results of every run of a "properly synchronized" program are guaranteed to be consistent with some sequentially consistent run of the program. We present these relaxed consistency models in roughly the chronological order in which they appear in the literature: Release Consistency (1990) in section 5.1, Lazy Release Consistency (1992) in section 5.2, Entry Consistency (1993) in section 5.3, and Dag Consistency (1996) in section 5.4. (Our past work on Location Consistency [8,9] appeared in the 1994-1995 time frame.)

5.1 Release Consistency

The goal of Release Consistency (RC) is "to exploit additional information about shared accesses to develop a memory consistency model that allows for more efficient implementations" [10]. The RC model distinguishes between ordinary shared memory accesses and synchronization instructions, and further distinguishes between *acquire* and *release* synchronization operations.

For example, in the RC protocol, a *release* triggers a write of a shared variable and an *acquire* triggers the read of a shared variable. The purpose of a *release* is to inform other processes that all accesses that precede it (in program order) have completed. Similarly, the purpose of an *acquire* is to await such a signal from another processor before initiating any further accesses.

RC guarantees sequential consistency for a specific class of programs which it classifies as "properly labeled". That is, properly labeled programs are guaranteed to contain enough synchronization to appear sequentially consistent on RC. Intuitively, properly labeled programs for RC are those which contain enough synchronization (such as acquire and release) to eliminate all data-races among ordinary accesses. That is, a program is properly labeled if there is enough synchronization so that for all legal interleaving of accesses, pairs of conflicting ordinary accesses are separated by a *release-acquire* chain i.e., the program has no data races.

The RC condition has imposed certain limitations on how memory operations should be ordered. In the DASH implementation of RC [14], each write operation must be explicitly acknowledged to ensure atomiticity, to satisfy the memory coherence condition. In addition, the release operation contains an implicit "fence" operation which ensures that all previous memory operations must be completed before subsequent operations can be issued — thus no data races may occur in a properly labeled program.

5.2 Lazy Release Consistency

Lazy Release Consistency (LRC) can be viewed as an extension to RC aimed at reducing the number of messages and the amount of data exchanged in a distributed shared-memory system implemented in software [12]. The basic motivation for the algorithm is the observation that cross-processor consistency information only needs to be propagated at acquire synchronization points, at which point RC requires *all ordinary accesses that precede the corresponding release to be performed with respect to the acquiring processor*. While the *eager* counterparts of LRC[1] make modifications globally visible at the time of a *release*, LRC exploits the intuition that only the processor that *acquires* a variable needs to see all modifications that *precede* the *acquire*. So, whereas in an eager implementation, consistency broadcasts are made at *release* points and may involve all processors, in the lazy implementation, the broadcasts are made at the points of *acquire*, and involve only the acquiring and releasing processors.

---

[1] For instance, the RC implementation on Munin [6].

In the DSM implementation of LRC [12], coherence actions only take place at the acquire points, while release operations involve no coherence actions. It is found that both the number of messages as well as the amount of data exchanged are generally smaller for LRC than an eager implementation of RC.

5.3 Entry Consistency and the Midway System

Entry Consistency (EC) [4] can also be viewed as a further relaxed extension of RC. The basic difference is the following: whereas in RC, a synchronization object (variable) protects access to all shared data, in EC, there is an explicit correspondence between synchronization variables and the shared data they guard. As in LRC, modifications are propagated at an acquiring synchronization, but now, only the shared data that the synchronization variable guards is guaranteed to be consistent at that point. This correspondence between shared and synchronization data, which are implicit in the structure of a parallel program, is required by EC to be made explicit to the compiler and run-time system. An aggressive implementation can make use of this information to reduce the number of consistency messages (e.g. cache invalidations and/or updates) flowing across the system.

As in previous models, programs that include all the necessary labeling information, and have no data races (i.e., are "properly synchronized") observe a sequentially consistent shared memory. Measurements made on the implementation of EC on Midway [4] show that a program written for EC requires substantially fewer consistency transactions than stronger models such as RC. However, the caching protocol is still essentially based on a "single-ownership" model.

5.4 Dag Consistency

Blumofe et al [5] defined the *dag consistency* memory model for deterministic spawn-sync multithreaded programs in which dynamic program execution is modeled as a computation dag (directed acyclic graph) of nonsuspensive "threads". Threads are created as follows. Main program execution begins in a single thread. A spawn (fork) statement creates a procedure call in a new thread that can execute concurrently with the caller. A sync (join) statement terminates the executing thread and creates a new thread for the computation that follows the sync statement. This new thread must wait for all threads spawned by the previous thread to terminate before it can start execution. Each vertex in the computation dag corresponds to a distinct thread. An edge in the computation dag represents a partial order constraint *i.e.,* an edge exists between the caller thread and the callee thread in a spawn statement and between each previously spawned thread and the continuation thread in a sync statement. Thus, the computation dag defines a partial order on threads. The computation dag is an abstraction used in defining dag consistency, and is not actually computed at runtime.

The shared memory of a multithreaded computation is said to be dag-consistent if the following two conditions hold:

1. When thread i reads a memory location, it receives a value that was written by some thread j such that $i \not\prec j$ in the computation dag.
2. For any three threads i, j, k such that $i \prec j \prec k$, if k reads a location written by both i and j then the value read by thread k is not the one written by thread i.

These conditions for dag consistency appear to be identical to the conditions for location consistency defined in [8, 9] when applied to the special case of deterministic spawn-sync programs considered in [5]. For deterministic programs, the set of most recent writes for a location in the LC model will always either be an empty set or a singleton set.

6 Discussions

In this section, we provide a brief outline of our views on future research directions and a list of open research areas.

6.1 Research directions

We believe that the research directions which will offer a quantum leap in providing an end-to-end memory consistency model for the next generation of computer systems are those that break away from the shackles of memory coherence. The fruit of these research directions will be seen in robust compiler optimization techniques for explicitly parallel programs, new (multiple-owner) cache consistency protocols that provide more efficient and scalable support for shared-memory multiprocessing, and an end-to-end view of memory consistency that is as simple to articulate as the semantics of an assignment statement.

The two important grand challenge problems for today's research in the area of memory consistency are: a) How to build a shared-memory multiprocessor that is effective at a scale of a thousand processors or more? b) How to compile and optimize an explicitly parallel program so that its single-processor performance is comparable to that of highly optimized uniprocessor program?

The major obstacle in solving these problems is the memory coherence assumption in today's memory consistency model that must be obeyed by all levels (software and hardware) of a computer system. Memory coherence leads to the use of single-writer cache consistency protocols which (by their very nature) have limited scalability. Memory coherence also restricts reordering of instructions and severely limits the optimizations (*e.g.*, register allocation, instruction scheduling) that can safely be performed on an explicitly parallel program.

We believe that defining memory consistency in the context of a partial order program execution model will lead to a new and simple paradigm for thinking about parallel programs and their behavior on shared data. Furthermore, there is a natural synergy between the areas of compilers and computer architecture in implementing a new end-to-end view of memory consistency. However, a synergistic relationship will also be built with other aspects of computer systems that need a semantics for sharing data with applications (*e.g.,* operating systems, graphics, networks, etc.)

6.2 Some open research areas

In our opinion, all the following research areas are important but we list them in an estimated order of diminishing pay-offs:

- (a) Sound but simple specification of memory consistency models based on partial order execution semantics.
- (b) Compiler optimization of explicitly parallel programs with shared memory for the memory consistency models in (a). The biggest source of explicitly parallel programs today are multithreaded programs that use popular thread libraries such as pthreads and Java threads. These optimization techniques may be employed either in static compilation or dynamic code generation contexts.
- (c) Design and implementation of (more) scalable cache consistency protocols for shared-memory multiprocessors.
- (d) Comparative studies of the performance obtained by (b) and (c) with the corresponding performance for traditional memory consistency models.
- (f) Special studies of (b) and (c) for multithreaded architectures with hardware support for fine-grain context switching.

7 Conclusions

In this paper, we argued that it is important to take an end-to-end view of memory consistency that is applicable at all levels of software and hardware in a computer system. In particular, we believe that the semantics of the memory coherence assumption in current memory consistency models is not meaningful in an end-to-end view of computer systems, and that the memory coherence assumption also imposes serious limitations on compiler optimization of parallel programs and scalability of shared-memory multiprocessors

We proposed that memory consistency models instead be based on the partial order execution semantics of parallel programs without relying on the memory coherence assumption. Location Consistency and Dag Consistency are two specific examples of memory consistency models based on a partial order execution semantics.

Acknowledgment

The authors gratefully acknowledge the support of the the National Science Foundation (NSF) of the United States. and the Natural Sciences and Engineering Research Council (NSERC) of Canada.

References

1. *Proceedings of the 17th Annual International Symposium on Computer Architecture*, Seattle, Washington, May 1990.
2. Sarita V. Adve and Mark D. Hill, "Weak Ordering—A New Definition," in *Proceedings of the 17th Annual International Symposium on Computer Architecture*, Seattle, Washington, pp. 2–14, May 1990.
3. Sarita V. Adve and Mark D. Hill, "A Unified Formalization of Four Shared-Memory Models," *IEEE Transactions on Parallel and Distributed Systems*, pp. 613–624, June 1993.
4. B. Bershad, M. Zekauskas, and W. Sawdon, "The Midway Distributed Shared Memory System," in *Proceedings of the IEEE COMPCON*, 1993.
5. Robert D. Blumofe, Matteo Frigo, Christopher F. Joerg, Charles E. Leiserson, and Keith H. Randall, "An Analysis of Dag-Consistent Distributed Shared-Memory Algorithms," in *Proceedings of the 8th Annual ACM Symposium on Parallel Algorithms and Architectures*, Padua, Italy, pp. 297–308, June 1996.
6. J.B. Carter, J.K. Bennett, and W. Zwaenepoel, "Implementation and Performance of Munin," in *Proceedings of the 13 ACM Symposium on Operating System Principles*, pp. 152–164, 1991.
7. Michel Dubois, Christoph Scheurich, and Faye Briggs, "Memory Access Buffering in Multiprocessors," in *Proceedings of the 13th Annual International Symposium on Computer Architecture*, Tokyo, Japan, pp. 434–442, June 1986.
8. Guang R. Gao and Vivek Sarkar, "Location Consistency: Stepping Beyond the Barriers of Memory Coherence and Serializability," ACAPS Technical Memo 78, School of Computer Science, McGill University, Montréal, Québec, December 1994. In ftp://ftp-acaps.cs.mcgill.ca/pub/doc/memos.
9. Guang R. Gao and Vivek Sarkar, "Location Consistency: Stepping Beyond Memory Coherence Barrier," in *Proceedings of the 1995 International Conference on Parallel Processing*, vol. II, Oconomowoc, Wisconsin, pp. 73–76, August 1995.
10. Kourosh Gharachorloo, Daniel Lenoski, James Laudon, Phillip Gibbons, Anoop Gupta, and John Hennessy, "Memory Consistency and Event Ordering in Scalable Shared-Memory Multiprocessors," in *Proceedings of the 17th Annual International Symposium on Computer Architecture*, Seattle, Washington, pp. 15–26, May 1990.
11. Alan H. Karp and Vivek Sarkar, "Data Merging for Shared-Memory Multiprocessors," *Proceedings of the 26th Hawaii International Conference on System Sciences, Wailea, Hawaii, Volume I (Architecture)*, pp. 244–256, January 1993.
12. Pete Keleher, Alan L. Cox, and Willy Zwaenepoel, "Lazy Release Consistency for Software Distributed Shared Memory," in *Proceedings of the 19th Annual International Symposium on Computer Architecture*, Gold Coast, Australia, pp. 13–21, May 1992.
13. Leslie Lamport, "How to Make a Multiprocessor Computer That Correctly Executes Multiprocess Programs," *IEEE Transactions on Computers*, 28(9):690–691, September 1979.
14. Daniel Lenoski, Kourosh Gharachorloo, James Laudon, Anoop Gupta, John Hennessy, Mark Horowitz, and Monica Lam, "Design of Scalable Shared-Memory Multiprocessors: The DASH Approach," in *Digest of Papers, 35th IEEE Computer Society International Conference, COMPCON Spring '90*, San Francisco, California, pp. 62–67, February–March 1990.
15. Shamir Merali, "Designing and Implementing Memory Consistency Models for Shared-Memory Multiprocessors," Master's thesis, McGill University, Montréal, Québec, April 1996.

16. Hisham Petry, "Comparison of SC-Derived Memory Models and Location Consistency on Shared Memory Architectures," Master's thesis, McGill University, Montréal, Québec, July 1997.
17. J. H. Saltzer, D. P. Reed, and D. D. Clark, "End-To-End Arguments in System Design," *ACM Transactions on Computer Systems*, 2(4):277–288, November 1984.

High Performance Distributed Object Systems

Dennis Gannon

Department of Computer Science, Indiana University, Bloomington, IN 47401

Abstract. This paper will provide a survey of current work on object oriented tools and techniques for metacomputing systems. More specifically, we consider the problem of designing a software component architecture that extends the current emerging desktop object composition models to the domain of high performance networks and massively parallel compute servers.

1 Introduction

Successful software development must now live by a new set of rules: A drive toward more end-user programmability. A movement away from monolithic applications and toward systems grown from the integration of many smaller components. Increased expectations of radically greater functionality, delivered under radically shorter schedules. *Grady Booch [2]*

The design of the current generation of desktop software technology differs from past generations in one fundamental way. The new design paradigm states that applications should be built by composing "off the shelf" components, much the same way that hardware designers build systems from integrated circuits. Furthermore, these components may be distributed across a wide area network of compute and data servers. Each component is defined by the *public interface* that specifies its function as well as the protocols with which other components may communicate with it. An application program in this model becomes a dynamic network of communicating components. This basic *distributed object* design philosophy is having a profound impact on all aspects of information processing technology. We are already seeing a shift in the software industry toward investment in software components and away from hand-crafted, stand-along applications and within the industry a technology war is being waged over the design of the component composition architecture.

High performance computing will not be immune from this paradigm shift. More specifically, as our current and future Internet continues to scale in both size and bandwidth, it is not unrealistic to think about applications that might incorporate 10,000 active components that are distributed over that many compute hosts. Furthermore, pressure from the desktop software industry will compel us to integrate the applications that run on supercomputer systems into distributed problem solving environments that use object technology. Metacomputing systems consisting of MPP servers, advanced, networked instruments, database servers and gigabit networks will require a robust and scalable object model that support high performance application design.

This paper will sketch the current software design ideas that make up the commercial desktop software environment. We will then explore where these ideas scale well and are applicable to scientific and engineering applications and where they fail to meet the needs of high performance computing. We conclude with a brief survey of several approaches to high performance object systems now in use in research settings.

2 Basic Concepts and Examples

The basic ideas behind Object Oriented software design that are important for this paper are as follows.

- Data, and the functions that operate on it, should be bound together into *objects*. These objects are each instances of an abstract data type called a *class*. The data associated with an object are called *data members* or *attributes* and the functions that are associated with a class of objects are called *member functions*.
- *Interfaces* describe a set of functions that can be used to interact with a family of objects. Those classes of objects that respond to a particular interface are said to *implement* that interface. A class of objects may implement more than one interface.
- A new class of objects may be built from an existing class by adding new data attributes or member functions. Instances of the new class each contain an instance of the original (*parent*) class and thus, still implement the same interfaces as the parent. This process of extending one class to build another is called *inheritance*: the extended class is said to inherit from the original class. It is also possible for the extended class to override the definition of a member function from a class that it extends. Consequently, it is possible to make an extended class specialize or modify the behavior of the parent, i.e. it responds to the same functions, but the action is different.
- It is also possible to create a new interface definition by extending the definition of one or more other interfaces by simply adding new functions.

Consider the following example. Suppose we are designing a new aircraft and we wish to simulate the aerodynamics of the new vehicle. The components of the design application might include a CAD design database which represents the new structure in a special polygon based format. The aerodynamic simulation will require a flow solver that requires a 3-D grid of the exterior of the vehicle as input and generate a flow field as an output. Another component would transform the CAD database description into the grid structure. A visualization system could be used to display the flow field. The interface to the CAD database might look as follows. It has a single member function that takes the name of the design and returns a stream of polygons. To describe this we can use an Interface Description Language (IDL) which might take the form

```
interface CAD_DB{
    sequence<polygon> fetch_design(in string vehicle_name);
    };
```

where "sequence" is a wrapper object used to describe a stream of some given data type. The polygon-to-mesh generator interface may be described by

```
interface Mesh_Gen{
    grid transform(in sequence<polygon> design);
};
```

where *grid* is a special data type to describe a set of 3-D arrays. The solver would take a grid object a return a sequence of vectors to represent the flow field.

```
interface Flow_Solver{
    sequence<vector> solve(in grid the_grid);
 };
```

The visualization tool would need to be able to render both the vehicle design and the flow field.

```
interface visualizer{
   void render(in sequence<polygon> vehicle);
   void render(in sequence<vector> field);
};
```

The simulation application is then created by connecting instances of object whose classes implement these interface together in a network as shown in the figure below.

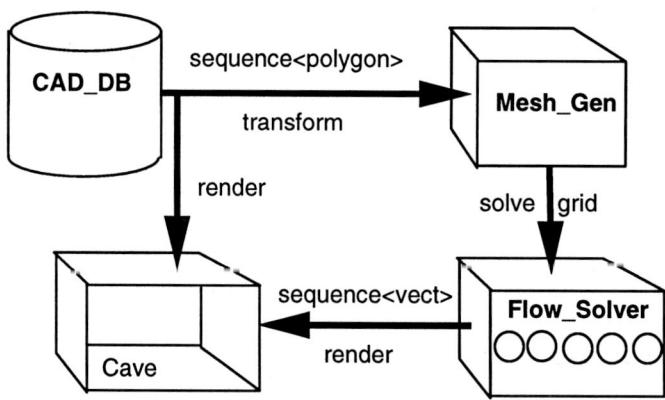

Fig. 1. The simulation application as a network of communicating components.

To accomplish the task of connecting the components together in this manner we must solve certain additional problems. First, the objects in the figure need to know about each other in order to be able to transmit the data and member function invocation as indicated. For example, the CAD database may be located in one city and the flow simulation may be running on a parallel processing system in another location. Furthermore, the visualization system may be an emmersive environment like a CAVE in another facility. A mechanism of *remote object references* must be created to allow one object know about another. Moreover, to obtain a remote object reference from the name or description of an object we need some type of directory service.

Second, our object interface descriptions need further refinement. In the tradition model of distributed computing, object represent *servers* and *client* programs make requests through member function calls and receive value results. In the application described here, rather than return a value result (which may be very large), we need an object to forward the result to one or more other components. There are several approaches to solving this problem. One method is to extend each component interface with application specific control methods that know about the overall application structure. For example, the CAD database can extended with attributes that represent the object references for the visualizer and mesh generator and a control method that only returns a value if there is an error. The standard mechanisms for dealing with errors is called *exception handling*. Our CAD control interface can be now described with the following IDL fragment.

```
enum Reasons { NO_SUCH_ITEM_IN_DB, VIS_FAILURE, MESH_GEN_FAILURE};
exception db_error{
    Reasons reason;
    };
interface CAD_Control: CAD_DB{
    attribute visualizer vis;
    attribute Mesh_Gen mg;
    void send_design(in string vehicle_name) raises(db_error);
    };
```

An implementation class for this interface would include the code for the *send_design* method which may take the form shown below.

```
void CAD_Control_Impl::send_design(char * vehicle_name){
  try{
    sequence<polygon> design = fetch_design(vehicle_name);
    }
   catch(Exception e){ throw db_error(NO_SUCH_ITEM_IN_DB) };
   try{ vis->render(design); }
    catch(Exception e){ throw db_error(VIS_FAILURE); }
   try{ vis->transform(design); }
    catch(Exception e){ throw db_error(MESH_GEN_FAILURE); }
}
```

The problem with this solution is that the implementation of the component must now know about the details of the interfaces of the other components that it will communicate with. A superior solution would be to use a visual programming system that allows a user to draw the application component graph and append the interconnection rules and exceptions with a scripting language. NAG Explorer [14]has some of these properties and scripting languages like Python [11] have been used control distributed object systems. Unfortunately NAG's type system is not very rich and it is not clear that graphical composition tools will scale to the networks of more than a few dozen objects. We would also like to describe networks that are dynamic and are able to incorporate new component resources "on the fly" as they are discovered.

A third problem is that our system must know how to transmit application specific objects over the network. This is called the *serialization* problem and a solution to it is an automatic protocol for packing and unpacking the components of data structures so that they may be reliably transmitted between different computer architectures in a heterogeneous environment.

3 The Components of an OO Metacomputing Model

The application in the preceding section only illustrates one distributed object design pattern. It also illustrates only a small subset of the important problems that must be addressed by a complete system. Most good object oriented systems must also include central mechanisms for the following additional problems.

- *Persistence, and storage management.* It is often the case that an object needs to be "frozen" so that its state is preserved on some storage device and then "thawed" later when it is needed again. An system with the ability to do this to objects is said to support persistence and it is closely related to serializability as described above.
- *Process and thread management.* Most instances of distributed objects are encapsulated within their own process, but it is possible that we may wish more than one object may belong to the same process. Also we may want the ability to have an object respond to different requests for the same method invocation concurrently. To do this the object system must be integrated with a thread system. There are many reasons that this can be a challenging problem [1].
- *Object distribution and object migration.* An object implementation may itself be distributed. This is important in the case of parallel programming, but it can also happen when part of a particular interface may need to be implemented on one system and another part on another. In addition, it is often important for an object to be able to migrate from one host to another. For example, when the first hosts compute resources become limiting, it is nice to be able to move the object to a second, more powerful host.
- *Networking.* For a good object system, the network protocol layer is transparent to the user. However, it is essential for high performance application to take advantage of high bandwidth connections when they are available. Furthermore, if an object implementation resides on a parallel compute server

and it is communicating with a similar object, large data sets like sequences should be moved in parallel if possible. Pardis [6] is one example system that supports this feature. In addition, support in the object model for multicast communication is very important. In addition, it should be possible to specify Quality of Service (QOS) requirements at the interface level.
- *Event Logging.* Debugging distributed systems is very hard. It is essential to have a mechanism that will allow the events associated with a set of distributed interactions to be logged in a way that will help identify what happened and when.
- *Fault tolerance.* An exception handling mechanism is the first step toward building reliable systems, but it falls far short of providing a mechanism where failure can be tolerated in a reliable manner. The system must be able to automatically restart applications and rollback transactions to a previous known state.
- *Authentication and Security.* Authentication allows us to identify which applications and users are allowed to access system components. Security means that these interactions can be accomplished with safety for the data as well as the implementations. It is an issue that goes far beyond the domain of the object system, but the object system must provide a way to allow the user access to the metacomputing authentication and security tools that are available.
- *Beyond Client/Server.* For high performance computation it is essential that the future distributed object systems support a greater variety of models than simple client/server schemes. As illustrated by the example in section 2, there are paradigms that include peer-to-peer object networks. In the future we can imagine massive networks of components and software agents that work without centralized control and dynamically respond to changing loads and requirements.
- *Support for Parallelism.* Beyond multi-treaded applications are those that involve the concurrent activity of many components. An object systems must allow both asynchronous as well as synchronous method calls. In addition, multicast communication and collective synchronization are essential for supporting parallel operation on very large numbers of concurrently executing objects.

4 The Current Desktop Distributed Object Models

Different object oriented programming languages and systems implement these concepts in different ways. For example, some combine the concept of class and interface. Others allow single inheritance for classes but not multiple inheritance. However, an distributed object system requires more than a programming language to support it.

The current commercial market for component systems is supported by three different distributed object models.

CORBA. The Object Management Group (OMG) is a consortium of several hundred institutions that have come together to define a platform independent standard for distributed object systems called the Common Object Request Broker Architecture (CORBA). In its current version, CORBA consists of a very large collection of technology specifications and it is still growing. CORBA IDL is an extension of the OSF DCE system IDL and was used in the examples in section 2.

OLE-DCOM-ActiveX. Microsoft has proposed a standard for Windows NT programming that has three components. Object Linking and Embedding (OLE) and ActiveX are the names of systems that provide user level composable components. DCOM is the Distributed Component Object Model.

Java RMI. Java brings something new to the table. By providing a common, virtual machine, execution environment for all machines, Java is able to create objects that are completely serializable. The Java Remote Method Invocation model use Java itself as an IDL to compile proxies for building remote object references. Java has thread support and security mechanisms built-in the language and virtual machine.

Of these three, CORBA has the greatest support for the features we need in a high performance, metacomputing object system. In the table below we list each system and grade it on its level of support for the feature list of the previous section.

In this table we have also listed three experimental systems that are designed to support parallelism and high performance metacomputing environments.

Pardis. Pardis [6] is an experimental extension of CORBA that supports a greater collection of method invocation methods, network support, distributed objects and parallel programming tools. Though it is designed as an extension of CORBA it does not yet support the full suite of CORBA services.

CC++ and HPC++. Composition C++ [7] (CC++) is an extension of C++ that supports multi-threaded computation, distributed objects and remote method invocation. HPC++ [8] is a collection of libraries and compiler technologies that provide similar features. Both HPC++ and CC++ are based on the Nexus run time system [12] and will operate in the Globus [13] metacomputing environment.

Legion. Legion [5] is a complete Metacomputing infrastructure that is based on the Mentat object model. It was designed from the ground up to support parallelism and scaling to many thousands of objects.

It should be noted that other experimental distributed object models exist that show great promise, but not all address the issues of parallelism and scale we are considering for high performance, metacomputing applications. In the complete version of this paper we will devote more attention to some of the important ideas they introduce.

System	Persist	Thread	Dist	Net	Event	Fault	Auth	Beyond C/S	Parallelism
CORBA	X	X				p		p	p
DCOM-ActiveX	X	X					p	p	
Java RMI	X	X	p	p			p	p	
PARDIS		X		X				X	X
CC++/HPC++	p	X	X	p			X	X	X
Legion	X	p		X	p	X	X	X	X

Table 1. Levels of support for distributed object system features for several commercial and experimental systems.

5 Conclusion

In this short, extended abstract we have sketched some of the issues that are important if we are to build a bridge between high performance parallel computing and the world of desktop object systems. As we have seen from the discussion, none of the current commercial distributed object environments meet the needs of speed, flexibility and scale required for large scale scientific and engineering applications. Furthermore, it is not likely that the economics of the commercial marketplace will be driven by the needs of high-end computing very soon. But if distributed object systems that address the issues of high performance are built, they will need to maintain a backward compatibility with the desktop environments. This form of interoperability will be essential for the design of integrated problem solving environments.

There are several sets of difficult problems that most be solved. First, we must find the right model of parallel distributed object system that breaks the mold of client server computing. It should be possible to compose large numbers of object components into hierarchies that allow us to manage the complexities of scale while still allowing for the greatest freedom in expressing and exploiting the natural parallelism in applications. Second, we must be able to design application components in a way that is independent from the context in which they execute. In other words, the model for component composition should be orthogonal to the model of component design. Third, our object systems must be adaptive. Component objects need to be able to migrate to appropriate execution environments and it should be possible to add new components to an application dynamically at runtime. As described in the preceding sections, there are a host of other technology problems that must be address.

However, this topic is important enough that we can expect a great deal of research activity in this area over the next few years. It will be interesting to see which solutions emerge.

References

1. Satoshi Matsuoka and Akinori Yonezawa, "Analysis of Inheritance Anomaly in Object-Oriented Concurrent Programming Languages," Research Directions in Concurrent Object-Oriented Programming, G. Agha, P. Wegner and A. Yonezawa (Ed.), MIT Press, 1993, pp. 107-150.
2. Grady Booch, "Object Solutions: Managing the Object-Oriented Project", Addison-Wesley, 1996.
3. JavaSoft, "RMI" in *The JDK 1.1 Specification*, 1997, see http:// javasoft.com/ products/ jdk/1.1/ docs/guide/ rmi/index.html.
4. Ken Arnold and James Gosling. *The Java Programming Language*. Addison Wesley, 1996.
5. Michael J. Lewis, Andrew Grimshaw, "The Core Legion Object Model", Proceedings of the Fifth IEEE International Symposium on High Performance Distributed Computing, IEEE Computer Society Press, Los Alamitos, California, August 1996.
6. Kate Keahey and Dennis Gannon, "PARDIS: A Parallel Approach to CORBA", 6th IEEE International Symposium on High Performance Distributed Computation, August 1997.
7. K. Mani Chandy and Carl Kesselman, "CC++: A Declarative Concurrent Object-Oriented Programming Notation", In *Research Directions in Concurrent Object Oriented Programming*, G. Agha, P. Wegner and A. Yonezawa (Ed.), MIT Press, 1993.
8. D. Gannon, P. Beckman, E. Johnson, E. and T. Green" "Chapter 3: HPC++ and the HPC++Lib Toolkit" in *Compilation Issues on Distributed Memory Systems*, Springer Verlag, 1997.
9. David Chappell, "Understanding ActiveX and OLE", Microsoft Press, 1997.
10. Randy Otte, Paul Patrick and Mark Roy, "Understanding CORBA", Prentice Hall, 1996.
11. M. Lutz, "Programming Python", O'Reilly and Associates, 1996.
12. I. Foster, C. Kesselman, S. Tuecke, "The Nexus Approach to Integrating Multithreading and Communication", J. Parallel and Distributed Computing, 37:70–82, 1996.
13. I. Foster, C. Kesselman, "Globus: A Metacomputing Infrastructure Toolkit." International Journal of Supercomputer Applications (to appear)
14. IRIS Explorer, see http://www.nag.co.uk/Welcome_IEC.html.

Instruction Cache Prefetching
Using Multilevel Branch Prediction

Alexander V. Veidenbaum

Dept. Of Electrical Engineering and Computer Science

University of Illinois at Chicago

alexv@eecs.uic.edu

Abstract

This paper presents an instruction cache prefetching mechanism capable of prefetching past branches in multiple-issue processors. Such processors at high clock rates often use small instruction caches which have significant miss rates. Prefetching from secondary cache can hide the instruction cache miss penalties but only if initiated sufficiently far ahead of the current program counter. Existing instruction cache prefetching methods are strictly sequential and cannot do that due to their inability to prefetch past branches. By keeping branch history and branch target addresses we predict a future PC several branches past the current branch. We describe a possible prefetching architecture and evaluate its accuracy, the impact of the instruction prefetching on performance, and its interaction with sequential prefetching. For a 4-issue processor and a cache architecture patterned after the DEC Alpha-21164 we show that our prefetching unit can be more effective than sequential prefetching. The two types of prefetching eliminate different types of misses and thus can be effectively combined to achieve better performance.

1. Introduction

Instruction-level parallelism is one of the main factors allowing the high performance delivered by state-of-the-art processors. Such a processor is designed to issue and execute K instructions every cycle. K=4 in today's typical processor. A processor organization consists of an instruction fetch unit followed by a decode and execution units. The fetch unit's task is to supply the decode unit with K instructions every cycle. This is accomplished by having a wide instruction cache (I-cache) supplying $\alpha*K$ instruction words every α cycles, where α is typically between one and three. This is a difficult task for a typical processor with a clock speed of 200+MHz and more so for a high-end processor with a 600MHz clock. It will become even more difficult with clock rates reaching 1GHz and beyond.

Two major problems limiting the instruction issue width K are branches and instruction cache misses. The former has received a lot of attention. Branch prediction has been used to allow execution to speculatively continue while a conditional branch is resolved. Overall, branch prediction has been very successful although conditional branches remain a major problem in further increasing the instruction issue rate. Branch target buffers and call/return stacks have been used to tackle the other types of branches, but again can be further improved.

The second problem, I-cache misses, is also a difficult one but with fewer solutions proposed for solving it. One brute-force solution is to increase the primary I-cache size. This may not always be possible or desirable for an on-chip I-cache because the cycle time of the cache is determined by its size [JoWi94] and is a major factor in determining the CPU clock speed. This limits a typical I-cache size to between 8 and 32KB in the current generation. Fast processors, like the DEC Alpha-21164 [ERPR95], are bound to have small I-caches in this range and thus higher miss rates. The Alpha-21164 8KB I-cache has miss rates as high as 7.4% for SPEC92 codes, as reported in [HePa96]. A cache hierarchy is used to reduce the miss penalties. For example, the DEC Alpha-21164 has a unified on-chip second-level cache (96KB) and an off-chip third-level cache (typically 2MB). A large L2 cache has a low (instruction fetch) miss rate, typically well under 1%.

The problem of high primary I-cache miss rates has been addressed via sequential instruction prefetching in the past. Sequential prefetching typically starts with an I-cache miss address and stops when a branch instruction or an I-cache hit are encountered. The key to successful prefetching is to issue a predicted future instruction addresses to the L2 cache at least T cycles before they are needed, where T is the L2 cache latency. Stated in a different way, instruction fetch prediction needs a lookahead of T*K instructions. T*K=24 for the 21164, which is representative of the L2 miss service time and issue width of a modern processor, and sequential prefetching cannot get far enough ahead of the fetch unit given average branch probabilities.

This paper presents an I-cache prefetch mechanism with a longer look-ahead. The term "prefetching" is used here to mean fetching from a second-level cache to the I-cache ahead of the current program counter (PC). Given typical branch frequencies in programs this calls for predicting a prefetch address across several branches. We use a "multilevel" branch prediction to get around this obstacle and predict and prefetch branch target instructions before they are requested. Sequential prefetching is also studied and its effectiveness and relationship with branch target prefetching is explored. Only blocking I-caches are considered, although lockup-free caches can help to combine branch prediction and prefetching.

This paper makes three major contributions. First, it defines a prediction and prefetching mechanism which can predict a likely future PC over several intervening branches and initiate the prefetch. Second, we analyze several benchmarks to understand the importance of different miss types, the predictor accuracy, and the effect on miss rate and CPU time. Third, we show the complimentary nature of sequential prefetching and the branch target prefetching and combine the two for best results. Our branch target prefetching gets ahead of the fetch unit while sequential prefetching initiates but often does not complete many of its requests in time.

2. The Approach

Our mechanism to predict a prefetch address with a sufficient time to complete the prefetch divides instruction addresses in two classes. First is a class of sequential addresses defined as M[PC + δ], where the range of δ addresses contains no transfer of control instructions. Sequential prefetching for this class has been widely used utilizing an instruction prefetch buffer. The second class are addresses of branch targets to which a transfer can occur from a given branch. Conditional branches, unconditional branches, and call/returns will be treated in the same since all require a new prefetching path.

For the second class, a branch has to trigger prefetching along one of the two possible paths as specified by a branch predictor. In this case an address of the branch target needs to be predicted in addition to a taken/not taken prediction. For an even longer lookahead several future branches need to be predicted and one of the possible target addresses predicted. This has been called multilevel branch prediction [YMP93]. In other words, if a branch B_i is currently executed we would like to predict branch outcome and target addresses for branches B_{i+1} through B_{i+K} that will follow B_i in the dynamic execution sequence. K is the lookahead distance in the number of branches past the current PC, e.g. K=3 predicts a target of a branch 3 (dynamic) branches past the current PC. K=0 is a standard branch prediction.

The approach combines the ideas of branch prediction and a BTB in the following way. The BTB concept is extended to look ahead K branches and return a prediction for the next prefetch address. We will call this a multilevel BTB (mBTB). A mBTB lookup is performed when a current instruction is a branch using its PC. The mBTB returns 2^K possible branch targets. A 2-bit history counter is stored with each target and is used to keep track of prediction success and select the target on lookup. The saturating up/down counters are used to select the most frequently encountered target as the most likely. A counter of a correctly predicted target is incremented while all other counters are decremented. Note that for K=0 our approach reverts to a standard BTB. A Branch History Register (BHR) maintains taken/not taken status, PC, and target address of the last K executed branch instructions. The history and target addresses in the BHR are used to update the mBTB. A new taken target, that of B_i, is added to a mBTB entry pointed to by the PC of B_{i-K} in the BHR.

I-cache prefetching architecture consists of K sequential prefetch buffers, each holding one or more cache lines. A prefetch buffer P_j is used to prefetch branch B_i such that (i mod K) = j. The prefetch buffers are thus used in a circular fashion. The next prefetch buffer in sequence is allocated and prefetching with a K-branch lookahead is initiated every time a branch is executed. Prefetch unit performs an I-cache lookup before issuing an address to the L2 cache. The K prefetch buffers are looked up in parallel with the I-cache and can return a line directly to the fetch unit.

The hardware complexity of the mBTB is $(2^K +1)*\log(Addr\_size)$ and grows rapidly with K. K also affects the latency: an associative lookup of K buffers needs to be performed in parallel with the I-cache lookup and one of the K+1 resutls selected. To keep the complexity down and table access time low only taken branch targets are predicted and stored in the BTB. This also relies on not taken branches being picked up by the sequential prefetching. Thus to predict over two branches, for instance, requires only 4 addresses and 4 two-bit counters. Multilevel branch prediction based prefetching (MLBP) can be combined with sequential prefetching which can pick up some of the fall-through paths, for instance by using multi-line prefetch buffers P_j. Overall, every time a branch is encountered by the CPU a new branch target will be prefetched.

3. Multilevel Branch Target Prediction Hardware

The number of target addresses to be associated with a branch address depends on the prediction level or lookahead. In general, one would expect to get a lower prediction accuracy with more levels, in addition to the higher implementation cost. We selected a two-level branch predicator in this work because of its natural balance between branch frequencies, average number of consecutive instructions between branches, and prediction accuracy. However, we will also investigate the 1- and 3-level prediction.

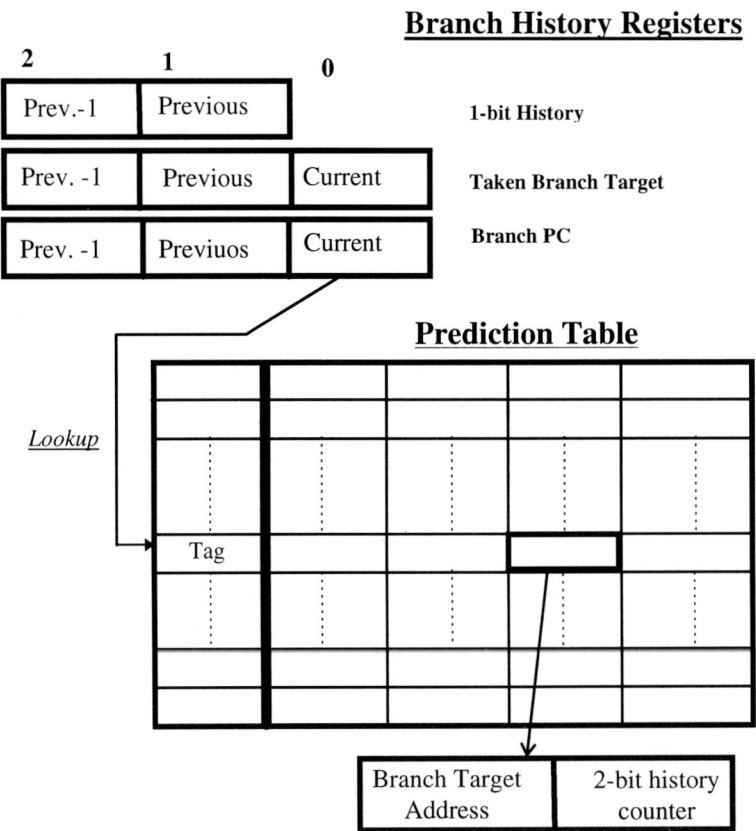

Figure 1. Direct-Mapped Two-Level Predictor Organization

Figure 1 shows a two-level predictor with a direct-mapped implementation. The fully-associative organization will also be investigated. The predictor consists of a Branch History Register (BHR), a Predictor Table (PT), and associated control logic (not shown). BHR holds PCs, target addresses, and taken/not taken history of previous K=2 branches plus the current branch. It is shifted left on each branch with current branch info shifting in. The information about the current branch (curr), the previous branch (prev), and the branch before the previous branch (prev.-1) are held in BHR positions 0, 1, and 2, respectively. 1-bit History, a 2-bit sub-register of BHT, holds the taken status of the previous two branches (global branch history). A PT entry holds four target address/2-bit "saturating" pairs for each possible branch target and a cache tag accessed with low-order bits of instruction address.

The predictor supports two operations: a lookup and an change/update. The lookup is performed when a new prediction is needed, i.e. when the current instruction is a branch. In this case the PC is used to access the PT entry, the counter with a maximum value among the four counters is identified, and the target address associated with the counter is returned as the prediction. On update, the PT entry is determined by the PC in BHR_2 and the addr/ctr pair in a slot pointed to by the $BHR_{2,1}$ History Bits is changed or updated. The pair is changed to the current PC and the counter set to a selected initial value if the target address is new, otherwise the counter is simply incremented while the other three counters in this entry are decremented.

4. System Organization

The focus of this paper is the instruction fetch logic of a processor and its I-cache. Either a static or a dynamic execution unit can follow the fetch unit. Ideal instruction issue logic, execution units, and primary D-cache are assumed and their stalls are not modeled in any detail because we are primarily interested in the effect of instruction fetch logic on availability of new instructions. To anchor our system in reality and support the claims of very high clock rates we base the processor on the DEC Alpha 21164. This means that the basic pipeline, its behavior and timing follow that of 21164. The I-cache organization and interface to on-chip, pipelined L2 cache, the branch predictor, and the sequential prefetcher are based on the description in [RPPR95]. Of course, our model is only an approximation. The L2 cache is assumed to have an extremely low miss rate and is modeled with simple timing in the first approximation.

The general system organization we study is shown in Figure 2. Four variants of this architecture are modeled to study instruction prefetching. An architecture may thus omit a particular prefetching unit, such as a stream buffer or MLBP prefetcher. The various system units are described first, followed by the description of the four systems.

4.1 Instruction Cache Organization

The instruction cache is a direct-mapped 8KB cache with 32B blocks and a 2-cycle latency. Thus a cache block contains eight instructions. A block is fetched into two 16Byte staging buffers for access by the fetch unit. Blocks have to aligned and a 32B fetch cannot cross cache block boundaries.

For comparison, we also present some results for 16KB I-caches, 2-way set associativity, and with 64B blocks. The 2-way set associative cache uses a random replacement policy.

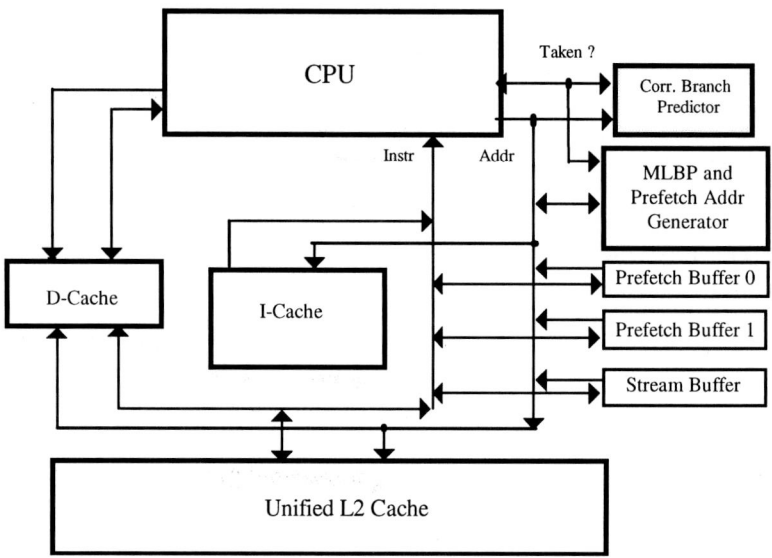

Figure 2. Overall System Organization

4.2 Processor Organization

The processor is a quad-issue superscalar processor. Every cycle an aligned 16-Byte set of four instructions is accessed from the I-cache staging buffer. The processor can issue a maximum four instructions per cycle. This corresponds to half a cache block and, in any cycle, the instruction issue will not cross the half block boundary.
The following are the only stages in the instruction pipeline we model:
S0 - cache access, delivers four instructions plus information for decoding and slotting
S1 - branch decoding, branch prediction, next PC computation
S2 - instruction slotting to execution units
S3 - instruction issue/register access
S4 - execute stage I
During S0 the instruction cache returns a naturally aligned block of four instructions along with necessary information for instruction decoding and slotting. During S1, the CPU decodes the four instructions, predicts branch outcome if one of the instructions is a branch, and uses branch prediction to generate the next I-cache half-block address to

be used. If one of the instructions is a branch and it is predicted taken the rest of the block is not executed. There is a 1-cycle stall for any taken branch to access the new block. The instructions flow through **S2** and **S3** without any stalls. The conditional branch outcome is known in **S4** resulting in a 5-cycle branch misprediction penalty.

4.3 Instruction Fetch Unit

The fetch unit issues a new address to the I-cache every other cycle. It also generates the next half-block address as either PC+16 or PC+Branch_displacement if there is a taken branch in the current half-block. The current half-block comes from one of the 2 staging buffers in the I-cache. If a half block contains a branch and the branch is predicted taken, the sequential instruction fetch is stalled for one cycle while the new PC is requested from I-cache. Note that a dedicated branch predictor is used here only for fetching from the I-cache and *not* for I-cache prefetching.

4.4 Branch Predictor

A (2,2) correlated branch predictor is used consisting of a 2-bit branch history register and a table of four 2-bit saturating counters in each entry. The table size is 1K entries. The low-order bits of the branch address are used to access the table for both updating and predicting. One of the four 2-bit counters in the accessed entry is selected based on the taken/not taken history of the previous two branches. The most significant bit of the selected counter predicts the direction. A counter is incremented or decremented based on the current branch outcome which is also shifted into the branch history register.

4.5 L2 Cache and Memory

A second-level cache with a 0% instruction miss rate is assumed. The L2 cache access is pipelined and has a latency of 6 cycles. A new fetch or prefetch can be issued to it every other cycle. In our system 3 units may simultaneously attempt to issue a request to the L2 cache: the instruction fetch, the sequential prefetch, and the MLBP-based prefetch units. An arbiter selects one of these with the following priorities: instruction fetch, MLBP prefetcher, sequential prefetcher, and stalls the other units.

4.6 Sequential Instruction Prefetcher

A sequential instruction prefetcher (or an instruction stream buffer) consists of an address register, an incrementor, and one 32Byte cache line buffer. The stream buffer is accessed in parallel with I-cache in one cycle. The stream buffer initiates a prefetch on an I-cache miss and stops when a branch or another cache miss are encountered. On a stream buffer hit the line is loaded into the I-cache.

4.7 Multilevel Branch Predictor

In this work, we study several different implementations of multilevel branch predictor. Multilevel branch predictor consists of a table in which each entry contains 2^{K-1} branch target addresses and frequency counters . The predictor receives the current branch address, branch direction, and branch target address from CPU, updates its internal prediction table, and predicts branch targets for prefetching. The direct-mapped and set-associative table organizations will be studied.

4.8 System Organizations under study

The effect of MLBP-based prefetching and its interaction with sequential prefetching are studied by analyzing performance of four systems described below. These systems add MLBP and/or sequential prefetch units to the base system and allow the miss rate and CPU time changes to be observed.

- Baseline Architecture (B)
  This architecture performs no prefetching and models a simple I-cache. It is used to as the basis for comparing the improvement from prefetching.
- Baseline plus Instruction Stream Buffer (BI)
  This architecture, adds an instruction stream buffer in parallel with the instruction cache. A CPU address is issued to both I-cache and stream buffer. This stream buffer targets sequential instruction prefetching.
- Baseline plus MLBP-based Prefetch Unit (BP)
  This architecture, adds a 2-level MLBP-based prefetch address generator and two 1-line buffers to store prefetched lines to the instruction cache. This architecture attempts to generate a prefetch request for *every* branch, conditional or unconditional, using an earlier branch as a trigger. It targets branch targets for prefetching.
  The two buffers are used cyclically and a new block overwrites the oldest block. A prefetch address is checked against the I-cache and then issued to the L2 cache. A CPU address is issued to both I-cache and the two prefetch buffers. If an address is present in a prefetch buffer but not in instruction cache, the block is loaded into the I-cache.
- Baseline plus Instruction Stream Buffer plus Prefetch Buffer (BIP)

  This architecture combines the I-cache with the two types of prefetch units. Figure 2 shows this system organization. A CPU addresses is checked in all three units. The instruction cache is loaded if a block is found in either the stream buffer or the MLBP-based buffers. A true instruction cache miss is considered to occur when a requested block is not present in any of the three units. The stream buffer will issue a prefetch for the next sequential block when either a true instruction cache miss occurs or when a hit occurs in either the stream buffer itself or the MLBP-based buffers.

5. Experimental methodology

We use trace-driven simulation to evaluate the effect of MLBP-based prefetching. Six benchmarks are used: five SPEC92 benchmarks with high I-cache miss rates and a widely-used UNIX data base manager. The SPEC benchmarks are compiled on an SGI system using SPEC scripts. Pixie software [CHKW86] is used to generate an instruction trace. The data base manager benchmark is a 10-minute sample of one data base manager process. This process is one of thirty such processes simultaneously executing. It represents several transactions of tpmC benchmark. We could not trace this process ourselves and relied on a trace supplied by others.

The programs were compiled and traced using used MIPS-I instruction set, compiler optimizations (-O3 flag) and statically linked libraries (non-shared flag). The one major deviation from the SPEC scripts was in compilation and tracing of gcc. We used only one copy of input files instead of five and "merged" these files into a single C program to avoid multiple startups. Both gcc and cc1 compilation were traced. Table 1 shows the basic benchmark statistics.

Program Name	Instruction Counts (Millions)	Percentage of Conditional Branches	Percentage of Unconditional Branches
doduc	1,350	7.312	0.993
fpppp	2,139	0.945	0.115
gcc	477	2.595	8.823
sc	72	19.284	1.279
xlisp	1,179	14.500	4.519
dbm	46	13.3	2.4

Table 1. Benchmark statistics.

6 Prediction Accuracy

Multi-level branch prediction accuracy is key to prefetching. We start by examining a correlated branch predictor used for "regular" branch prediction and the effect of the table size on its performance. Next, a direct-mapped 2-level branch predictor is studied while varying the size of its table, followed by a fully-associative predictor. The size of a direct-mapped table ranges from 32 to 4K entries and from 64 to 512 entries for a fully-associative table. The latter are kept smaller to maintain approximately the same access time since associative lookup is slower. Finally, the 1- and 3-level predictors are investigated in addition to the 2-level predictor.

6.1 Correlated Branch Predictor Accuracy

The effect of table size on prediction accuracy for each benchmark program is summarized in
Figure 3. The prediction accuracy reaches 90% and above for all benchmarks once a table size of around 1K entries is reached. A 1K entry predictor is used for CPU branch prediction in the rest of experiments.

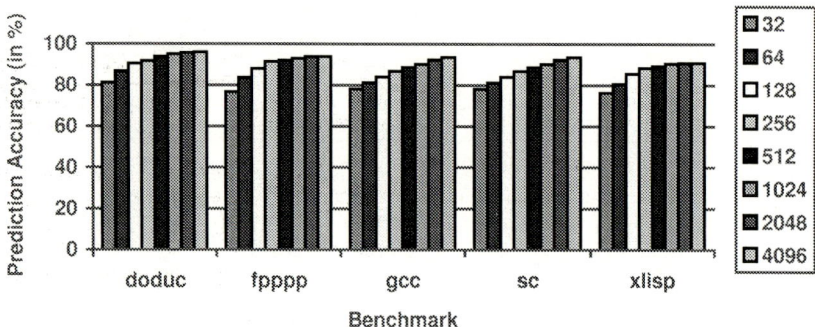

Figure 3. Prediction accuracy for different size correlated predictors

6.2 Direct-mapped multi-level prediction

Figure 4 summarizes the overall prediction accuracy of selected benchmark programs relative to the total number of taken branches (since only taken branches are predicted). The prediction accuracy increases smoothly with the increase in table sizes in most cases and ends up in the range of 55 and 85% for all program except gcc. The problem with gcc is a large number of branches based on a target address in a register which makes it hard to predict based on the branch address. While the results are not as high as for correlated branch prediction (Figure 3), they may be sufficient for prefetching since there is a high probability of finding data in the cache.

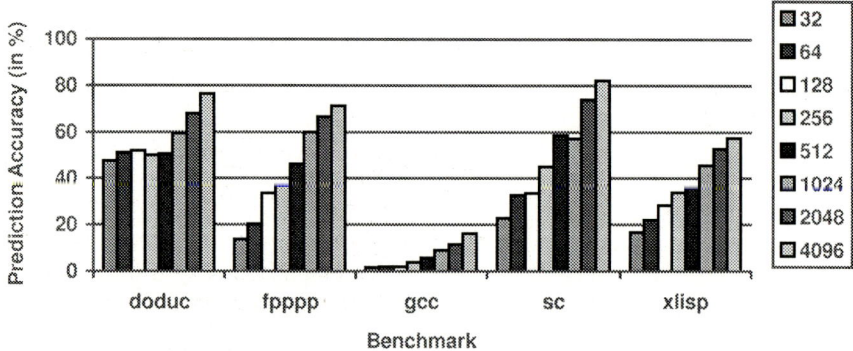

Figure 4 Direct-mapped prediction accuracy for various table sizes

Prediction accuracy for 1- and 3-level direct mapped predictors is not shown to save space. It generally follows the pattern seen for the fully-associative case below.

6.3 Fully-associative prediction

Next we analyze the performance of a fully-associative predictor with table sizes of 64, 256 and 512, and prediction levels of 1, 2 and 3. All entries in the mBTB's fully associative table are compared in parallel to the address, just as in a cache or standard BTB. The prediction accuracy for the fully-associative predictor is shown in Figure 5 as a function of table size and predictor level (see size-level caption). It is clear that prediction accuracy drops with decrease in table size and increase in the number of levels. Overall, however, a 256-entry, 2-level predictor does quite well and the results are competitive with larger, direct-mapped predictors.

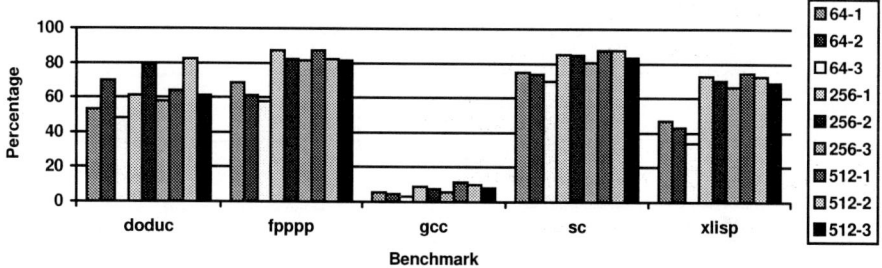

Figure 5 Fully-associative prediction accuracy for 1,2,3-level prediction

7 Prefetching Results

As shown above, multi-level branch prediction works fairly well in predicting the branch target addresses. Next we use MLBP to provide an address prediction which will be used to initiate instruction prefetch. We start by simulating several I-cache organizations to assess the baseline performance, followed by the I-cache prefetching variants described above. Finally, the effect of instruction fetch on execution time and issue rate is investigated.

7.1 Instruction Cache Miss Rates

I-cache sizes of 8 and 16KBytes and line sizes of 32 and 64Bytes are studied. The size and associativity of the I-cache are kept small to guarantee high clock rates. Table 2 shows the results. The miss rates range from below 1 to 10%. While doubling the associativity or line size leads to a miss rate reduction, overall the miss rates remain unacceptably high.

The distribution of cache misses caused by branch targets is shown in Table 3. These misses will be targeted by MLBP prefetching. It shows that the branch target misses can be a very high fraction of I-cache misses. But they are not very sensitive to associativity and cache size, except for fpppp. Fpppp has approximately 2% of instruction cache misses caused by branch targets.

Cache Size	Line Size	Associativity	Miss Rate	Cache Size	Line Size	Associativity	Miss Rate
GCC							
8K	32	1	7.26	16K	32	1	4.86
8K	32	2	6.65	16K	32	2	4.87
8K	64	1	3.89	16K	64	1	2.59
8K	64	2	3.57	16K	64	2	2.59
XLISP							
8K	32	1	1.24	16K	32	1	1.15
8K	32	2	0.40	16K	32	2	0.12
8K	64	1	1.05	16K	64	1	0.73
8K	64	2	0.36	16K	64	2	0.28
SC							
8K	32	1	1.71	16K	32	1	1.07
8K	32	2	1.14	16K	32	2	0.58
8K	64	1	1.27	16K	64	1	0.80
8K	64	2	0.81	16K	64	2	0.42
FPPPP							
8K	32	1	10.51	16K	32	1	6.92
8K	32	2	10.30	16K	32	2	6.66
8K	64	1	5.35	16K	64	1	3.53
8K	64	2	5.24	16K	64	2	3.40
DODUC							
8K	32	1	3.36	16K	32	1	1.48
8K	32	2	2.75	16K	32	2	1.31
8K	64	1	1.96	16K	64	1	0.85
8K	64	2	1.59	16K	64	2	0.73
DBM							
8K	32	1	10.9	16K	32	1	9.9
8K	32	2	10.7	16K	32	2	9.5

Table 2 Instruction cache miss rates

Organization	doduc	fpppp	gcc	sc	xlisp
8K, Direct-mapped	14.5	2.38	9.86	47.7	42.7
8K, 2-way assoc.	12.4	2.21	9.44	48.1	44.3
16K, Direct-mapped	14.9	2.05	9.25	46.3	42.8
16K, 2-way assoc.	13.0	1.72	7.50	49.4	45.4

Table 3 Percentage of branch target misses for 32Byte line

7.2 The Performance Impact of Prefetching

We analyze the effect of prefetching on system performance by measuring its effect on I-cache miss rate and the reduction in CPU stall cycles caused by cache misses. The baseline architecture (B) does no instruction prefetching, the second architecture (BI) adds a one-line instruction stream buffer, the third architecture (BP) adds a 2-level MLBP-based prefetch generator and two 1-line prefetch buffers, and the last architecture (BIP) adds both a stream and a MLBP-based prefetch units.

First, results for the direct-mapped, 4K-entry predictor are presented, followed by a fully-associative, 256-entry MLBP. The results are shown for an 8KB, direct mapped instruction cache organization with 32Byte lines unless otherwise specified. A cache configuration is represented by a triplet (cache size, line size, associativity). First, we look at the miss rate change followed by the CPU time analysis for the four architectures.

7.3 Direct-mapped MLBP Implementation

7.3.1 Prefetch Effect On I-Cache Miss Rate

Recall that with prefetching, a miss occurs when neither the I-cache or a prefetch buffer contains the requested line. The case when a prefetch buffer has issued a request but a line is not yet available is counted as a hit. The stall cycles spent waiting in this case will be shown in the next section. Figure 6 summarizes the I-cache miss rate for the four architectures and Figure 7 shows the I-cache miss rate change.

For doduc, fpppp, and gcc, the stream buffer is more efficient than multilevel branch prediction in removing cache misses. For sc and xlisp, the effect of the two organization is quite close. The stream buffer sequential prefetching reduces the miss rate by 80 to 97% for 3 benchmarks, but for the other three it only removes 40% or fewer of the misses. Branch target prefetching removes close to an additional 40% of the misses in the latter two benchmarks, while giving 3 to 11% improvement in the former three benchmarks. When the two prefetching methods are used together, the effect is very close to purely additive. For dbm the miss rate reduction is not very large in all the cases, but very important as we will show in the next section.

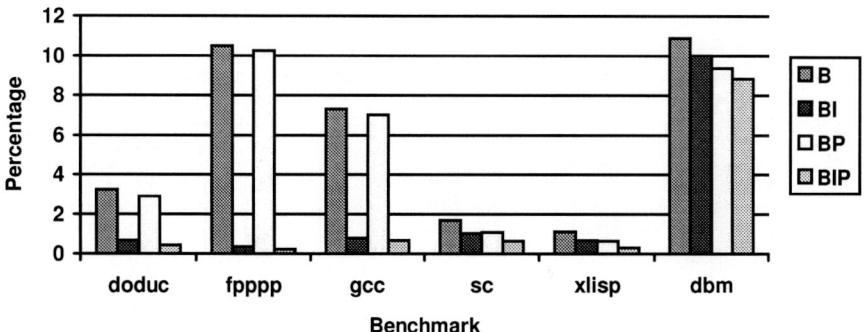

Figure 6. I-cache miss rates

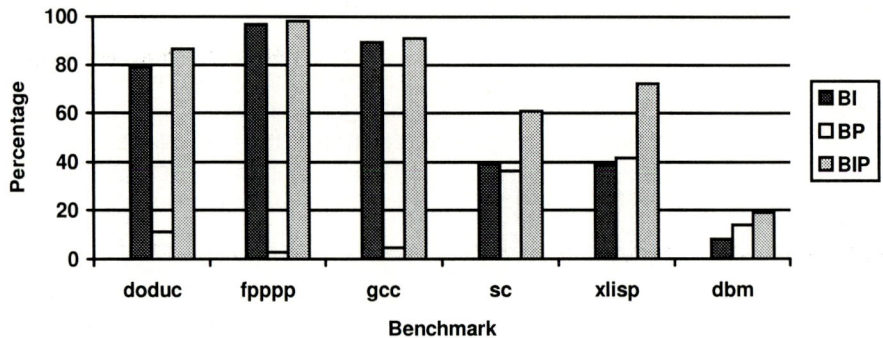

Figure 7. Percentage of baseline architecture (B) misses removed

7.4.2 The Effect on Execution Time

We have seen that the stream buffer significantly reduced the instruction cache miss rates for doduc, fpppp, and gcc while MLBP-based prefetching significantly helped sc and xlisp, and the combination of two methods produced additive results. Next the actual execution time change from prefetching is evaluated. Figure 8 summarize the base execution time reduction due to prefetching. The range of improvement from stream buffer prefetching is 1.7 to 10.3%, 1 to 8% from MLBP prefetching, and 5.8 to 16% for combined prefetching.

The effect is smaller than one might expect from the miss rate reduction. As shown in the next section, there is still a significant stall component in stream buffer prefetching indicating it may not be started early enough and may need more bandwidth. The relative performance of stream and MLBP-based prefetching also changes. For sc and xlisp, multilevel branch prediction based prefetching becomes more efficient than stream buffer, for dbm they are very close and completely additive, and for doduc the difference is significantly reduced. This reflects two facts. First, in fpppp and gcc sequential misses dominate, while in sc, xlisp and dbm a large fraction of instruction cache misses, up to 40%, are caused by branch targets (see Table 3). Second, MLBP allows prefetches to be issued earlier than in stream buffer prefetching.

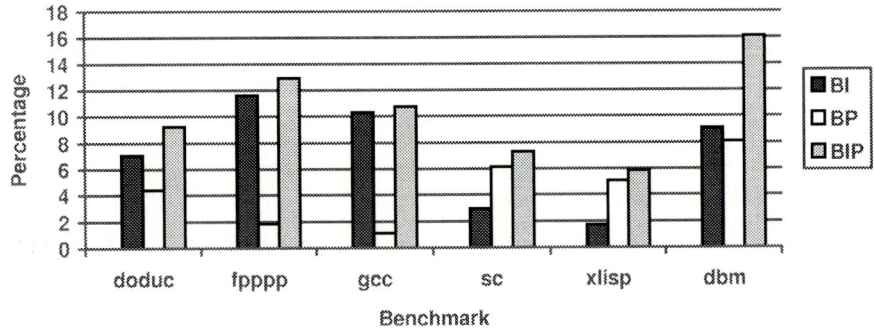

Figure 8. Relative CPU time decrease

7.4.3 CPU Cycle Breakdown

The following approximation is used to derive benchmark execution time for our systems. The processor stalls on an I-cache miss (6 cycles), any taken branch (1 cycle), and a branch misprediction (5-cycles). We do not model any other stalls, such as L2 cache misses, conflicts with primary D-cache misses, or execution unit stalls.

(Equation 1) gives a total benchmark execution time accounting for the stalls we model.

$$T_{cpu} = T_{hb} + T_{tb} + T_{cm} + T_{mp} + T_{pr} - T_{ovlp} \quad \text{(Equation 1.)}$$

Where:

T_{hb} - time to fetch all four-instruction blocks (not all are executed due to branches)

T_{tb} - taken branch stall time when the branch predictor predicts branch taken

T_{cm} - stall time for servicing instruction cache misses from the L2 cache

T_{mp} - branch missprediction stalls

T_{pr} - stall on an issued prefetch which is not serviced when a I-cache miss occurs

T_{ovlp} - cache miss and misspredicted branch overlap

Event counts for the T_{cpu} components are collected during simulation, multiplied by the corresponding delay time, and added up to obtain the total. T_{ovlp} is the time over-charged for a misprediction and a cache miss at the same time, thus T_{ovlp} is subtracted. Table 4 summarizes the results, showing the T_{cpu} and T_{pr} directly while showing event counts for other categories. For an event 'xx', the stall time T_{xx} is found by multiplying the 'xx' column of the table by the stall duration. For example, the cache miss stall time T_{cm} can be found by multiplying the 'cm' column of the table by the L2 latency of 6 cycles.

The results clearly demonstrate the need to reduce I-cache misses. The two components dominating the CPU time are the half-block fetches followed by cache miss or taken branch stalls. Recall that we used a 6-cycle cache miss stall, a 8-cycle stall would make cache misses the second largest CPU time component. At the same time, the programs where stream buffer prefetching had a large effect on miss rate show a large amount of prefetch stalls. The MLBP-based prefetching alone has no prefetch stalls and demonstrates the effectiveness of prefetching across branches, even in the data base manager.

DODUC

System	Tcpu	hb	tb	cm	mp	Tpr	ovlp
B	736124	391027	67422	43997	4385	0	1282
BI	692759	391027	67422	9201	4384	165244	1217
BP	709123	391027	67422	39101	4384	0	1235
BIP	668144	391027	67422	5948	4384	159132	1176

FPPPP

B	1907158	547121	12639	224268	1248	0	782
BI	1685369	547121	12639	7279	1248	1080541	1080
BP	1878085	547121	12639	219030	1248	0	726
BIP	1660889	547121	12639	4464	1248	1072525	1072

GCC

B	375184	146921	3070	34815	4182	0	943
BI	336562	146921	3070	3749	4182	147525	989
BP	371776	146921	3070	33472	4182	0	849
BIP	334838	146921	3070	3230	4182	148192	913

SC

B	44442	26355	9378	1212	458	0	152
BI	43343	26355	9378	737	458	1750	147
BP	42165	26355	9378	772	458	0	147
BIP	41737	26355	9378	473	458	1755	144

XLISP

B	657086	424711	95440	12994	13444	0	721
BI	647899	424711	95440	7941	13444	20334	623
BP	629181	424711	95440	7593	13444	0	684
BIP	625054	424711	95440	3613	13444	25543	615

DBM

B	46691	13785	3352	4658	1075	0	737
BI	42439	13785	3352	4281	1075	7559	726
BP	42896	13785	3352	4009	1075	0	714
BIP	39184	13785	3352	3769	1075	7390	707

Table 4. CPU time components for direct-mapped predictor (in thousands)

7.5 Fully-Associative Implementation

A fully-associative MLBP implementation uses a tagged table of 256 entries, otherwise it is identical to the direct-mapped case. Figure 9 and Figure 10 show the percent reduction in the miss rate and CPU time, respectively.

The results are very close to the CPU times for the 4K-entry direct-mapped implementation (usually within 1%), except for the dbm benchmark. The dbm benchmark suffers a significant reduction in its ability to perform MLBP prediction and address generation. This indicates shows that a tagless table may work better as it will make a prediction even if the history belongs to another reference. Insufficient size to hold the predictions is another reason. The miss rate reduction is also seriously affected in sc but it still generates a noticeable decrease in the CPU time.

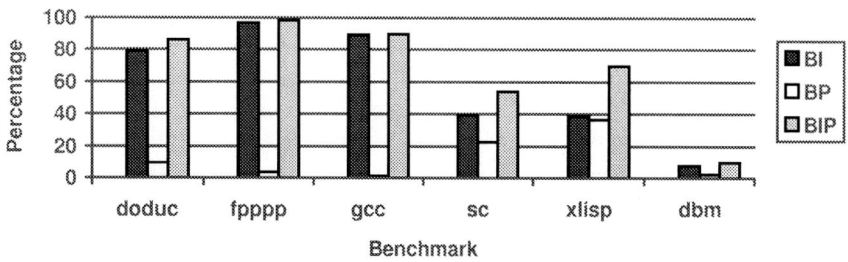

Figure 9. Misses removed by a fully-associative predictor

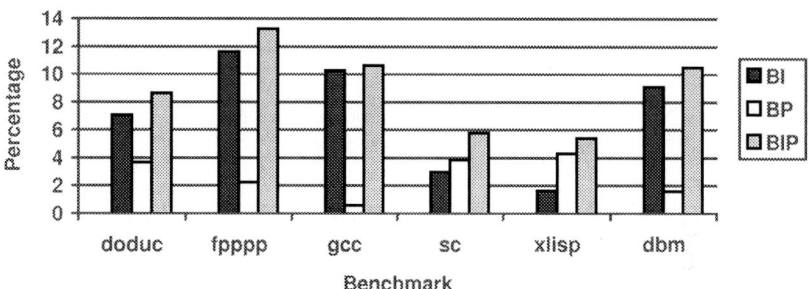

Figure 10. CPU time decrease with a fully-associative predictor

8. Related work

Instruction prefetching has been addressed in the past primarily through sequential prefetch or code layout techniques [Smit82, DEC82, SmHs92, HwCh89, McFa89, ERPR95, UNMS95, XiTo96, LBCG95 Intel93]. Sometimes instruction prefetch was initiated along both possible branch paths [Intel93]. Compiler assistance can help by code layout or by identifying the end of a basic block to stop prefetching [HwCh89, McFa89, XiTo96]. The main improvement comes from adding a sequential prefetcher as has been done in many existing machines. The problem in existing approaches is that prefetching stops when a branch instruction is encountered and the predicted address is non-sequential.

An approach to prefetch speculatively along both paths while waiting for a branch to be resolved has been used (Intel Pentium™), but was aimed at getting the instructions from the I-cache to the Decode unit. [ULMS95] consider sequential prefetching for small I-caches. "Optimistic" prefetch policy is used in [LBCG95] to predict a branch and prefetch down the predicted path.

Multilevel instruction prefetching can be accomplished using the Lookahead Program Counter [BaCh91], which advances forward one instruction per cycle using a standard predictor. True multilevel branch prediction has been proposed in order to speculatively fetch instructions along the most likely path and to "collapse" them into a contiguous sequence [YMP93, DuFr95, CMMP95]. Our multi-level brediction follows these techniques. An interesting solution using a history-based predictor proposed in [RoBS96] uses multiple history table lookups for 3 levels of prediction.

Finally, a related approach [SJSM96] was independently developed to fetch the next two cache line even if they contain branches. Our work differs in its intenet to prefetch to the I-cache and in that it allows, in theory, any lookahead distance to be used.

8.1 Branch Prediction

This subject has been widely researched and is still an active area of research. Branch prediction algorithms try to utilize the past and surrounding information to predict its outcome as accurate as possible. A branch direction can be predicted as soon as a PC of a branch is known, even the branch target address can be predicted at the same time using a branch target buffer (BTB). High prediction accuracy has been achieved using many innovative ideas.

Smith [Smit81] proposed using a table of 2-bit saturating up-down counters to keep track of dynamic branch information. A Two-Level Branch Predictor proposed by and Patt [YePa91] uses two levels of branch history to predict branch direction. Hybrid branch predictors composed of several single scheme predictors and a way to select one of them at a particular time have been proposed [McFa89, ChHP95].

9. Conclusions

In this work, the concept of instruction prefetching using multi-level branch target prediction (MLBP) is developed and its effectiveness studied. It predicts branch direction and branch target across K branches, a K-level prediction. We concentrate on a 2-level prediction to balance hardware complexity and performance. The behavior of five SPEC92 benchmark programs with highest instruction cache miss rates and of a data base manager is analyzed. Integer and data base management benchmarks showed the largest improvements, up to 15% CPU time reduction, from prefetching. It benefits small, fast caches the most.

The MLBP prediction accuracy was found to be quite close for a fully-associative 256-entry and direct-mapped, 4K-entry predictors, about 70% on average. This is low by branch prediction standards but may be sufficient for prefetching. Part of the reason for low accuracy is our implementation choice of tracking only taken branches. It was done to make the MLBP implementation simple in order for it to keep up with the CPU and require less hardware.

The average fraction of cache misses caused by branch targets is about 25% for an 8K instruction cache with 32 byte line size in the benchmarks studied. This can lead to a significant miss rate reduction when MLBP is used. The effect on CPU performance is smaller, around 4%. For a larger I-cache miss service time the effect on CPU performance will increase.

In some programs sequential prefetching works well and our additional prefetching hardware provides only a small improvement. In other benchmarks the

MLBP-based prefetching produces a larger effect than sequential prefetching. It success is due in part to the lookahead afforded by the multi-level prediction.

The MLBP and sequential instruction prefetching, such as an instruction stream buffer, are complementary. They can be used together to effectively remove both sequential and branch target instruction cache misses. The sequential prefetch is often only partly effective because it is not initiated early enough. A larger stream buffer can help but will consume a lot of L2 cache bandwidth and given the frequency of branches may not help after all. MLBP does not have this problem.

Overall, the results of MLBP-based prefetching are encouraging and need further study. There are several areas where they can be improved, the most important being the predictor accuracy. Multi-level predictor design we used does not distinguish between conditional and all other types of branches. Given limited predictor table space, jumps (unconditional branches), calls/returns, and "indirect" jumps (via a register address) do not use the space efficiently. The latter two cannot be easily predicted using PC-based predictor while the latter does not need the number of taken branch entries we used.

We are currently exploring two techniques to solve these problems. First, call/returns may be filtered out using a separate, stack-based predictor, as in the actual DEC Alpha 21164 implementation. The other solution is to switch to history-based multi-level branch prediction and to re-organize the mBTB table to contain just a single entry. This will lead to a longer lookahead and much better space utilization and prediciton accuracy.

Acknowledgements

The author would like to thank Qingbo Zhao for many fruitful discussions and for collecting most of the data presented here.

References

[BaCh91] Jean-Loup Baer and Tien-Fu Chen. "An effective on-chip preloading scheme to reduce data access penalty", Supercomputing'91, pp. 176--186. November 1991.

[CaGr95] B. Calder and D. Grunwald, "Next Cache Line and Set Prediction", International Symposium on Computer Architecture, pp.287--296, May 1995.

[ChHP95] P. -Y. Chang, E. Hao, and Y. N. Patt.: Alternative Implementations of Hybrid Branch Predictors. In: 28th ACM/IEEE International Symposium on Microarchitecture, Nov. 1995.

[CHKW86] Fred Chow, A. M. Himelstein, Earl Killian and L. Weber, "Engineering a RISC Compiler System," IEEE COMPCON, March 1986.

[CMMP95] T.M. Conte, K. N. Menezes, P.M. Mills, and B.A. Patel, "Optimization of Instruction Fetch Mechanism for High Issue Rates", International Symposium on Computer Architecture, pp.333--344, May 1995.

[DuFr95] Simonjit Dutta and Manoj Franklin, "Control Flow Prediction with Tree-Like Subgraphs for Superscalar Processors". International Symposium on Microarchitecture (Micro-28), pp. 258--263, November 1995.

[ERPR95] J. H. Edmondson, P. R. Rubinfeld, Ronald Predton, and Vidya Rajagopalan. "Superscalar Instruction Execution in the 21164 Alpha Microprocessor". IEEE Micro, Vol. 15, No. 2, April 1995

[DEC82] VAX Hardware Handbook, Digital Equipment Coporation, 1982.

[EvCP96] M. Evers, P-Y Chang, and Y. N. Patt, "Using Hybrid Branch predictors to Improve Branch Prediction Accuracy in The Presence of Context Switches", International Symposium on Computer Architecture, pp. 3--13, May 1996.

[HePa96] John L. Hennessy and David A. Patterson, "Computer Architecture, a Quantative Approach", 2nd edition, pp. 465, 1996.

[HwCh89] W.-M. Hwu and P. Chang, "Achieving High Instruction Cache Performance with an Optimizing Compiler", International Symposium on Computer Architecture, pp. 242-251, May 1989.

[Inte93] Pentium Processor User's Mannual, Vol.1: Pentium Processor Data Book. Intel, 1993.

[Joup90] Norman P. Jouppi. "Improving direct-mapped cache performance by the addition of a small fully-associative cache and prefetch buffers", International Symposium on Computer Architecture, pp. 364--373, May 1990.

[JoWi94] Norman P. Jouppi and Steven J.E. Wilton, "Trade-offs in Two-level On-chip caching", International Symposium on Computer Architecture, pp. 34-45, April 1994.

[LBCG95] D. Lee, J.-L. Baer, B. Calder, D. Grunwald "Instruction Cache Fetch Policies for Speculative Execution", International Symposium on Computer Architecture, pp. 357-367, May 1995.

[McFa89] S. McFarling, "Program Optimization for Instruction Caches", International Conference on Architectural Support for Programming Languages and Operating Systems, pp. 183-191, 1992,

[PaSR92] S-T Pan, K. So, and J.T. Rameh, Improving the Accuracy of Dynamic Branch Prediction Using Branch Correlation", International Conference on Architectural Support for Programming Languages and Operating Systems, pp. 76-84, October 1992.

[RoBS96] Eric Rotenberg, Steve Bennett, and James E. Smith, "Trace Cache: a Low Latency Approach to High Bandwidth Instruction Fetching", 29th Annual International Symposium on Microarchitecture, pp. 24-34, December 1996.

[SaPN96] Ashley Saulsbury, Fong Pong and Andreas Nowatzyk. "Missing the Memory Wall: the Case for Processor/Memory Integration". Computer Architecture News, Vol. 24, No. 2, pp.90-101, May, 1996.

[SeLM96] S. Sechrest , C-C Lee and T. Mudge. "Correlation and Aliasing in Dynamic Branch Prediction ", International Symposium on Computer Architecture, pp. 22--32, May 1996.

[SJSM96] A. Seznec, S. Jourdan, P. Sainrat, P. Michaud. "Multiple-Block Ahead Branch Prediction", International Symposium on Computer Architecture, pp. 116-127, May 1996.

[Smit81] J. E. Smith. "A Study of Branch Prediciton Strategies." Proceedings of the 8th International Symposium on Computer Architecture, pp.135-148, May, 1981.

[SmHS92] J.E. Smith and W.-C. Hsu, "Prefetching in Supercomputer Instruction Caches", International Supercomputing Conference, pp. 588-597, July 1992

[UNMS95] R. Uhlig, D. Nagle, T. Mudge, S. Sechrest, and J. Emer, "Instruction Fetching: Coping with Code Bloat", International Symposium on Computer Architecture, pp. 348--356, May 1995.

[YePa91] T.-Y. Yeh and Y. N. Patt. "Two Level Adaptive Branch Prediction." 24th ACM/IEEE International Symposium on Microarchitecture, Nov. 1991.

[YeMP93] T-Y Yeh, D.T. Marr, and Y. N. Patt, "Increasing Instruction Fetch Rate via Multiple Branch Predictions and a Branch Address Cache", International Conference on Supercomputing, pp. 67-76, July 1993.

[XiTo96] C. Xia and J. Torrrellas, "Instruction Prefetching of Systems Codes with Optimized Layout for Reduced Cache Misses", International Symposium on Computer Architecture, pp. 271--283, May 1996.

[Zhao96] Q. Zhao, "Performance evaluation of instruciton prefetching using multi-level branch prediction", M.S. Thesis, EECS Dept., University of Illinois at Chicago, October 1996.

High Performance Wireless Computing

George Cybenko

Dartmouth College
Thayer School of Engineering
8000 Cummings Hall, Hanover, NH, 03755-8000, USA
george.cybenko@dartmouth.edu

Abstract

This session deals with fundamental issues arising in wireless networking and their implications for high performance business, consumer and military computing applications. We will describe the physical basis for wireless networking, mobile IP solutions and challenges as well as ongoing research efforts to make high performance and high confidence computing over wireless networks possible.

One of the major technical challenges for wireless computing is the volatility of the network links and nodes. Bandwidth and latency vary much more rapidly than in wired networks and network connectivity is highly intermittent because of transmission anomolies and power conservation, among other factors. Moreover, the network topology is constantly changing because the nodes are mobile.

These challenges can be met using a variety of novel network management ideas and applications software. We are developing support for scalable mobile IP, dynamic network sensing and predictive routing, proxy servers and docking systems. Many of these developments use mobile agents as a building block. The mobile agent system we are using is based on Agent Tcl which was developed at Dartmouth (see http://www.cs.dartmouth.edu/ agent).

Part of the research presented is funded by the US Department of Defense through a Multidisciplinary University Research Initiative at Dartmouth with subcontracts to Harvard, University of Illinois, RPI, Lockheed Martin and Alphatech. Industrial partners include Merrill Lynch, Northern Telecom, Digital Equipment, Lucent and AT&T.

High-Performance Computing and Applications in Image Processing and Computer Vision

Hamid R. Arabnia

University of Georgia
Department of Computer Science
Graduate Studies Research Center
Athens, Georgia 30602-7404, USA hra@cs.uga.edu

Abstract

A variety of parallel computer architectures are being used today to cope with the computationally intensive tasks in the areas of image processing and computer vision. Most image processing algorithms can readily exploit SIMD (Single Instruction, Multiple Data Stream) machine architectures. The mapping of these algorithms to such machines is rather straightforward. The fine granularity parallelism and regular data units are inherent in the nature of these algorithms. The basic disadvantage of the SIMD systems is their inadequacy to handle problems where the data involved is high level and irregular and the operations defined on them are complex. The MIMD (Multiple Instruction, Multiple Data Stream) machine architectures have the potential to deal with this kind of problem, common in computer vision. Our studies show that a reconfigurable system termed as the Reconfigurable Multi-Ring Network (RMRN) supports image processing as well as computer vision algorithms within the same architecture framework. We show the RMRN to be a viable architecture for image processing and computer vision prob- lems by demonstrating the parallel computation of a set of imaging algorithms on the SIMD, SPMD (Single Program, Multiple Data Stream), or cluster of workstations. In particular, we will address the problems of stereo image reconstruction, image classification, and image segmentation used in digital mammography.

Present and Future of HPC Technologies

Tadashi Watanabe

Supercomputers Marketing Promotion Division
NEC Corporation
watanabe@sxsmd.ho.nec.co.jp

Abstract. From the beginning of computer history, there has been strong demand for faster speed of computations, while at the same time there is growing demand for lower cost including lower operating cost and ease-of-use. In order to respond to these demands, there are three major points from the technological aspects though those are not limited to high performance computing area. The state-of-art technologies , however, are always required to achieve the highest speed particularly in high performance computing.
The following three technologies are discussed in this talk.
- Architecture
- Hardware Technology
- Software Technology

1 Architecture

First of all, systems architecture, i.e. systems configuration, must be considered. It is needless to say that the parallel processing architecture is indispensable to achieve higher processing capability. The major configurations are the shared memory processors (SMP), distributed memory processors (DMP) and distributed /shared memory architecture (or cc-NUMA:cache coherency Non-Uniform Access Architecture). The SMP has greater advantages from the view point of ease-of-use over other configurations, but has some limitations t o achieve faster speed. Furthermore, we must consider about the network topology connecting between processors and memories. Pros and cons of those architectural configurations, and the future trend will be presented. The next architectural consideration is a processor architecture where the major topics are cache, RISC, Super-Scalar, VLIW, and Vector architecture.

2 Hardware Technology

Hardware technologies to be talked here include chip and packaging technologies which are keys for achieving faster cycle time, smaller physical size and lower cost. Most of HPC vendors already employed CMOS chip as key devices that has various advantages over ECL such as high density , low power consumption and low cost. Those features of CMOS will be constantly enhanced in these ten years. By extrapolating the technology improvements, the current vector type processor of GFLOPS level speed may become one chip within ten years. Memory technology must be also considered. There are two typical devices in memory, DRAM and SRAM. DRAM is the most popular memory device used in all types of computer memory while DRAM is slower in speed than SRAM. SRAM is sometimes used in the main memory in supercomputers due to its faster speed while it is less in density than DRAM.

3 Software Technology

UNIX is the only operating system for HPC systems currently. But the UNIX for HPC must handle large memory, large files , high speed I/O, and parallel processing function. Languages and tools for debugging and optimization must also be considered. Although Fortran is still a major language in HPC, even Fortran continues to be enhanced and improved by incorporating various functions particularly demanded by HPC community. As the compiler for HPC, the most of the current Fortran have automatic vectorization and automatic parallelization functions, and are reaching to a mature level. The recent topics in compiler for HPC is how to handle the distributed memory architecture. To support this architecture, there are two approaches, data distribution and process distribution. The data distribution is supported by HPF (High Perfomance Fortran) which is a kind of the extension of Fortran90. On the other hand, the process distribution is typically handled by MPI (Message Passing Interface) which is a collection of libraries called by Fortran and C where each process on the processor is executed coordinating one another.

Evaluation of Multithreaded Processors and Thread-Switch Policies

Richard J. Eickemeyer[1], Ross E. Johnson[1], Steven R. Kunkel[1],
Beng-Hong Lim[2], Mark S. Squillante[2], C. Eric Wu[2]

[1] IBM AS/400 Division, Rochester MN 55901, USA
[2] IBM T. J. Watson Research Center, Yorktown Heights NY 10598, USA

Abstract. This paper examines the use of coarse-grained multithreading to lessen the negative impact of memory access latencies on the performance of uniprocessor on-line transaction processing systems. It considers the effect of switching threads on cache misses in a two-level cache system. It also examines several different thread-switch policies. The results suggest that multithreading with a small number (3–5) of active threads can significantly improve the performance of such commercial environments.

1 Introduction

As processor speeds continue to increase at a higher rate than memory speeds [4, 14], memory access latencies become increasingly significant. Moreover, the first-level cache miss latency in a two-level cache memory system also contributes significantly to execution time. Commercial applications have poor cache behavior, magnifying the impact of memory and cache latencies on execution time [6, 12]. The trend towards object-oriented programming and micro-kernel based operating systems in commercial environments are expected to further increase these effects [5, 3]. It is imminent that uniprocessors executing commercial application workloads will be limited by cache-miss delays unless the system provides methods to tolerate or avoid long and frequent cache-miss delays.

Multithreaded architectures attempt to decrease idle time by maintaining multiple thread contexts in the processor. There are three basic forms of multithreading. *Fine-grained* multithreading interleaves different threads on a cycle-by-cycle basis, thus eliminating most pipeline dependencies by separating the instructions in a single thread. *Coarse-grained* multithreading interleaves the instructions of different threads on some long-latency events. *Simultaneous* multithreading [13] assumes a superscalar architecture and improves upon fine-grained multithreading by allowing instructions from multiple threads to be arbitrarily interleaved in the pipeline.

This paper evaluates the effectiveness of coarse-grained multithreaded architectures for improving the performance of single-processor on-line transaction processing systems where there is a natural, coarse-grained parallelism among the tasks (resulting from the concurrently executed transactions) and no application software modifications is required. When a running thread waits for a long-latency event, the processor switches to a new thread to increase processor utilization. This research is a continuation of our previous study [8] that considered thread switches solely on second-level cache misses.

The focus here is on switching threads on either a first-level or a second-level cache miss. When a thread switch takes less time than a first-level cache miss, there is a benefit to switching earlier, should a second-level cache miss also occur.

Most of the results are obtained from detailed trace-driven simulations of a multi-threaded processor architecture, based on commercial application traces from an IBM AS/400 system. Several different policies for switching threads on first-level cache misses are evaluated, together with a policy of switching on second-level cache misses. The results show that multithreading can significantly improve the performance of uniprocessors in commercial computing environments, decreasing mean response time and increasing throughput. Larger performance benefits are observed for switching on first-level misses than on second-level misses. The number of threads that provide the best performance depends upon a number of factors, including processor speed, memory latency, cache miss rates and thread-switch policy. The results suggest that 3 to 5 threads are best for the thread-switch policies and architectures considered. A larger number of threads may be required for future designs with longer memory latencies.

The next section describes the system environment assumed in this study. Sections 3 and 4 present the results. Section 5 provides a brief summary of related work. Section 6 summarizes the conclusions and discusses the implications of the results on the usefulness of multithreading in uniprocessor commercial systems.

2 Architecture, Workload and Methodology

This section presents the commercial workload and the multithreaded microarchitecture used in this study. It also describes the different thread switch policies evaluated, as well as the methodology used to obtain the performance results.

2.1 Application Workload and Processor Microarchitecture

The application workload and processor microarchitecture are essentially the same as in our previous study [8]. We therefore summarize here their key features and refer the interested reader to this previous study for more details.

The workload is based on traces from a hardware monitor attached to an AS/400 model D70 executing the TPC-C benchmark, which is representative of an on-line transaction processing environment with medium-weight transactions. These traces capture the interaction of the many tasks and interrupts of a fully-loaded, multi-tasking, multi-user commercial system. To simulate the traces for a multithreaded processor, we parse the traces into *trace segments* that each represent a task executing for a single scheduling quantum. Addresses are taken from one active trace segment until a thread switch occurs, at which point addresses are taken from another active segment. The number of active segments is equal to the number of threads, denoted by N. When a trace segment is exhausted, another segment from the trace is made active. The simulations are representative of a different task being executed in each processor thread.

The base microarchitecture includes first-level on-chip caches (L1) and a second-level off-chip cache (L2), all of which are non-blocking to avoid degrading multithreading performance [8]. The states of the N threads are maintained in hardware and all N

threads share the L1 and L2 caches. An L1/L2 cache miss causes a thread switch that is performed by the hardware. The thread is blocked (*i.e.*, switched out) until its cache miss is satisfied. Meanwhile, other threads may use the caches.

The base configuration comprises 1-way, 16KB L1 caches with 32B lines and 8-cycle miss latencies (L1M), and a 1-way, 512KB L2 cache with a 64B line and 60-cycle latency (L2M). The base thread switch time[3] (TS) is 4 cycles. Since the probability of all threads waiting for a cache miss also depends upon the rate at which references are made to the caches, we use an infinite-cache processor speed of 1.3 cycles per reference (CPR). A $4ns$ cycle time is assumed throughout.

We vary several microarchitecture design parameters that span a range of processor designs to examine their effect on our results. We examine L1 cache sizes of 8KB and 32KB, line sizes of 16B and 64B, and L1M of 4 and 12 cycles. 2-way and 4-way associativity is further used to represent processors with a less aggressive cycle time, shorter pipelines, or a 2-cycle cache access. The base L1M=8 represents a processor with an off-chip L2 cache and a moderate cycle time, whereas L1M=4 is fast and represents processors with less aggressive cycle times, expensive L2 caches, expensive packaging (a multi-chip module), or an on-chip L2 cache. L1M=12 is slow and represents processors with an aggressive cycle time or a low cost, off-chip cache.

L2 cache sizes of 256KB and 1MB, line sizes of 32B and 128B, L2M of 50 and 70 cycles, and 2-way and 4-way associativity are also examined. The L2 cache sizes were chosen because they show the knee of the curve where the effect of multithreading on the L2 miss rate becomes negligible; the effect of the higher miss rates with a RISC architecture and object-oriented applications also suggest these sizes.

The base infinite-cache processor speed of 1.3 CPR approximates the reference rate in a 4-way superscalar processor running a transaction processing workload. To study a more aggressive design, CPR=1 is also evaluated. A number of design features affect TS, including the length of the pipelines, whether registers from several threads can be accessed at once, the size of instruction queues and/or the number of reservation stations, and whether instruction dispatch buffers exist for more than one thread. The base TS=4 represents short pipelines with little instruction queueing. We also examine TS of 1 and 8 cycles. TS=8 represents a long pipeline with much queueing and register files that only allow access to one thread at a time, while TS=1 represents a design in which the state of the pipelines is duplicated for each thread.

2.2 Thread Switch Policies

The thread switch policy, which determines the thread to be executed next, is an important issue for switching on L1 misses. This section presents four different switch policies. Note that the performance of the single-thread case is independent of the policy.

OPT On an L1 miss, the *optimal* policy (OPT) switches to the thread that had or will have its miss satisfied earliest. If there are ready threads, the earliest ready

---

[3] The thread switch time is defined to be the time from the detection of a miss until an instruction from the next thread reaches the pipeline stage in which an L1 miss could be detected. Note that, when switching on L2 misses, the L1 miss stalls the pipeline until the L2 miss is detected. This definition of TS counts all non-productive cycles caused by the thread switch.

thread will be scheduled immediately after a delay of TS. Otherwise, OPT chooses the thread whose miss will be satisfied earliest and the 'clock' is advanced to max{current_time + TS, time_miss_satisfied} if it switches to another thread, or simply to time_miss_satisfied if it switches to itself. Ties are broken in a round-robin manner. There is no penalty for determining which thread gets scheduled next. Although OPT cannot be implemented as it requires knowledge of future events, it provides an upper bound on performance.

NT On a miss for any thread, the *next-thread* policy (NT) immediately switches to the next thread in a round-robin fashion and the 'clock' is advanced to max{current_time + TS, time_miss_satisfied}. If the next thread is not ready to execute, the processor stalls until the thread is ready to run. This is the simplest policy that we consider.

NRT The *next-ready-thread* policy (NRT) is an extension of NT. On a miss for thread α, NRT immediately switches to the next thread (in a round-robin order), incurring the TS penalty for each switch until it finds a ready thread or it returns to thread α and waits.

SWR The *switch-when-ready* policy (SWR) relies on hardware to detect ready threads before switching. If there are ready threads, then SWR is equal to OPT with the exception that the thread switch costs an extra cycle to determine which thread to execute. When no threads are ready, SWR waits for the first miss to be satisfied, pays a 1 cycle penalty to detect this, and then pays a normal TS overhead to switch to this thread. SWR relies on additional hardware to determine the next thread to activate. A simple scheme to provide this hardware support, which requires minimal additional storage and logic, is presented in [7]. A good design should be able to eliminate the extra 1-cycle penalty assumed for SWR in our simulations.

2.3 Simulation and Models

The performance evaluation combines trace-driven simulation and a general analytic computer system model. The multithreaded processor microarchitecture of Section 2.1 is simulated under the commercial application workload described there. From the simulation data, we obtain detailed characterizations of the processor performance and cache miss behavior. This is used to parameterize an analytic model of the computer system to obtain mean job response time. Due to space limitations, we briefly summarize these aspects of our study and refer the interested reader to [7, 8] for more details.

Multithreaded Processor Simulation. A trace-driven simulation models in detail the multithreaded microarchitecture at the behavioral level, including the internal states of the L1-I, L1-D and L2 caches. Miss rates are obtained for each of the caches, under the assumption that multi-level inclusion is not enforced. The simulator first reads the binary address traces as a mapped file and parses the trace into segments. It then assigns the first N segments as the N threads. Experiments with different orderings of the trace segments suggest that this does not affect the performance trends of our results. The processor reads the next reference for the current thread and the clock is appropriately advanced based on the reference. The reference is then passed to the corresponding L1 cache. If the reference hits in the L1 cache, the processor remains active and processes the next reference from the current thread. Otherwise, if an L1 miss occurs and $N = 1$, then the clock and the idle-cycles sum are both increased by the miss latency. For $N \geq 2$, the processor switches to the next thread as defined by the thread switch policy, and the clock and the idle-cycles sum are both advanced by the cycles to perform the thread switch under this policy. If the (new) current thread is still waiting for its previous cache request to complete, then the clock is set to the cycle when this miss will be satisfied and the idle-cycles sum is increased by this difference.

When a thread returns from a cache miss, it is possible that the referenced cache line has been replaced by references from other threads. The simulator checks whether the last reference of each thread is still in the cache before proceeding with its execution. If the line has been replaced, the processor issues the request again and stalls until the reference is returned to the cache. This simulates a common hardware technique for avoiding livelock.

Computer System Model. We develop a general model for a computer system that is based on a multithreaded processor with N threads. An open model is used since this is representative of the large-scale, multi-user commercial systems of interest, although our approach is easily extended to handle a closed system model.

Jobs are submitted to the computer system according to a general probability distribution $\mathcal{A}(\cdot)$. When the system contains i jobs, $i \geq 1$, the computing demands of each of these jobs follow a general probability distribution $\mathcal{B}_i(\cdot)$, where $\mathcal{B}_{N+k}(\cdot) \equiv \mathcal{B}_N(\cdot)$, $k \geq 0$. The load-dependent distributions $\mathcal{B}_i(\cdot)$ are fitted to match the detailed statistics from the trace-driven simulation of the multithreaded processor. In this paper we solely consider the case where the processor throughputs are used to parameterize the load-dependent service rates.

The state space of this model is obtained by representing in detail the states of the computer system. By exploiting a general class of probability distributions, this model formulation leads to a well-structured Markov chain, the properties of which we exploit to obtain an efficient solution for steady-state measures. The mean job response time can be written as a closed-form expression in terms of these measures, from which it can be computed in an efficient manner.

3 Measurements from Simulation Experiments

This section presents a subset of the experimental results. Table 1 lists the miss rates and throughput for the OPT policy under various cache configurations with 1 – 6 threads. Miss rates are in terms of misses per *total* number of processor references when there are N threads. These statistics are used to obtain the time between cache misses that cause thread switches, which we term *intermiss time* (in nanoseconds). Both the mean and coefficient of variation[4] (CV) are given. The throughput is the fraction of time when the processor is not waiting for cache misses or performing thread switches.

---

[4] The coefficient of variation is the ratio of the standard deviation to the mean.

The first configuration in the table represents the base cache and latency configuration. Several parameters are then varied around the base configuration to explore the effect of multithreading on different design points. The full set of results covering all the thread-switch policies is reported in [7], and the results for switching on L2-cache misses are reported in [8].

3.1 Impact on Cache Miss Rates

We first address the effect of multithreading on cache miss rates, where more threads may increase the miss rate. In the base configuration in Table 1, I-cache miss rates increase by 4.1% from 1 to 2 threads, and increase by 6.5% for 6 threads. The increase in miss rates is magnified for a smaller L1-I cache and large number of threads. An 8KB L1-I cache encounters a 4% increase in miss rates from 1 to 2 threads, and a 15% increase for 6 threads. This increase is in contrast to L1-I miss rates that remain largely unchanged when switching only on L2 misses [8]. However, a larger L1-I cache or a higher associativity mitigates the increase in cache misses. The direct-mapped 32KB L1-I cache and the 2-way (or greater) set-associative 16KB L1-I cache handles a larger number of threads without a significant increase in miss rates.

The L1-D caches encounter a more significant change in miss rates as N increases. There are 57% more misses for the base configuration with 6 threads than for 1 thread. This increases to 70% for an 8KB L1-D cache and reduces to 41% for a 32KB L1-D cache. Comparing the two-thread miss rate for the 16KB cache (0.0629) with the one-thread miss rate of an 8KB data cache (0.0667) shows that the miss rate with $N = 2$ is better than physically partitioning a 16-KB cache between two threads. Although the miss-rate trends under different L1-D cache sizes are the same as when switching on L2 misses, the miss rates increase more when switching on L1 misses.

The effect of multithreading is less for L2 caches. The base configuration shows a 24% increase in L2 misses for $N = 6$. Comparing different L2 sizes and combined L1 cache miss rates shows that a larger cache reduces the effect of multithreading on miss rates. Since a trace of a multitasking operating system is used and there are relatively frequent context switches, a larger cache will contain multiple contexts even without multithreading. The more frequent task switching induced by multithreading does not have a significant effect.

Other cache parameters also affect miss rates. 2-way associativity yields a smaller increase in the miss rate for the L1-D and L2 caches. For the L1-I cache, associativity can make the effect of multithreading negligible. There is negligible change in L2 miss rates when associativity is at least 4. Associativity has more of an effect when switching on L1 instead of L2 misses. For the L1-I and L2 caches, associativity almost eliminates all effect of multithreading on miss rates. Since the effect is larger when switching on L1 misses, the improvement with associativity is more important. This trend is observable for the L1-D cache, but less dramatic.

Larger caches and higher associativity reduce miss rates and are less sensitive to multithreading effects. The opposite is true for larger lines that result in reduced miss rates (for the range shown), but are more sensitive to multithreading effects. In each of the throughput results, however, a larger line size is preferable. The lower miss-rate is

N	L1-I/L1-D/L2 Caches				Intermiss Time		Throughput	
	total size	set assoc	line size	misses per processor reference	avg	CV	act	rel
1	16K/16K/512K	1/1/1	32/32/64	0.0556/0.0538/0.0195	122.3	1.1	0.39	1.00
2	16K/16K/512K	1/1/1	32/32/64	0.0579/0.0629/0.0207	73.0	1.3	0.59	1.52
3	16K/16K/512K	1/1/1	32/32/64	0.0586/0.0705/0.0218	60.5	1.2	0.67	1.71
4	16K/16K/512K	1/1/1	32/32/64	0.0571/0.0745/0.0226	57.4	1.2	0.69	1.77
5	16K/16K/512K	1/1/1	32/32/64	0.0594/0.0789/0.0235	55.3	1.2	0.68	1.75
6	16K/16K/512K	1/1/1	32/32/64	0.0603/0.0843/0.0244	53.9	1.1	0.67	1.72
1	8K/8K/256K	1/1/1	32/32/64	0.0679/0.0667/0.0303	124.7	1.0	0.31	1.00
2	8K/8K/256K	1/1/1	32/32/64	0.0706/0.0808/0.0326	69.5	1.0	0.49	1.60
3	8K/8K/256K	1/1/1	32/32/64	0.0727/0.0916/0.0350	54.3	1.0	0.58	1.88
4	8K/8K/256K	1/1/1	32/32/64	0.0722/0.0985/0.0365	49.5	1.0	0.62	1.99
5	8K/8K/256K	1/1/1	32/32/64	0.0744/0.1057/0.0378	47.0	1.0	0.61	1.98
6	8K/8K/256K	1/1/1	32/32/64	0.0776/0.1140/0.0396	45.6	0.9	0.60	1.92
1	16K/16K/512K	1/1/1	16/16/32	0.0938/0.0709/0.0285	105.1	1.2	0.30	1.00
2	16K/16K/512K	1/1/1	16/16/32	0.0965/0.0793/0.0299	61.0	1.4	0.48	1.61
3	16K/16K/512K	1/1/1	16/16/32	0.0978/0.0862/0.0311	49.2	1.4	0.57	1.91
4	16K/16K/512K	1/1/1	16/16/32	0.0949/0.0898/0.0321	46.1	1.4	0.61	2.03
5	16K/16K/512K	1/1/1	16/16/32	0.0946/0.0932/0.0330	44.7	1.4	0.62	2.06
6	16K/16K/512K	1/1/1	16/16/32	0.0952/0.0976/0.0341	43.8	1.3	0.62	2.05
1	8K/8K/256K	1/1/1	16/16/32	0.1137/0.0844/0.0436	111.0	1.1	0.24	1.00
2	8K/8K/256K	1/1/1	16/16/32	0.1165/0.0972/0.0461	61.1	1.2	0.40	1.69
3	8K/8K/256K	1/1/1	16/16/32	0.1184/0.1073/0.0480	46.6	1.2	0.49	2.09
4	8K/8K/256K	1/1/1	16/16/32	0.1201/0.1131/0.0498	41.3	1.2	0.54	2.28
5	8K/8K/256K	1/1/1	16/16/32	0.1214/0.1186/0.0514	39.1	1.1	0.55	2.35
6	8K/8K/256K	1/1/1	16/16/32	0.1244/0.1265/0.0525	37.8	1.1	0.55	2.32
1	16K/16K/512K	1/1/1	64/64/128	0.0345/0.0433/0.0146	143.9	1.1	0.46	1.00
2	16K/16K/512K	1/1/1	64/64/128	0.0368/0.0536/0.0158	86.8	1.3	0.66	1.43
3	16K/16K/512K	1/1/1	64/64/128	0.0370/0.0616/0.0170	73.1	1.3	0.72	1.55
4	16K/16K/512K	1/1/1	64/64/128	0.0375/0.0668/0.0179	68.9	1.3	0.72	1.56
5	16K/16K/512K	1/1/1	64/64/128	0.0387/0.0720/0.0189	66.5	1.2	0.71	1.52
6	16K/16K/512K	1/1/1	64/64/128	0.0409/0.0788/0.0196	63.4	1.2	0.68	1.47
1	32K/32K/1024K	1/1/1	32/32/64	0.0433/0.0435/0.0127	127.1	1.4	0.47	1.00
2	32K/32K/1024K	1/1/1	32/32/64	0.0440/0.0491/0.0132	82.6	1.7	0.68	1.44
3	32K/32K/1024K	1/1/1	32/32/64	0.0445/0.0538/0.0138	71.8	1.6	0.74	1.56
4	32K/32K/1024K	1/1/1	32/32/64	0.0433/0.0560/0.0140	69.8	1.6	0.75	1.59
5	32K/32K/1024K	1/1/1	32/32/64	0.0447/0.0595/0.0147	67.2	1.5	0.74	1.58
6	32K/32K/1024K	1/1/1	32/32/64	0.0448/0.0612/0.0148	66.5	1.5	0.74	1.56
1	16K/16K/512K	2/2/2	32/32/64	0.0533/0.0446/0.0137	118.7	1.7	0.45	1.00
2	16K/16K/512K	2/2/2	32/32/64	0.0541/0.0509/0.0140	75.8	2.1	0.65	1.46
3	16K/16K/512K	2/2/2	32/32/64	0.0530/0.0558/0.0145	66.3	2.0	0.72	1.61
4	16K/16K/512K	2/2/2	32/32/64	0.0528/0.0593/0.0146	62.9	2.0	0.74	1.65
5	16K/16K/512K	2/2/2	32/32/64	0.0547/0.0617/0.0147	60.8	1.9	0.73	1.64
6	16K/16K/512K	2/2/2	32/32/64	0.0551/0.0662/0.0151	58.9	1.8	0.73	1.63

Table 1. Multithreading results for switching N threads on L1 misses under the optimal thread-switch policy (OPT) with L1M=8, L2M=60, TS=4, CPR=1.3, and 4ns cycle time.

more important than the increase due to multithreading. This is observed for both L1 and L2 caches, as well as for switching on L1 and L2 misses.

Another important cache effect is the number of additional misses (not included in Table 1) resulting from the livelock detection and recovery mechanism described in Section 2.3, which we summarize for the base case with $N = 2$. OPT exhibits the smallest percentage per processor reference of such misses in the L1 caches (0.019% for L1-I, and 0.055% for L1-D), whereas SWR yields the worst (0.036% and 0.081%). NRT (0.024%, 0.062%) and NT (0.031%, 0.081%) lie in between OPT and SWR. The smallest percentage of livelock-detection events in the L2 cache were observed for NT (0.0008%), followed by OPT (0.0010%), SWR (0.0020%) and NRT (0.0023%).

The impact of the four thread-switch policies on cache misses are shown in Table 2. It also provides the absolute and relative throughputs for these switch policies.

N	Policy	L1-I/L1-D/L2 Caches misses per processor reference	Throughput act	rel	N	Policy	L1-I/L1-D/L2 Caches misses per processor reference	Throughput act	rel
1	OPT	0.0556/0.0538/0.0195	0.39	1.00	1	SWR	0.0556/0.0538/0.0195	0.39	1.00
2	OPT	0.0579/0.0629/0.0207	0.59	1.52	2	SWR	0.0584/0.0631/0.0206	0.53	1.37
3	OPT	0.0586/0.0705/0.0218	0.67	1.71	3	SWR	0.0581/0.0704/0.0218	0.61	1.58
4	OPT	0.0571/0.0745/0.0226	0.69	1.77	4	SWR	0.0573/0.0746/0.0226	0.64	1.64
5	OPT	0.0594/0.0789/0.0235	0.68	1.75	5	SWR	0.0576/0.0781/0.0237	0.64	1.64
6	OPT	0.0603/0.0843/0.0244	0.67	1.72	6	SWR	0.0581/0.0831/0.0242	0.63	1.61
1	NRT	0.0556/0.0538/0.0195	0.39	1.00	1	NT	0.0556/0.0538/0.0195	0.39	1.00
2	NRT	0.0579/0.0629/0.0207	0.59	1.51	2	NT	0.0574/0.0631/0.0210	0.48	1.24
3	NRT	0.0582/0.0706/0.0219	0.63	1.62	3	NT	0.0575/0.0703/0.0218	0.53	1.36
4	NRT	0.0571/0.0745/0.0227	0.65	1.67	4	NT	0.0589/0.0755/0.0227	0.57	1.46
5	NRT	0.0581/0.0780/0.0237	0.66	1.69	5	NT	0.0593/0.0786/0.0238	0.60	1.55
6	NRT	0.0598/0.0841/0.0243	0.65	1.67	6	NT	0.0616/0.0849/0.0244	0.62	1.60

Table 2. Multithreading results of the four thread-switch policies, for L1M=8, L2M=60, TS=4, CPR=1.3 and direct-mapped cache sizes 16K/16K/512K with line sizes 32/32/64.

3.2 Impact on Throughput

The first row of Table 1 presents the contribution of memory latency to the total execution time. Fig. 1 plots the system throughputs for various cache sizes, line sizes and set-associativities.

A processor execution rate of 1.3 CPR, plus the L1 miss component of (0.0556+0.0538) multiplied by the 8-cycle L1M (0.8752 CPR), plus 0.0195 multiplied by the 60-cycle L2M (1.17 CPR), yields 3.35 cycles of execution time per reference. Dividing 1.3 by 3.35 gives the reported throughput of 0.39. In other words, L1 and L2 cache misses make up 61% of the execution time. If all cache misses could be

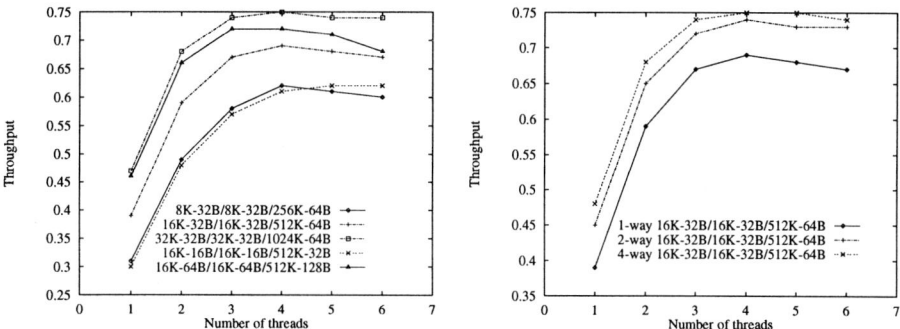

Fig. 1. Absolute throughputs achieved under various cache sizes, line sizes and set associativities, for L1M=8, L2M=60, TS=4 and CPR=1.3. Caches are direct-mapped, unless noted otherwise.

eliminated, throughput would increase by 2.57 times, giving an actual throughput of 1.00. With TS=4, a 1.92 times improvement is an upper bound on relative throughput, computed by removing the L2 CPR component and replacing the 8-cycle L1M with the 4-cycle TS. Overall, multithreading reduces the average cache-miss penalties incurred. These bounds change for each different cache configuration and switch policy.

While multithreading reduces the impact of miss latencies, it also increases the number of misses. Our results show, however, that latency hiding is more important and that over a factor of 2 increase in throughput is possible. The base configuration achieves a 77% throughput improvement (0.69 actual throughput) with 4 threads. There are two reasons why this is less than the 92% upper bound computed above. First, the miss rates increase with N; repeating the calculation for 4 threads with the miss rates in Table 1 shows an 83% upper bound. Second, when all threads are waiting for an L1 cache miss the effective L1M is greater than the 4-cycle TS.

Throughput declines when $N > 4$. Since nearly all L1 miss latency is hidden with 4 threads, there is little improvement from adding more threads. In fact, more threads increase the miss rates so there are more frequent thread switches and less time spent computing. In Table 1, 4 or 5 threads achieves the maximum throughput.

When switching only on L2 misses, the numbers decrease: an upper bound on improvement is only 48%, actual improvement peaks at 38% with $N = 3$, where actual throughput is 0.54. The performance improvement is less than when switching on L1 misses because there is less cache-miss latency that can be hidden. The optimal number of threads is less because L2 misses occur less often and there is less miss time to cover, making it less likely that all threads will be waiting for misses that can be covered by switching. Although the quantitative results are lower, the trends are similar in that multithreading improves performance up to some number of threads, at which point the increased utilization of the processor does not compensate for the increase in cache-miss rates.

3.3 Impact on Intermiss Statistics

The remaining columns to be discussed in Table 1 are for intermiss times. As the number of threads increases, the intermiss time decreases since it is more likely to have a ready thread waiting to execute. The decrease is related to the increase in throughput and miss rates. The smaller intermiss time increases request rates and bandwidth requirements to the memory system. The CV of intermiss times increases for the $N = 1$ case, as cache size or associativity increase. There is a smaller CV in intermiss times when switching on L1 misses than when switching on L2 misses. Furthermore, the CV increases more slowly with multithreading in the L1-switch case. As the associativity increases, the CV for the L1-switch case becomes similar to that of the L2-switch case.

3.4 Impact on Response Time

The final set of results considers the mean job response time in multithreaded computer systems. Intuitively, one might expect that the above improvements in processor throughput with multithreading is obtained at the expense of an increase in mean job response time. We use the model of Section 2.3 to demonstrate and explain why this is not the case. In each of our modeling experiments, average response time measures were obtained for each of the multithreaded microarchitectures simulated. The performance trends were quite similar for each of these cases under a fixed set of parameter distributions. To illustrate a representative sample of these trends, we plot in Fig. 2 the ratio of the mean response time for $N = 1$ to that for $N = 2$ and $N = 4$, under the case where jobs are submitted according to a Poisson process and the computing demands of these jobs have an exponential distribution.

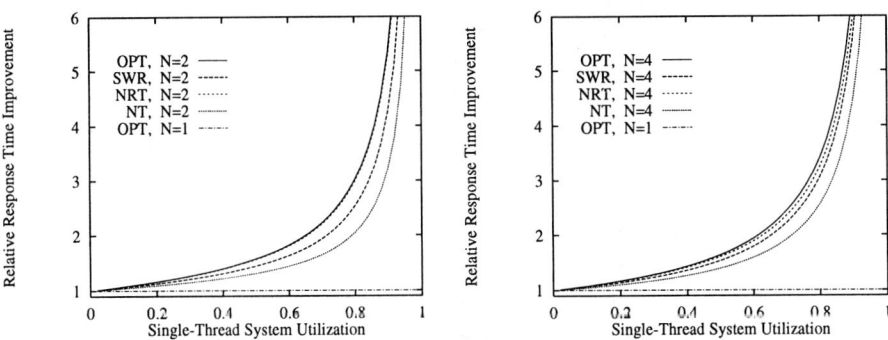

Fig. 2. Ratio of Single-Threaded Mean Response Time to Multithreaded Mean Response Time with N=2 and N=4, as a Function of Single-Thread System Utilization.

Our results show that the mean job response time is significantly smaller in the multithreaded system under all thread-switch policies than in the single-thread system, with a proportionally greater reduction in mean response time than the increases in processor throughput. We also observe that the mean response time improvements in the multithreaded system increase considerably with the system load. To explain these characteristics, we first note that job response time is the sum of the job queueing delay and job execution time. Although multithreading can increase the job execution time, it reduces processor idle time and the queueing delays experienced by these jobs. Under very light loads, the system rarely has more than one job so that both systems provide equivalent performance. As the system load increases, however, the reduced queueing delays and idle times outweigh the increase in execution times. Thus, the multithreaded system provides smaller mean response times. The improvement in response time becomes even more significant at heavier system loads where queueing delays represent an increasingly dominant fraction of the mean job response time. The improvement also results from the increasing performance impact of processor idle times as the load rises, which causes the single-thread system to saturate (*i.e.*, response times grow unbounded) well before the multithreaded system. Finally, we note that our model shows larger response time improvements under more variable job arrival and/or service time distributions, and vice versa.

4 Performance Implications of System Design

4.1 Effect of Thread-Switch Policy

This section considers the effect of the different thread-switch policies on performance. The results in Section 3 show that different thread-switch policies do not affect cache miss rates significantly. Using these measures, Fig. 3 plots the relative throughput achieved by each of the policies, varying N and the machine parameters. Most of the performance difference between the policies is due to the different amounts of time spent waiting for cache misses to complete. As expected, OPT yields the best performance since it always switches to the thread with the shortest waiting time. NT yields the worst performance, sometimes as much as 30% worse than OPT. SWR and NRT perform reasonably well, and are almost always within 9% of OPT.

The most interesting performance differences occur at small numbers of threads. As N increases, the performance difference among the policies decreases. The performance gain from adding more threads also decreases. A similar effect occurs when increasing TS. The knee of the curve occurs at 3 or 4 threads in almost all cases for OPT, NRT and SWR. A shorter switch time brings the knee further out towards more threads, while the opposite is true for longer TS.

NRT is usually better than SWR for $N = 2$ because predicting which thread will be ready first is easier with only two threads to choose. This makes a lower bound of 50% prediction accuracy, and helps NRT. In comparison, SWR avoids making a prediction and always incurs a TS penalty. However, in two cases, NRT performs worse than SWR on two threads for reasons that we will describe below. For larger N, NRT and SWR perform almost equally well. When there are ready threads, the most likely thread to be ready is the next thread in the round-robin ordering. This makes NRT's prediction accuracy high, and both NRT and SWR incur a TS penalty. SWR incurs an additional cycle for determining which thread is ready while NRT occasionally incorrectly predicts which thread will be ready first. When there are no ready threads, it is usually because all threads are waiting on an L2 miss. In this case, NRT ends up switching to the next

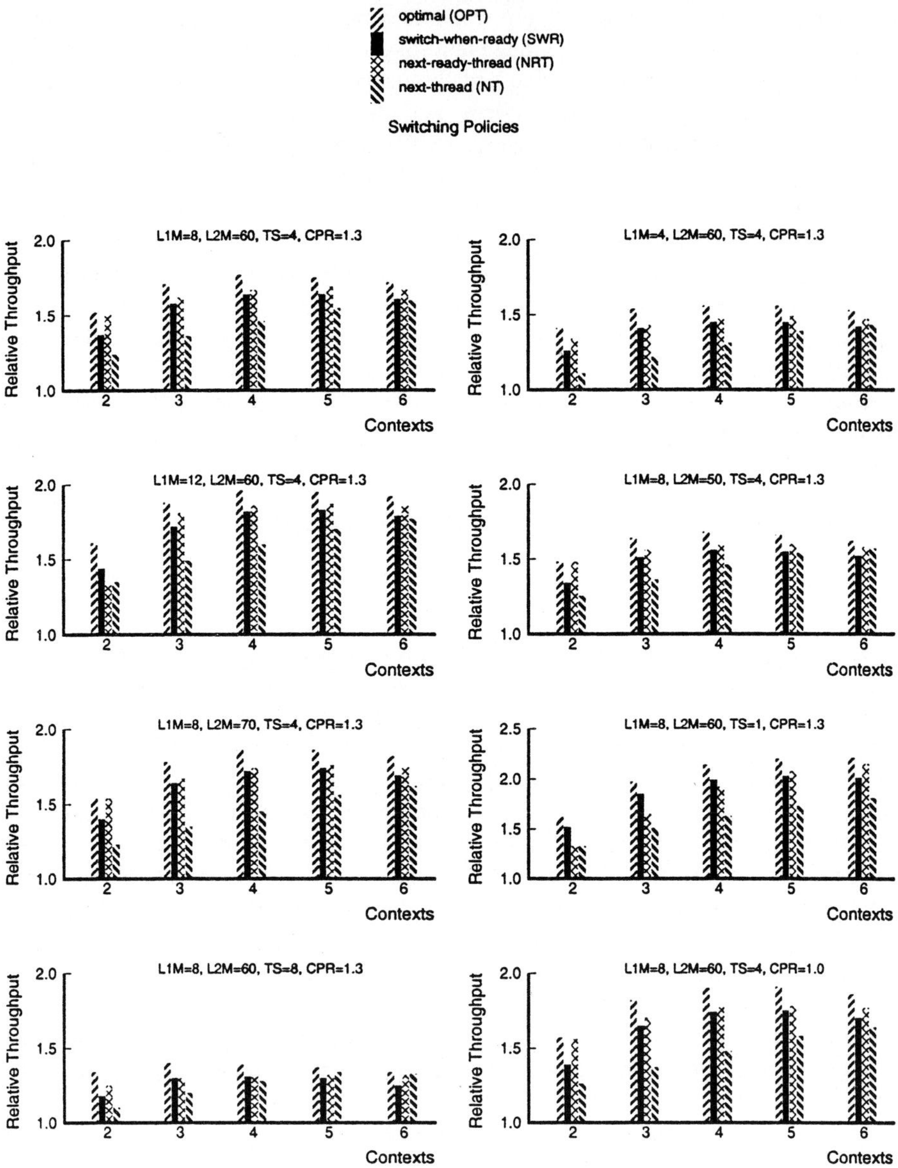

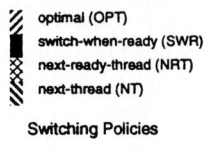

Fig. 3. Relative throughput achieved by various thread switching policies (relative to single-thread throughput), varying the number of threads and machine parameters. Cache parameters are: L1/L2 line size = 32/64 bytes, L1/L2 cache size = 16KB/512KB, associativity = 1.

thread in the round robin order and waiting for it to becomes ready after visiting all of the other threads. In this case, the prediction accuracy of NRT is good and SWR incurs a thread switch penalty while NRT usually does not. However, note that it is rare to have all threads waiting when N is large, so NRT has only a slight advantage.

There is a quirk in NRT that occurs whenever the ratio between L1M and TS is larger than N. Consider the $N = 2$ case where one of the threads is waiting for an L2 miss. The second thread is likely to incur an L1 miss, causing the processor to switch back to the thread waiting on the L2 miss, which is probably still pending. The processor then switches back to the thread that had the L1 miss. Because its L1 miss time is longer than twice TS, the L1 miss is still pending. Finally, the processor switches to the thread with the L2 miss and waits for the L2 miss to complete because it has already visited all the threads, even though the L1 miss will likely complete first. Thus, the processor ends up waiting unnecessarily. This situation occurs frequently and limits the performance gains of NRT. This performance quirk can be seen in Figure 3 when L1M increases from 8 to 12 cycles or when TS decreases from 4 to 1 cycle.

The naive NT policy yields the worst performance. Although it consistently shows a performance gain with increasing threads, the gain is significantly smaller than the other policies. Therefore, statically predicting that the next thread in the round-robin ordering will be the first thread ready to execute is not a good predictor. For example, consider an L1 miss that causes a switch to the next thread that had incurred an L2 miss. NT waits for the L2 miss to complete even though the L1 miss completes first. This occurs frequently, limiting the performance gain of this policy. As N increases, the probability of waiting on an L2 miss while another thread has a satisfied L1 miss decreases, and the performance gap between NT and the other policies narrows. Also, NT is more sensitive to L2M because NT frequently waits for L2 latencies.

4.2 System Design Effects

The results in Section 3 show the effect of changing the system parameters, namely latencies, under various thread-switch policies for the base configuration. The cache miss rates change since different latencies result in a different order of thread execution.

When L1M increases, the throughput decreases but the gain from multithreading increases. The larger the L1M, the greater is the benefit possible by switching threads. This result (i.e., relative throughput) contrasts with thread switching on L2 misses where the multithreading benefit was larger for smaller L1M since the L1 miss latency was not covered by thread switching. In terms of absolute throughput, smaller L1M results in better throughput regardless of switching policy. This is true for thread switching on either L1 or L2 misses. For L2-miss switching, however, throughput is significantly better for smaller L1M.

When L2M increases, the throughput decreases but the gain from multithreading increases, as for L1M. At 4 threads the degradation from a larger L2M is negligible for thread switching on L2 misses. For thread switching on L1 misses under NT, the throughput degradation is significant. If the next thread has an L2 miss, the processor will wait regardless of the other thread's status. When the latency is smaller, this case happens less often. Throughput degradation for other L1 thread-switch policies is somewhat visible, but not as significant.

The effect of TS is considered next, and the results are related to the effect of changing the cache miss latencies. Since L1 misses are considerably shorter than L2 misses, and switch times are comparable to L1-miss times, there is an increased sensitivity to TS when switching on an L1 miss. When L1M equals TS, the benefit is similar to that of switching on L2 misses; there is some improvement because switching on L1 misses that become L2 misses is better than waiting for the L2 miss to happen. As the ratio of L1M to TS increases, the potential benefit of multithreading grows. For example, there is about a 25% performance gain for OPT from TS=8 to TS=4, and another 25% from TS=4 to TS=1. For NRT with $TS \ll L1M$ (such as the $TS = 1$ and $L1M = 8$ cases) the performance (i.e., throughput) is not as good as that in OPT or SWR, and NRT behaves like NT especially when N is small. Similar degradation for NRT is found when $N = 2$, $TS = 4$, and $L1M = 12$. The number of threads to achieve peak performance also tends to increase compared to L2-miss switching.

Finally, with a smaller CPR, multithreading provides a larger relative gain. The processor portion of the single-thread execution time is smaller, resulting in a greater portion of the total time available to be covered by multithreading.

5 Related Work

Multithreading has been proposed and studied in several different contexts over a number of years. One such area consists of fine-grained multithreaded processors, such as Tera [2], where a thread switch occurs at each cycle in a round robin fashion. Since each stage of the processor pipeline contains a different thread, this approach avoids pipeline stalls and eliminates the logic to detect such stalls. A potential drawback of this approach is poor single-thread performance, since a single thread cannot occupy more than one pipeline slot at any time. In this paper, coarse-grained multithreading is of most relevance because we focus on commercial systems where response time is important and there is coarse-grained parallelism among the tasks.

Coarse-grained multithreaded processors are largely based on commodity processor designs, and switch threads on long-latency events. Single-thread performance is better because a single thread can occupy the entire pipeline. An example is the Sparcle processor in the MIT Alewife machine [1] that maps multiple threads into multiple register windows of a modified SPARC processor. A number of architectural techniques in addition to multithreading are considered for reducing and tolerating long latencies by Gupta *et al.* [10].

Laudon *et al.* [11] propose a pipeline interleaving of multiple threads and compare the performance of fine-grained and coarse-grained multithreading. They show that uniprocessor improvements for coarse-grained multithreading are small under a 34-cycle minimum L2M and a 7-cycle TS on data-cache misses, which is significant compared to their 9-cycle L1M. In contrast, we examine coarse-grained multithreading in the context of a commercial workload with few limits on parallelism and a relatively short pipeline with 4-cycle TS. And, we find significant performance improvements for coarse-grained multithreading with only two threads. The different assumptions are the basis for our different conclusions. We believe that interleaving of threads will improve throughput over what we are currently reporting.

To improve the utilization of the functional units of a superscalar processor, Tullsen *et al.* [13] dispatch instructions from several threads at each cycle, eliminating the notion of switching threads. Farrens and Pleszkun [9] measure multithreading performance on a processor with two threads and compare several thread switching policies.

6 Conclusions

The rate of improvement in processor speeds clearly exceed the rate of improvement in memory speeds. The characteristics of commercial application environments magnify these effects to cause a significant fraction of idle processor cycles due to cache misses. Future commercial systems will therefore be limited by local memory access times unless the designs incorporate methods to tolerate or avoid long memory latencies. This paper examined the effectiveness of coarse-grained multithreaded architectures that switch on a first-level cache miss in a two-level cache system for improving the performance of uniprocessor on-line transaction processing environments.

The performance benefits over a variety of L1 and L2 cache configurations and latencies show significant throughput improvements over a single-threaded processor, with a proportionally greater reduction in mean response times. Multithreading has a relatively small effect on cache miss rates. Our results also demonstrate that the potential benefits of switching on L1 misses are significantly larger than for switching on L2 misses. The thread-switch policy has a large impact on multithreading performance when switching on L1 misses. In particular, our results show that NRT performs relatively close to OPT in most cases, whereas NT and naive hardware support for choosing the next thread are not as effective. NRT performs worst, relative to OPT, when N is less than the ratio of L1M to TS. In this case it is possible for the processor to be stalled on the L2 miss of one thread while other threads are ready to execute since their L1 misses have been satisfied. A pure round-robin strategy for selecting the next thread to execute will eliminate these problems, but could potentially introduce other problems. We are currently investigating schemes that provide the benefits of these two approaches while eliminating their shortcomings.

Our simulations were largely based on a CISC environment. We now consider the implications of our results on RISC-based architectures. High-end RISC processors have increasingly small cycle times, placing a severe constraint on the size of L1 caches. The trend in RISC architectures is towards multiple cache levels with small, low-associativity L1 caches and a larger, slower L2 cache. This trend implies that L1 cache misses will occur more frequently with a higher miss penalty on RISC processors. Also, RISC instruction footprints are larger than CISC due to a larger number of instructions, further increasing L1-I misses. This makes it more important to switch on L1 misses as well as L2 misses in RISC processors. Another RISC trend is towards longer cache line sizes to take advantage of memory bandwidths and locality of reference. Comparing configurations one, three and five of Table 1 shows that longer cache lines tends to reduce the cache miss rate. This implies that there is significant locality in commercial workloads that should favor RISC processors with longer cache lines.

Multithreading architecturally presents itself to the software as multiple processors. Absolute throughput is maximized with about 4 threads in most cases. This confirms our previous findings that tens or hundreds of threads per processor are not required for the environments considered herein. The low sensitivity of the throughput gains to TS suggests that the pipeline state need not be duplicated, thus avoiding a potential impact to cycle time and chip area. Moreover, the ability of multithreading to reduce sensitivity to memory latency allows less expensive system costs without significant performance degradation. Multithreading increases performance significantly relative to the chip area required for implementation. The predominant trends in computer architecture and software indicate that multithreading should become more important for commercial systems.

References

1. A. Agarwal, J. Kubiatowicz, D. Kranz, B.-H. Lim, D. Yeung, G. D'Souza, and M. Parkin. Sparcle: An evolutionary design for large-scale multiprocessors. *IEEE Micro*, 13(3):48–60, 1993.
2. R. Alverson, D. Callahan, D. Cummings, B. Koblenz, A. Porterfield, and B. Smith. The Tera computer system. In *Proc. of the Intl. Conf. on Supercomputing*, pp. 1–6, Jun. 1990.
3. T. Anderson, H. Levy, B. Bershad, and E. D. Lazowska. The interaction of architecture and operating system design. In *Proc. of the Intl. Conf. on Arch. Support for Prog. Lang. and Operating Systems*, 1991.
4. K. Boland and A. Dollas. Predicting and precluding problems with memory latency. *IEEE Micro*, 14(4):59–67, 1994.
5. B. Calder, D. Grunwald, and B. Zorn. Quantifying behavioral differences between C and C++ programs. Technical Report CU-CS-698-94, University of Colorado, Jan. 1994.
6. Z. Cvetanovic and C. Bhandarkar. Characterization of Alpha performance using TP and SPEC workloads. In *Proc. of the Intl. Symp. on Comp. Arch.*, pp. 60–70, Apr. 1994.
7. R. J. Eickemeyer, R. E. Johnson, S. R. Kunkel, B.-H. Lim, M. S. Squillante, and C. E. Wu. Evaluation of multithreaded processors and thread-switch policies. Technical report, IBM Research Division, Jul. 1996.
8. R. J. Eickemeyer, R. E. Johnson, S. R. Kunkel, M. S. Squillante, and S. Liu. Evaluation of multithreaded uniprocessors for commercial application environments. In *Proc. of the Intl. Symp. on Comp. Arch.*, pp. 203–212, May 1996.
9. M. K. Farrens and A. R. Pleszkun. Strategies for achieving improved processor throughput. In *Proc. of the Intl. Symp. on Comp. Arch.*, pp. 362–369, May 1991.
10. A. Gupta, J. Hennessy, K. Gharachorloo, T. Mowry, and W.-D. Weber. Comparative evaluation of latency reducing and tolerating techniques. In *Proc. of the Intl. Symp. on Comp. Arch.*, May 1991.
11. J. Laudon, A. Gupta, and M. Horowitz. Interleaving: A multithreading technique targeting multiprocessors and workstations. In *Proc. of the Intl. Conf. on Arch. Support for Prog. Lang. and Operating Systems*, pp. 308–318, Oct. 1994.
12. A. M. Maynard, C. M. Donnelly, and B. R. Olszewski. Contrasting characteristics and cache performance of technical and multi-user commercial workloads. In *Proc. of the Intl. Conf. on Arch. Support for Prog. Lang. and Operating Systems*, pp. 145–156, Oct. 1994.
13. D. M. Tullsen, S. J. Eggers, and H. Levy. Simultaneous multithreading: Maximizing on-chip parallelism. In *Proc. of the Intl. Symp. on Comp. Arch.*, pp. 392–403, Jun. 1995.
14. W. Wulf and S. McKee. Hitting the memory wall: Implications of the obvious. *Comp. Arch. News*, 23(1):20–24, 1995.

A Multithreaded Implementation Concept of Prolog on Datarol-II Machine

Peter Kacsuk*   Makoto Amamiya**

*MTA-MSZKI Research Institute
of the Hungarian Academy of Sciences
H-1525 Budapest, P.O.Box 49, Hungary
kacsuk@sunserv.kfki.hu

**Kyushu University
6-1 Kasugakoen Kasuga-shi
Fukuoka 816 Japan
amamiya@is.kyushu-u.ac.jp

Abstract

The paper presents a massively parallel implementation method of Prolog on the multithreaded parallel machine, Datarol-II. First the Logicflow model is introduced which was developed for implementing Prolog on massively parallel computers. The Logicflow is a dataflow-like graph in which nodes are macro dataflow nodes and tokens represent macrothreads. The Datarol-II architecture efficiently supports both the management of macrothreads derived from Logicflow Model and management of microthreads created when remote loads are necessary. The architecture of the Datarol-II machine and the macrothread/microthread management of Prolog programs are described in detail.

1. Introduction

The multithreading approach is promising for hiding remote memory access and synchronisation latencies in massively parallel machine. However, the application of this new architecture requires new compilation techniques and execution models for language implementations, too. One of the most accepted AI language is Prolog which requires large computational power for solving complex AI problems.

Multithreaded architectures are often considered as hybrid dataflow architectures. The Datarol-II machine also belongs to this class of multithreaded machines [1][9]. A natural idea is to use data-driven approaches for programming these machines. A number of research works have been pursued in order to implement Prolog based on the dataflow execution model [6][8][7][3][4][2][17], however, only one of them aims at implementing Prolog on multithreaded machines [17].

In the current study we investigate the possibility of implementing Prolog on the multithreaded Datarol-II parallel computer based on the Logicflow execution model of Prolog. The Logicflow Model inherits the dataflow principles, and has been developed for implementing Prolog on massively parallel computers. A parallel Prolog abstract Machine, called 3DPAM (Distributed Data Driven Prolog Abstract Machine) has been defined based on the Logicflow Model. The 3DPAM abstract machine has been implemented on several multi-transputer machines including a 34-processor Supernode machine. Though the multi-transputer implementations have justified that Prolog can efficiently be implemented on distributed memory computers based on the Logicflow Model, they also revealed some weaknesses of such implementations. Namely, the memory consumption was extremely high due to the structure copying nature of the closed binding environment concept [5] applied in the 3DPAM. The other time consuming bottle-neck was the necessary loading and unloading mechanism that was frequently applied to load machine registers from tokens and vice versa.

All of these drawbacks can be eliminated on a multithreaded architecture where special hardware mechanism supports fast access to registers and where due to the memory latency hiding mechanism intensive structure copying is not necessary anymore. Such a multithreaded architecture is Datarol-II that was designed at the Kyushu University and is now under construction. The implicit load/store mechanism of Datarol-II will significantly accelerate the machine register access mechanism of 3DPAM eliminating of the main drawbacks of its current massively parallel implementation. The continuation based split-phase execution mechanism of Datarol-II makes it unnecessary to use the closed binding environment concept and as a result the frequent copying of compound terms can be avoided.

In the current paper we show how to apply the Logicflow execution model and how to modify the 3DPAM abstract machine model in order to implement Prolog on the Datarol-II multithreaded parallel computer. Section 2 briefly summarises the main features of the Logicflow Model, and the 3DPAM abstract machine, while section 3 overviews those properties of the Datarol-II architecture which are relevant from the viewpoint of the Prolog implementation. Section 4 describes how the multithreaded computation can be exploited in the 3DPAM abstract machine. Section 5 explains the mapping of the 3DPAM on the Datarol-II architecture highlighting the benefits of register and memory management.

2. LOGFLOW and 3DPAM

LOGFLOW is a parallel Prolog system which is able to exploit inherent OR- and pipeline AND-parallelism of Prolog programs. LOGFLOW is based on the Logicflow Model which is a generalised version of the dataflow execution model. Computation in the Logicflow Model is based on the Logicflow Graph (LG) similar to a dataflow graph. Nodes of the Logicflow Model are specialised towards Prolog oriented activities like unification, OR-connections, CUT, etc. Detailed description of the nodes of the Logicflow Model can be found in [15]. Every Prolog program is translated into the

Logicflow Graph that represents the static nature of the Prolog database. A simple Prolog program and its Logicflow Graph is shown in Fig. 1.

The dynamic behaviour of the Prolog program is driven by tokens. Tokens moving

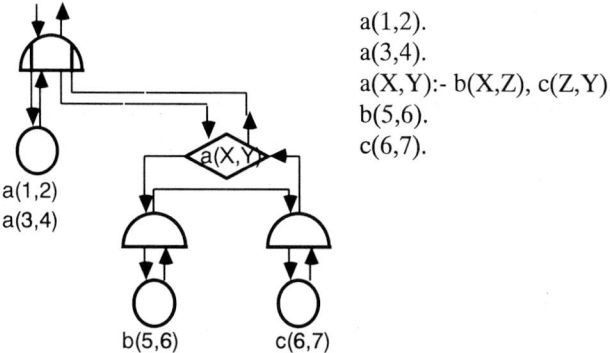

Fig. 1. LG of the example program

on the arcs of the Logicflow Graph represent queries (DO and SUB tokens) or results of the resolution (SUCC and FAIL tokens). LOGFLOW delivers all possible solutions for a query. These solutions are collected and transferred in the Logicflow Graph by token streams (SUCC or SUB streams). The last element of a token stream is always a FAIL token indicating that there is no more possible solution for the query.

3DPAM (Distributed Data Driven Prolog Abstract Machine) is the abstract machine that implements Prolog based on the Logicflow Model. The structure of the 3DPAM is depicted in Fig. 2. The rectangles represent the basic data structures while the circles show the basic processes executed in parallel. Two queues for waiting tokens are applied: LTQ (Local Token Queue) contains tokens to be processed by the local processor and RTQ (Remote Token Queue) temporarily stores tokens targeted to other processors.

The 3DPAM code of a Prolog program is logically equivalent to the Logicflow Graph of the program. In other words the LG is represented by the 3DPAM code area based on the 3DPAM instructions. (Details of the 3DPAM instruction set can be found in [11].) The 3DP Engine realises the generalised dataflow execution scheme of LOGFLOW. It takes a token from the LTQ, loads the token registers based on the contents of the selected token and executes the 3DPAM code representing the LG node where the token is targeted. Executing the node program, the 3DP Engines either creates no new token, or it generates one or more tokens. In the former case, it fetches a new token from the LTQ. If it generates exactly one token it keeps the token and continues the execution based on that token. If several tokens were generated by the engine, one of them is kept by the engine and the others are passed through the RTQ to the Output Token Manager which realises a granularity control and load balancing strategy [14] and decides the target of the tokens. The Input Manager process receives tokens from other processors and places them into the LTQ. Detailed explanation of the 3DPAM can be found in [12].

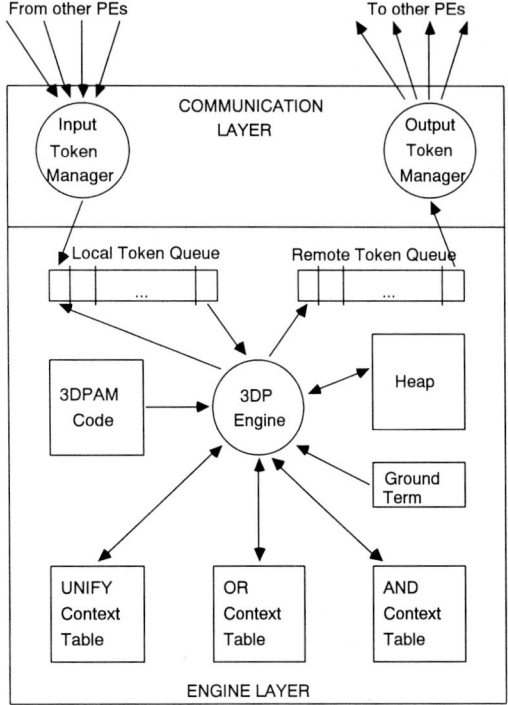

Fig 2. Structure of a 3DPAM processor

3. Datarol-II Architecture

The Datarol-II architecture [9] was designed to eliminate the drawbacks of the original Datarol machine [1]. These drawbacks (unable to use high speed registers and pipeline technique of conventional RISC processors) severely limited the performance of the Datarol machine and hence, an optimised redesign became necessary. This new version, called Datarol-II extracts fine-grain threads from a dataflow graph (in case of our Prolog implementation from the Logicflow Graph), and executes these threads by means of a program-counter-based pipeline equipped with high-speed registers similar to that of the conventional RISC processors. However, the continuation-based thread activation control, which includes thread synchronisation and split-phase operations, is still executed in a circular pipeline, as in the original Datarol machine.

In Datarol-II, since several flow controls are performed by a program-counter-based control, the number of tokens in the circular pipeline is reduced. Reducing the number of tokens is very important since the throughput of the circular pipeline, especially that of the synchronisation unit, is likely to be a bottleneck in the processor.

In addition to this execution control, an implicit register-loading mechanism is embedded in the execution pipeline in order to reduce the context switching overhead. This mechanism overlaps memory accesses with instruction executions. A two-level hierarchical memory system and a load control mechanism are introduced in order to reduce the local memory access latency.

The structure of the Datarol-II processor is shown in Figure 3. The Datarol-II processor consists of a Function Unit(FU), a Memory Unit(MU), a Communication Unit(CU), an Activation Controller(AC), a Ready Queue(RQ), a Structure Memory Unit(SMU), and a Network Interface(Net-IF).

FU is the main processing unit in the processor. MU holds the logical register values for each instance. (An instance in Datarol corresponds to a process.) CU handles the input packets for the processor, and AC controls the activation of threads. RQ is a queue for waiting threads. SMU controls the I-structure-based memory access.

The Datarol-II processor has two execution control mechanisms: a program-

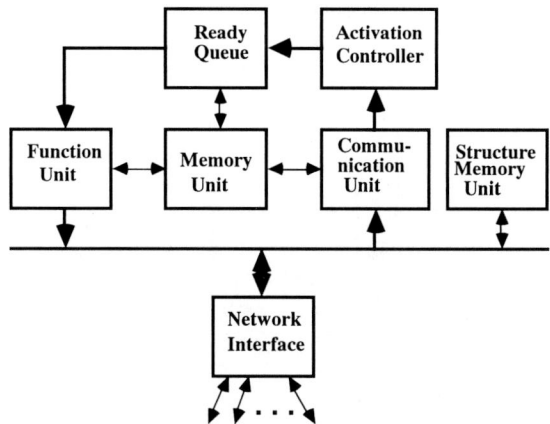

Fig. 3. Structure of a Datarol-II processing element

counter-based one for sequential threads execution, which is performed in FU, and a continuation-based one for split-phase execution and synchronisation, which are performed in a circular pipeline consisting of FU, CU, AC and RQ.

In Datarol-II, threads are invoked by packets. Each instance has logical registers, and matching counters (currently eight) which are used for synchronisation. Using these logical registers and matching counters, threads are activated by a data-driven mechanism.

4. Multithreaded Computation in the 3DPAM Prolog Abstract Machine

In applying the Logicflow Model for the Datarol-II architecture we have to distinguish two kinds of threads:
   1. A macrothread (instance) represents a logical thread derived from the Logicflow computation model.
   2. Microthreads are created at any time when remote memory loads are necessary.

A macrothread can be dynamically divided into microthreads when remote memory load occurs. Creating microthreads is a way of hiding remote memory latency (see details in [9]). On the contrary, macrothreads are logical entities to organise parallel activities based on the Logicflow Model.

There are two sources of creating new macrothreads in the Logicflow Model:
1. OR node that creates activation-thread
2.   UNIT node that creates solution-threads and one fail-thread

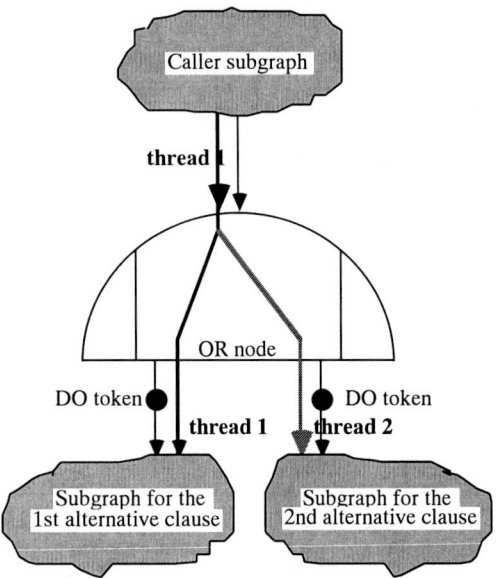

Fig. 4. Creating activation-threads by the OR node

When a thread execution reaches an OR node activation, the OR node executes a fork-like operation by copying the incoming DO token into two copies. The left one represents the continuation of the original thread, while the right one initiates a new thread that can be allocated to any processor of the machine based on load measuring information. After allocating the second DO token for a processor the corresponding thread is started on the selected processor. From that moment the two threads can progress in parallel on two different processors. Figure 4 shows the thread creation mechanism of the OR node. Notice that the OR-thread represents the source of OR-parallelism exploited by the parallel system.

A UNIT node can represent several unit clauses which all are candidates of releasing a new solution-thread after a successful unification. The UNIT node takes the input DO token and executes unfication with each clauses it represents. Whenever a successful unification is performed, a new solution-thread is created in the form of a SUCC token. This SUCC token should be returned to the caller AND node which transforms it to a SUB token and directs the SUB token to the next node of the UNIFY/AND ring (see Fig. 5.). If the next node is an AND node, the SUB token is transformed to a DO token that can be again freely allocated to any processor providing physical parallelism in the exploitation of pipeline AND-parallelism. Figure 5 depicts the thread creation mechanism of the UNIT node and its application for pipeline AND-parallelism. It can be seen from the figure that each thread creates a different instance of the subgraph belonging to the second goal of the rule body. In this way, the producer goal and the different instances of the consumer goal can work in parallel based on the Logicflow Model.

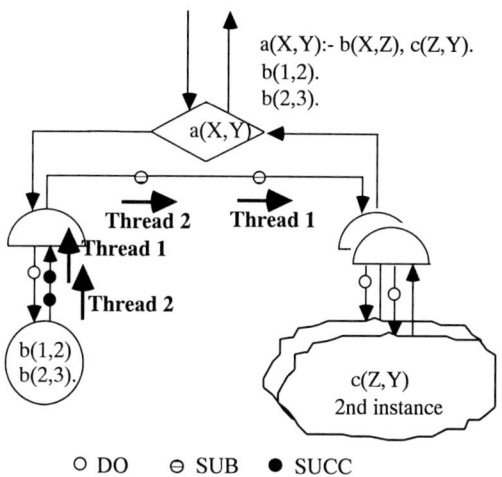

Fig. 5. Creating solution-threads by the AND node

If the UNIT node has completed unifications with all the represented clauses, a fail-thread is created by the UNIT node in the form of a FAIL token. The fail-threads do not represent any Prolog like activity, their only role is to co-ordinate the work of solution-threads as described by the Logicflow Model.

Notice that the other tokens of the Logicflow Model are not used to initiate threads, they represent parts of already existing threads. However, from the point of view of migrating threads from one processor to another one, they are valuable units of moving threads among processors. As it was shown in [14] request tokens (DO and SUB) can be migrated among processors at any time and this is the way how load balancing can be achieved in the massively parallel architecture. However, reply tokens (SUCC and FAIL) are always obliged to return to the caller node of the Logicflow Graph, i.e. to the processor that executed the code of the caller node. This return information is provided by the return address fields of the caller DO tokens and transformed to be the target field of the reply tokens.

Prolog programs are translated to the Logicflow Graph represented in a textual format in the code memory by applying the instruction set of the 3DPAM [11]. The code generator of the 3DPAM compiler is based on the thread concept, i.e. it generates code segments according to the thread concept. Here we show a typical example of thread code generation.

The following series of activities in the Logicflow Model can be represented by one thread:

1. The UNIFY node consumes a DO token, executes a successful unification and creates a SUB token on its request.out arc.
2. The first AND node of the UNIFY/AND ring consumes the SUB token and creates a DO token.

The DO token could belong to the same thread but since, it can be used to create a new instance of the same Prolog predicate, it is used as the initiator of a new macrothread. Notice however, that the number of threads remained unchanged in the system since, the original thread is terminated. Figure 6 shows how the DO and SUB tokens are merged into a single macrothread and how this macrothread is replaced with the new macrothread at the output arc of the AND node. The figure also demonstrates the 3DPAM code of the macrothreads for the following Prolog predicate:

a(g(X,Y), f(Y,1),S) :- b(S,U), c(X,U).

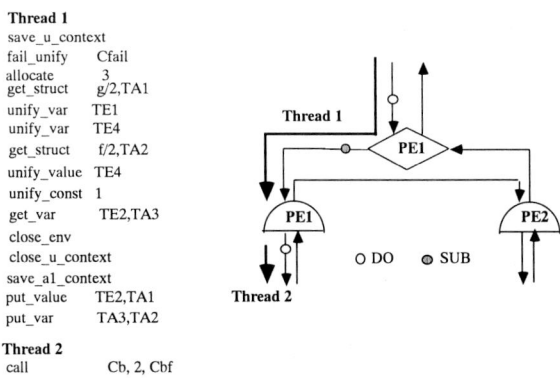

Fig. 6. Relationship between threads and tokens and the corresponding 3DPAM code

5. 3DPAM on Datarol-II

The structure of the 3DPAM machine is shown in Figure 2. In the current section we show how to map the 3DPAM abstract machine to the physical Datarol-II machine and how Datarol-II will execute Prolog programs based on the 3DPAM code.

5.1. Mapping 3DPAM to Datarol-II

If we compare the structure of 3DPAM and Datarol-II the similarity between the two architectures is obvious. The 3DPAM code is loaded into the Memory Unit of Datarol-II. The Function Unit of Datarol-II will play the role of the 3DP engine performing the 3DPAM instructions fetched from the Memory Unit. The Local Token queue of the 3DPAM is mapped to the Ready Queue of Datarol-II. Notice that only those tokens of the Logicflow Model will be stored in the ready queue that can initiate new threads or can be used to migrate threads among processors. Other tokens of the Logicflow Model implicitly occur in the token (activation) registers. The Input Token

Manager of the 3DPAM is realised by the Communication Unit of Datarol-II. The functions of the Remote Token Queue and Output Token Manager are played by the Network Interface of Datarol-II. The context tables of the 3DPAM are placed in the Memory Unit of Datarol-II.

The Heap of the 3DPAM is mapped into the Structure Memory Unit of the Datarol-II which is a distributed shared memory. One of the main differences of implementing the 3DPAM on distributed memory multicomputers and on the Datarol-II derives from the organisation of the Heap. In the distributed memory multicomputer implementation of 3DPAM, the Heap was distributed among local memories that could not be accessed by other processors and hence, intensive message passing and structure copying was necessary. In Datarol-II, the distributed shared memory concept dramatically reduces the need of copying data structures between processing elements. Even in the case when a remote access is necessary, the split-phase execution mechanism and the large number of available threads will hide the remote memory access latency. At this point we should emphasise the benefit of the Logicflow Model that can exploit a large amount of fine grain parallelism from logic programs providing the necessary number of threads even in a massively parallel computer. The multi-transputer implementations of the 3DPAM proved that even from small size Prolog programs like the queens problem large number of macrothreads (in the order of magnitude 1000) are generated by the Logicflow Model.

5.2. Executing 3DPAM on Datarol-II

Each executable 3DPAM thread is represented by a token which is placed in one of the Ready Queues of the Datarol-II in the form of a continuation. A 3DPAM token consists of the following registers:
    a/ token context registers:
        TID   Token Id (colour)
        TADDR Token Address (Address of the target node's code)
        TRPE   Token Return Processing Element
        TSR   Token Succ Return (Address of the sender CALL node)
        TFR   Token Fail Return
    b/ token environment registers (TEN,TE1,TE2,...)
    c/ token argument registers (TAN, TA1, TA2, ...)

Notice that TID corresponds to the colour, TADDR to the target node of Logicflow Graph and the return field is represented by <TRPE,TSR> in case of SUCC token and by <TRPE,TFR> in case of FAIL token. The environment field of the Logicflow Model is realised by the token environment registers, while the argument field by the token argument registers.

The token context registers play similar role as the control registers in the WAM and the token environment and argument registers correspond to the Ai registers of the WAM. The efficiency of the WAM comes from the optimised use of the WAM registers. Similarly, the optimised use of token registers plays a central role in the definition of the 3DPAM instruction set. More than that the efficient loading and storing of these registers represent a crucial architecture design issue.

The implicit load/store mechanism of Datarol-II plays a key role in realising efficient register access in the 3DPAM implementation. In the Datarol architecture, each macrothread (instance) has its own logical register set which corresponds to the token registers of the 3DPAM. The Memory Unit holds these logical register sets for each macrothread in the Operand Memory. In fine-grain program execution, like the parallel Prolog execution mechanism based on Logicflow, memory accesses are issued so frequently that memory access latency should be kept very low. In order to achieve fast memory accesses from FU (3DP Engine), a high speed memory is introduced in the Datarol-II processor. This high speed memory is called Register Buffer(RB). Figure 7 shows the structure of MU. RB is accessed by a page number and an offset address. Each page holds the register values of one macrothread, i.e. the token registers of the token that represents the macrothread. The offset address corresponds a token register inside the page holding the token register set. The token register values of a new macrothread are loaded into RB by the loading request from the Ready Queue before FU (3DP Engine) starts the new macrothread execution.

FU can have several register sets (currently four) and each register has a presence bit which indicates whether the register value is valid or not. FU also has temporary registers used for data passing from instructions to instructions within the same thread. When FU starts to execute a new thread, one of the register sets is allocated to the new thread. In this register set allocation, if there is a register set already allocated to the same macrothread, FU reuses it. Otherwise, one of the register sets is selected and the presence bits of the register set are cleared. FU uses this register set as a current token register set. As the token register sets are stored in MU, FU has to read the token register values from MU before they can be accessed. The implicit load mechanism performs this register loading automatically by a hardware means.

A similar automatic store mechanism is available in the five-stage pipeline execution mechanism of FU. The five stages of the pipeline and their data access mechanisms are shown in Figure 8. The five stages are:

1. Instruction Fetch (IF)
2. Instruction Decode and Register Check (ID&RC)
3. Operand Fetch (OF)
4. Execution and Write Back 1 (EX&WB1)
5. Write Back 2 (WB2)

The IF stage fetches an Instruction Memory (IM) cell, whose address is specified by the Instruction Pointer (IP), and passes the fetched instruction to the next stage. If the termination bit of the fetched instruction is on, FU gets a new thread ID from RQ and sets it into IP. Otherwise, FU increments IP, i.e. processing the current thread continues. The ID&RC stage decodes the instruction and determines source and destination register addresses. Then the presence bits of the source registers are checked. In the OF stage, source register values are read from the register-set or MU according to the RC stage result. The EX&WB1 stage executes the instruction. When the OF stage reads the source register value from MU, the read data is stored in the corresponding register. In the WB2 stage, the result data is stored into both the FU register and the MU cell at the same time.

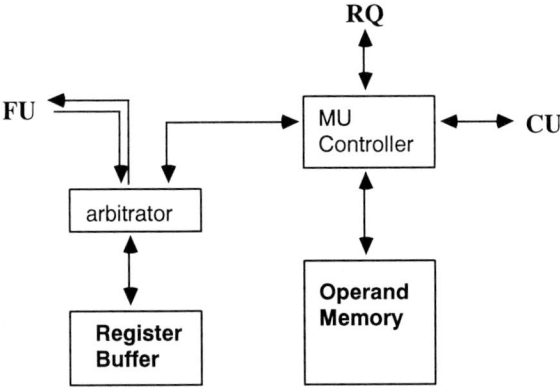

Fig. 7. Memory Unit of the Datarol-II

By means of this pipeline, MU access is achieved implicitly, and no explicit load/store instructions are needed. As the data access to MU is achieved in parallel with execution, load/store overhead is hidden. Notice that the load and store of token registers in the multi-transputer implementation of 3DPAM has been executed by time consuming software operations leading to large overhead in case of token switching. It means that the Logicflow Model needs a computer architecture like Datarol-II in order to eliminate the thread switching overhead. On the other hand Datarol requires a parallel Prolog execution model like the Logicflow Model since, the utilisation of the five-stage pipeline can be optimal only if large number of threads are available. The Logicflow Model can exploit a large amount of parallelism in logic programs and therefore represent an ideal computation model to implement Prolog on Datarol-II.

5.3. Memory Management of 3DPAM on Datarol-II

In order to exploit the fine-grain data driven nature of Datarol-II, a parallel logic programming computational model is needed that can produce the necessary large number of tokens. This approach needs a radically different organisation of the Prolog binding environment. In Logicflow we have solved this problem by replacing the conventional 4-stack solution with a combination of context and token tables (Figure 9). UNIFY, UNIT and AND Context tables store the state of the UNIFY, UNIT and AND nodes of the Logicflow Graph and they are logically equivalent with the Local Stack of the WAM machine. OR Context table holds information that makes this table logically similar to the Choice Point Stack of the WAM[1], while the role of Heap is preserved in the 3DPAM abstract machine.

---

[1] Since 3DPAM is an all-solution engine, i.e. it generates all the possible solutions without backtracking, the role of OR context table is similar but not equivalent to Choice Point Stack in WAM. For the same reason the Trail information appears in the appropriate fields of tokens.

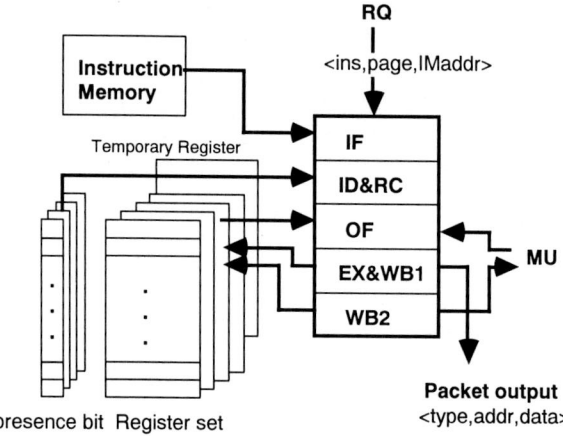

Fig. 8. Function Unit of the Datarol-II processor

The binding environment was redesigned in order to reduce the intraprocessor overhead [21]. The new binding environment tries to combine the benefits of the closed and other non-closed environments. The idea is that while computation remains on a single PE, variables occurring in structures are handled according to one of the well-known global addressing methods, and they form a closed environment (local addressing scheme) when they are migrated between processing elements [22]. In such a way, the most expensive part of the closing procedure, the structure scanning and copying, can be discarded, more precisely, they can be postponed until the task is really migrated between processors.

In distributed memory implementations the Heap was distributed among local memories that could not be accessed by other processors. These implementations showed, that the memory consumption, especially the Heap utilisation of 3DPAM is extremely high. It is due to two reasons:

- 3DPAM uses a closed binding environment, where structures are copied quite often both at unification and at back-unification (intraprocessor overhead)
- when tokens are migrated between processing elements, structures which occur in either the arguments or the environment registers, must be copied to the memory of the new PE (interprocessor overhead)

Both reasons are caused by the fact that in distributed memory systems the access to other PE's memory is not supported at hardware level.

The global storage is a hash table (Figure 10), not equivalent to [23]. The organisation of hash table offers an easy way to duplicate tokens and thus, an effective and fast way to reach variables and due to its implementation, the unused cells can be reclaimed easily [21]. Global variables are detected at run-time by modifying some abstract instructions, e.g. unify_var, write_var, etc.

Detailed performance tests [24] showed (see some examples in Figure 11.), that the new environment reduced the execution time and the heap utilisation on a single processor (intraprocessor overhead.) On the other hand, the multiprocessor execution did not become significantly better and in some cases a slow-down was achieved due to the interprocessor communication. As other measurements showed, it happens due to a large amount of structure copying which occur at token migration (interprocessor overhead). Notice that the new binding scheme was not planned to eliminate this type of overhead. The Datarol-II offers a hardware solution for this problem.

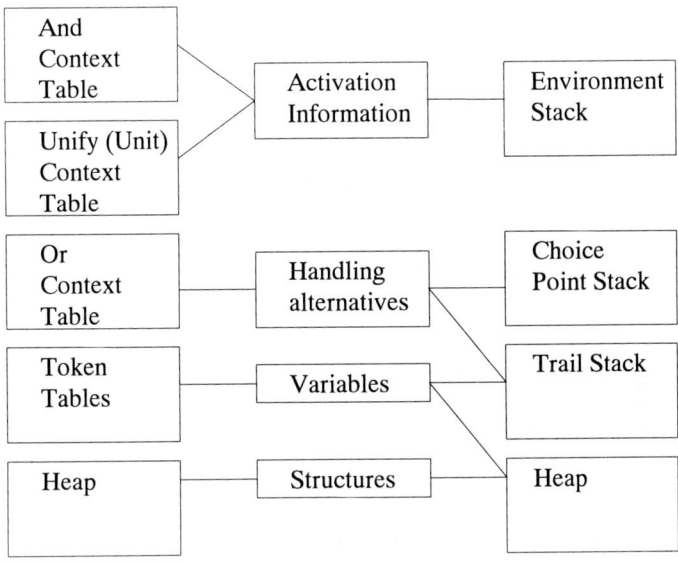

Fig. 9. Analogy between 3DPAM and WAM data structures

The Datarol-II has a distributed shared memory called Structure Memory which is an ideal place to hold the Prolog structures. In such a way structures need not be copied anymore when the corresponding token is executed on another PE. The multithreaded property of Datarol-II can hide the memory latency when accessing the Structure Memory.

According to the previous multi-transputer implementation of 3DPAM an the test results of the closed and hybrid binding environments, we can conclude, that such a hybrid environment would fit optimally the Datarol architecture:

1. It scraps the usual 4-stack implementation of Prolog abstract machine which either assumes a shared memory or yields in an intensive stack copying. From this point it fits the long-cycle execution of Datarol based on dataflow principle.
2. It handles the structures in a way which resembles the usual WAM implementation and requires a shared memory area. If structures were handled in a dataflow way, they would be copied each time they migrate to a different processing element thus jeopardising the overall performance.

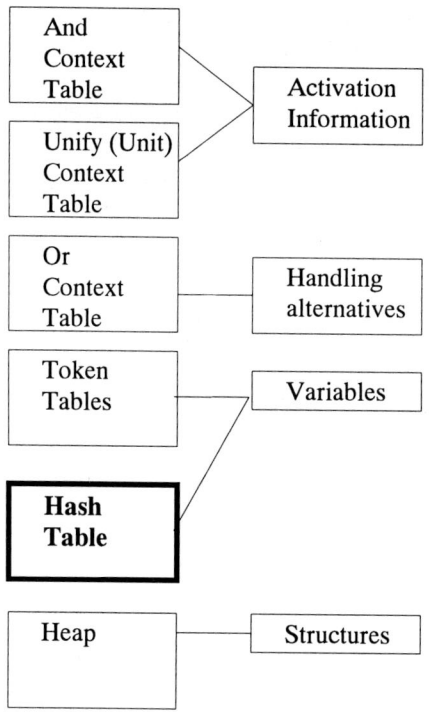

Fig. 10. The data structure of 3DPAM with the new binding environment

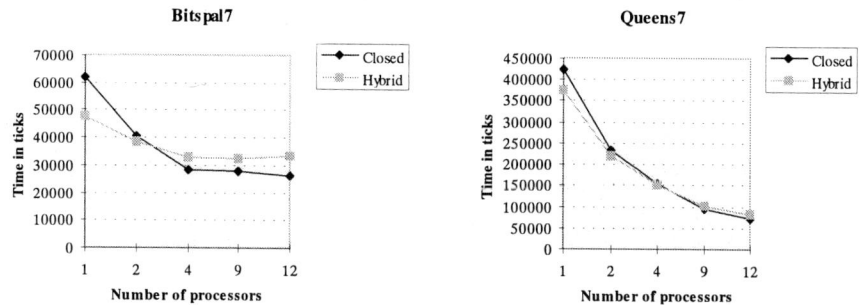

Fig. 11. Comparison of the closed and hybrid binding environment in case of 7-long bit palindromes and 7-Queens

Conclusions

We have shown in the paper that the Logicflow Model and the Datarol-II massively parallel multithreaded architecture are perfect match to implement Prolog on a massively parallel architecture. Based on the Logicflow Model both OR-parallelism and pipeline AND-parallelism can be exploited in an all-solution Prolog system.

The paper also demonstrated that the implicit load/store mechanism of Datarol-II can reduce significantly the administration overhead of the 3DPAM machine concerning the activation of threads. Another advantageous feature of Datarol-II is the split-phase execution control mechanism that eliminates the intensive structure copying that was necessary in the distributed memory multicomputer implementations of the 3DPAM.

The hybrid binding environment concept of LOGFLOW fits well the distributed shared memory architecture of Datarol-II. According to the performance tests in distributed memory environment, holding the structures in a shared memory - meanwhile the memory latency is hidden - could reduce the interprocessor communication overhead thus, the hybrid binding scheme and the shared memory together give a possible solution to the problem of structure handling. It must be emphasised, that Datarol-II behaves like a dataflow machine in its long-cycle executions and it is like a sequential von Neumann machine in its short-cycle execution. Neither the closed environment, nor the conventional 4-stack implementation but the hybrid environment described above can fit best this type of architecture.

Acknowledgement

The work described in the current paper was supported by the Monbusho International Scientific Research Program: Joint Research of Japan under the grant number 09044174, and the National Scientific Foundation (OTKA) of Hungary under the grant number T022106.

References

[1] Amamiya,M. and Taniguchi,R. Datarol: A Massively Parallel Architecture for Functional Language, Proc. Second IEEE Symposium on Parallel and Distributed Processing, 1990, 726-735
[2] Baiardi, F., Candelieri, A. and Ricci, L. *A Data-Driven Static Model for the Execution of Logic programs on Distributed Memory Systems*, Proc. of JICSLP'92 Post-Conf. Joint Workshop on Distr. and Parallel Impl. of Logic Prog. Sys., 1992
[3] Bic,L and Lee,C. *A Data-Driven Model for a Subset of Logic Programming*, ACM Trans. on Prog. Lang. Syst., vol. 9, no. 4. 1987, 618-645
[4] Biswas, P. and Tseng C. *LogDf: A Data-Driven Abstract Machine Model for Parallel Execution of Logic Programs*, Proc. of the Int. Conf. on Fifth Gen. Comp. Sys., 1988, 1059-1070
[5] Conery, J. S. *Binding Environments for Parallel Logic Programs in Non-shared Memory Multiprocessors*, Proc. of the 1987 Symp. on Logic Prog. Systems, San Francisco, 1987, 69-79
[6] Hasegawa,R. Amamiya,M. *Parallel Execution of Logic Programs based on Dataflow*, Proc. of the ICOT Conference 1984, 507-516
[7] Halim,Z. *A Data Driven Machine for OR-Parallel Evaluation of Logic Programs*, New Generation Computing, 1986, 5-33

[8] Ito,N. et al, *Data-flow based Execution Mechanism of Parallel and Concurrent Prolog*, New Generation Computing, 1985, no. 3, 15-41
[9] Kawano,T., Kusakabe,S., Taniguchi,R. and Amamiya,M. *Fine-grain Multi-thread Processor Architecture for Massively Parallel Processing*, in Proc. First IEEE Symp. High-Performance Computer Architecture, 1995, 308-317
[10] Kacsuk,P. *Execution Models of Prolog for Parallel Computers*, Pitman Publishing and the MIT Press, 1990
[11] Kacsuk,P. *Distributed Data Driven Prolog Abstract Machine*, in Implementations of Distributed Prolog, Edited by P. Kacsuk and M.J.Wise, John Wiley,1992, 89-118
[12] Kacsuk,P. *LOGFLOW-2: A Transputer Based Data Driven Parallel Prolog Machine*, in Proc. of the Transputer World Congress, Aachen, 1993, 1154-1169
[13] Kacsuk,P. *Memory Management in LOGFLOW*, in Proc. of the EUROMICRO Conf., Barcelona, 1993
[14] Kacsuk,P. *Wavefront Scheduling in LOGFLOW*, in Proc. of the 2nd EUROMICRO Workshop on Parallel and Distributed Processing, Malaga, 1994, 503-510
[15] Kacsuk,P. *Dataflow and Logicflow Models for Defining a Parallel Prolog Abstract Machine*, in Proc. of the PACT'94 Conf., Montreal, 1994, 289-298
[16] Kacsuk,P. *Execution Models for a Massively Parallel Prolog Implementation*, Keynote speech, Int. Workshop on Computational Models and their Applications, Cesme, Turkey, 1994
[17] Kim,H. and Gaudiot,J-L. *Exploitation of Fine-grain Parallelism in Logic Languages on Massively Parallel Architectures*, Proc. of PACT'94 Conf., Montreal, 1994
[18] Taylor,S. Safra,S. Shapiro,E. *A Parallel Implementation of Flat Concurrent Prolog*, in Concurrent Prolog Collected Papers, Edited by E. Shapiro, The MIT Press, 1987, 575-604
[19] Warren,D.H.D. *An Abstract Prolog Instruction Set*, Technical Note 309, SRI International, 1983
[20] Warren,D.H.D. *The SRI Model for OR-Parallel Execution of Prolog- Abstract Design and Implementation Issues*, Proc. of the 1987 Symp. on Logic Prog., 1987, 92-102
[21] Nemeth, Zs and Kacsuk, P. *Analysis and Improvement of the Variable Binding Scheme in LOGFLOW*, Technical report, 1996.
[22] Kim, H. and Gaudiot, J-L. *A Binding Environment for Processing Logic Programs on Large-Scale Parallel Architectures*, Technical Report, 1994.
[23] Borgwardt, P. *Parallel Prolog Using Stack Segments on Shared-Memory Multiprocessors*, Proceedings of the 1984 Symposium on Logic Programming.
[24] Nemeth, Zs and Kacsuk, P. *Evaluation of the hybrid binding scheme in LOGFLOW*, Technical report, 1997.

Thread Synchronization Unit (TSU): A Building Block for High Performance Computers[*]

Paraskevas Evripidou

Department of Computer Science University of Cyprus
P.O. Box 537 1678 Nicosia, CYPRUS
Tel: +357-2-338705, skevos@turing.cs.ucy.ac.cy

Abstract. The Thread Synchronization Unit (TSU) is a hardware mechanism that provides data-driven thread synchronization and data consistency for multi-threaded architectures built with control-flow (i.e. commodity) microprocessors. The TSU design is based on the Decoupled Data-Driven model of execution. This model decouples the synchronization from the computation portions of a program and allows them to execute asynchronously. At compile time a program is partitioned into a number of threads of variable granularity and the Data-Driven thread synchronization graph is also constructed. The TSU is responsible for maintaining the synchronization graph implicitly, it determines when a thread is ready for execution without interruption and then feeds it to the microprocessor for execution. The TSU-based machines exhibit the tolerance to long memory and communication latencies, of the data-driven model, with very little overhead and also exploits short-term optimal cache placement and replacement policies.

1 Introduction

It has been evident for some time now that the design of Parallel machines will be based on commodity microprocessors [1]. However, the synchronization capabilities of these commodity components cannot efficiently support large scale parallel processing. The TSU design fullfils the need for scalable efficient synchronization. The TSU provides data-driven synchronization to existing microprocessors and extend their memory management to facilitate large scale parallel processing. The proposed TSU brings the tolerance to long memory and communication latencies, of the data-flow model, to multi-threaded machines that utilize control-flow multiprocessors. The TSU-based processor design is the evolution of our previous work on the prototypical implementation of the Decoupled Data-Driven (D^3)-machine [1, 2].

---

[*] This material is based upon work supported in part by the National Science Foundation under Grant No. CCR-93-095572 and in part by the University of Cyprus Research Council

The Decoupled Data-Driven model of execution decouples (separates) each actor (thread[2]) into two parts: the graph (or synchronization) portion and the computation portion. The computation portion of each actor is a collection of conventional instructions (load/store, add, etc). The graph portion contains information about the executability of the actor and its consumers. Thus, a decoupled data-driven graph can be viewed as a partially ordered conventional program with a data-dependency graph superimposed on it. The graph synchronization is handled by the Data Flow Graph Engine (DFGE) and the Computation Engine (CE) executes the computation actors. The two engines execute in a decoupled i.e., asynchronous mode. Several simulation experiments have been conducted in order to test the ability of the D^3-machine to tolerate latency and exploit parallelism and locality. The results have shown that the D^3-machine can tolerate communication latency. Increasing the communication latency from 1 to 5 (500%) cycles results to an increase in execution of as little as 25%, going from 5 to 15 cycles increases the execution by a factor of 37%. Furthermore, it does exploit locality: the experimental results have shown that increasing thread length does indeed reduce execution time. Finally the experiments have shown that the D^3-machine in the most part neutralizes the overhead associated with the data-driven synchronization. A five-fold increase in the processing time per actor in the DFGE resulted in an increase of the overall execution time of as little as 15%.

The TSU is a hardware mechanism that allows us to develop architectures that achieve efficient, scalable synchronization. At compile time, an application program is partitioned into a number of threads of variable granularity. During program execution, the TSU schedules each thread based on data availability. Scheduling based on data availability provides tolerance to long memory and communications latencies inherent in large-scale multiprocessors, thus making the proposed architecture truly scalable and easily programmed. The TSU also provides for significant performance improvements by independently managing the flow of data to a processing node's cache based on data availability. An overall development goal will be to eliminate cache misses in each of the processing nodes through the use of the TSU design. The TSU will enable the design of scalable Multi-threaded machines with hybrid data-driven/control-driven synchronization with mostly commodity components and thus overcome the most serious obstacle in the proliferation of such machines. Hybrid multiprocessor[3, 2, 4, 5] exhibit tolerance to communication and memory latencies of the data-driven model of execution. In addition, they benefit from the efficient instruction scheduling and locality the control-driven model.

2 Design of a Thread Synchronization Unit

In the Decoupled Data-Driven Architecture [2] (depicted in Figure 1.a), the Data Flow Graph Engine (DFGE) executes all graph operations (determination of actor executability) and the Computation Engine (CE) executes all computation

---

[2] we use the terms thread and actor interchangeably in this paper

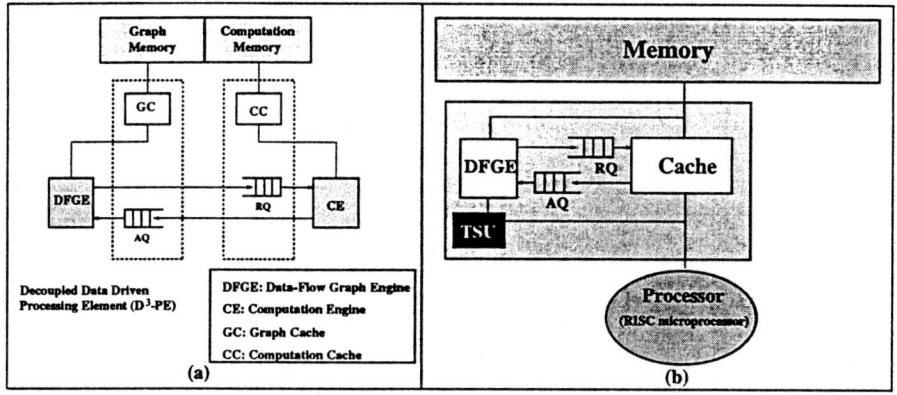

Fig. 1. (a) Decoupled Processing Element (b) Processing Element with TSU

operations (code fetching and execution). The two engines execute in an asynchronous manner, i.e., the CE does not have to execute the computation portions of actors in the same order as the DFGE executes the graph portions. A key feature of the D^3-machine is that it maps the entire dynamic data-flow graph onto the virtual space of the machine.

The goal of the TSU is to provide hardware support for data-driven thread synchronization on conventional microprocessors. The TSU is designed as a plug-in board for existing multiprocessor designs. The TSU integrates the functions of the DFGE, RQ, and AQ of the Decoupled architecture with the off-the-chip cache and memory management unit of the target microprocessor (Figure 1.b). The TSU is placed between the memory and the microprocessor. A schematic diagram of the TSU is depicted in Figure 2, its major components are:

Graph Cache (CC): A block in the GC holds the graph sub-template for one thread: [Consumer1, Consumer2, Inst-Frame-Pntr, Data-Frame-Pntr]

Synchronization Memory Cache (SMC): Stores the status-word of the threads. The status-word of a thread is decremented whenever some other thread generates one of its inputs. Whenever the status word reaches zero the thread is ready to execute. The SM-cache has hardware to perform the *decrement-and-check-if-zero* operation.

AQ: The acknowledgment queue holds the executed threads awaiting post-execution synchronization: notification of consumer threads that the completed thread has produce some (or all) of their inputs.

RQ: The ready queue stores all threads deemed executable. The RQ is implemented in two stages, the waiting and firing stage. When a thread-identification enters (< address of the first instruction in the thread, Context >) the RQ it is also send to the Graph-Cache. As seen in Figure 2 the slots of the GC that hold the Instruction-Frame-Pointer, Data-Frame-Pointer of a thread are wired to query input Q of the instruction and data caches respectively. When both computation caches signal that the requested blocks resides in the caches the thread identification is moved from the waiting stage to the firing stage of the ready queue.

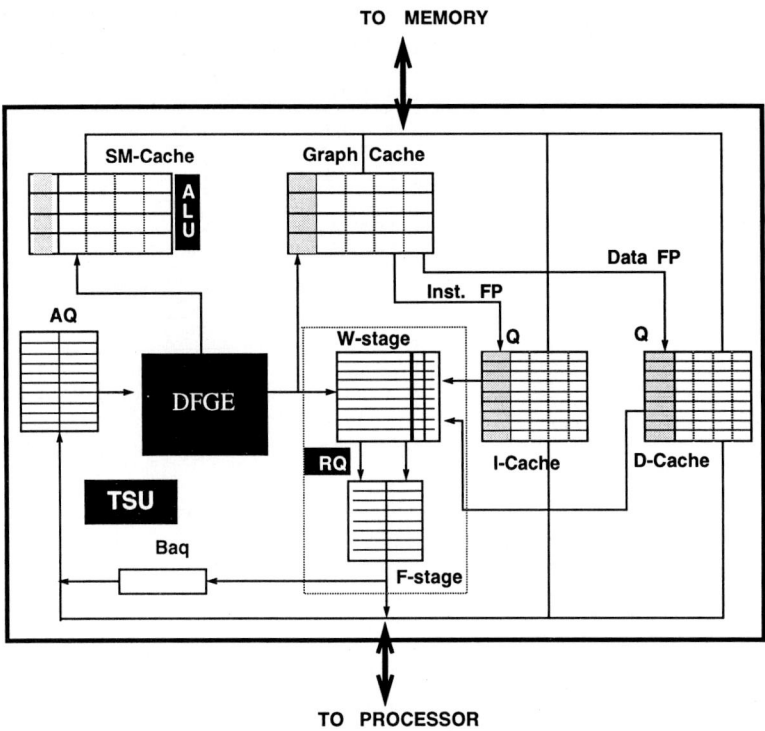

Fig. 2. Schematic diagram of the TSU design

Data-Flow Graph Engine: The DFGE is basically the control unit of the TSU. Its scope is similar to the DFGE of the D^3-machine.

Computation Instruction & Data Cache (CIC & CDC): The CIC stores computation instructions and the CDC the data. Both the CIC and CDC differ from conventional caches design in the fact that they have a query input Q. When a tag of block is applied at the query input Q the CIC & CDC check if the requested block is in the cache. If there is a miss they initiate the placement of the block. When the block is in the cache the CIC signals to the RQ of that effect. To implement this we need to provide dual-ported RAM for storing the cache tag bits and we also need to duplicate the hardware for comparing tags.

The TSU uses dynamic data-frames for implementing the data-driven principles of execution. Each dynamic instance of an actor reads its inputs form and writes its output into unique memory locations. Application programs written in SISAL or FORTRAN [6] compiled into decoupled data-driven graphs (D^3-graph). The decoupled data-driven graph of the inner product example is shown in Figure 3. The shaded lines represent data dependencies with no actual movement of data; the solid lines represent actual data movement. This graph has one static value (n) and five dynamic values (i, l-mul, r-mul, l-add, psum). The static values remain constant throughout the execution of the block, as opposed

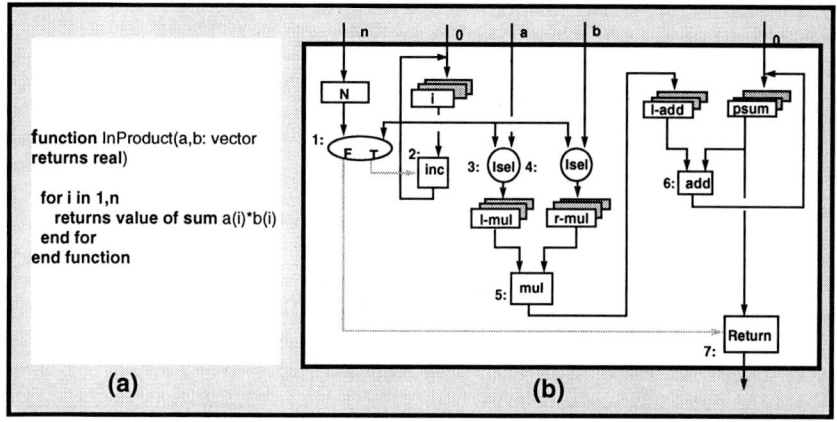

Fig. 3. The vector inner product example (a) SISAL code and (b) D^3-graph

to the dynamic values, for which a new value is created at each iteration. Figure 4 depicts the contents of graph and computation memories for the inner product example. Each actor is represented by an actor template, which is made up of a graph and a computation sub-template. The graph sub-template contains the following: (1) the address of the corresponding computation code; (2) the actor number within the block; and (3) the consumer list. A line of the Graph-Cache holds an entire graph sub-template. The computation sub-template is a thread of computation.

The basic component of the D^3-graphs is the *context-block*. The *context-blocks* mark the boundaries of a loop that changes the iteration context part of the tag. We are using the U-Interpreter[7] tag structure: c is the context, s is source line and i the iteration identifier. The inner product example of Figure 3 and 4 represents one *context-block*. For each invocation of the *context-block* a new graph and computation base address must be assigned. Thus the context in a *context-block* is synonymous with the iteration number. The computation data-frame is composed of all the dynamic values of the block. The graph synchronization frame is composed of the status-words (number of tokens required for the actor to fire) of each thread in the block.

The contents of the Ready Queue is the 2-tuple actor $<$ *actor address, context* $>$. The CE uses the actor address to fetch the first instruction of the actor and also loads the new context in the context register. The dynamic values belonging to certain context are accessed by using the context register (Rc) as index register. The instruction: `load i(Rc),R1` causes register R1 to load the dynamic instantiation of the value i that belongs to the context pointed to by the Rc.

The first actor of Figure 4 is the implementation of the *switch* actor. In the decoupled data-driven graph, the *switch* actor does not transfer any values but instead activates the "true" or the "false" block according to the evaluation of its predicate. To achieve this the switch actor is assigned two graph templates with consecutive address. The first one contains the consumers of the true block and the second one the consumers of the false block.

The last actor of each context-block is the *Return* actor, which stores the result of the context-block in the parent's data-frame. It also triggers garbage collection for the data frames used by the context-block. In Figure 4, the RESULT label points to the location where the result of the context-block must be stored.

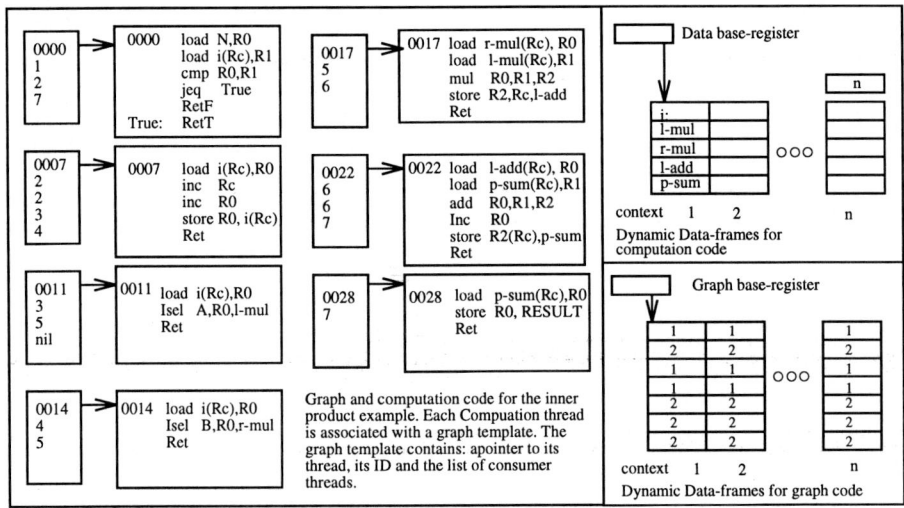

Fig. 4. Contents of the graph and computation memories for the inner product example

When a block is activated, the addresses of all ready actors are placed in the Ready Queue. The CE removes one actor address at a time, executes all instructions of that actor, and places the actor address into the AQ. The DFGE removes the address of an "executed" actor from the AQ and reads its consumer list from its graph sub-template. It then decrements the status word of each consumer. The triplet $<base\ address,\ actor\ number,\ context>$ makes up the address of the status word of each actor. When the status word reaches zero, the actor is deemed executable, and the DFGE places the address of its computation code in the RQ.

Nested contexts or nested loops can be handled in two ways. The change in context due to nested loops can be recorded either the context identifier c or the iteration identifier i of the tag $[c.s.i]$ If the context field c of the tag is used for recording changes in the context due to nested loops, then loops and function invocations are handled in the same fashion. Under this scheme, a special actor *new_context* allocates new data frames for each invocation of the innermost loop. Each data frame for the innermost loop has a frame location for the corresponding j.

The instruction load R1,(i)Rc (R1 ←M[i + Rc × 4]) uses index addressing to fetch the corresponding value of i from the current data frame pointed to by the context register Rc. No index addressing is needed for fetching the corresponding j because the each data frame stores the corresponding j. Thus, the instruction load R2,j fetches the corresponding j.

In the second case, the iteration field is used for recording changes in the context due to nested loops. In this case a single data frame is allocated for all nested loops. The iteration identifier is not reset with each invocation of the innermost loop but instead it is incremented. In the general case the starting value of the tag of the inner n^{th} loop is given by $\sum_{k=1}^{n-1} i_n * range_k$ where $range_n$ is max range at n^{th} level and i_n is the value of the "iteration" identifier of the tag at the nth level. In order to fetch the j corresponding to a particular context, the context register must be divided by the range of the innermost loop. div R3,Rc,range1 After determining the frame location of j, index addressing is used for fetching the appropriate value of j, load R2,(j)R3.

2.1 Cacheflow: Data-Driven Cache Placement and Replacement Techniques

The data-driven nature of the Decoupled model, at the thread level, makes it possible to tailor placement and replacement policies for the Decoupled architecture. The contents of the two queues (RQ and AQ) represent the *near-future* execution patterns for the two engines respectively.

The Ready Queue is divided into two stages the waiting stage and the firing stage. The RQ/Computation Cache (CIC and CDC). placement algorithm is made up of the following steps:

1. The DFGE places a token (2-tuple of context and address) in the RQ.
2. The address part of the token is used to determine if the corresponding block is in the cache.
   - If in cache the token is placed in the firing stage of the RQ.
   - If not in cache the token is waiting in the waiting stage of the RQ until the block is moved in the cache.
3. At the firing stage of the RQ tokens are sorted according to the graph-level priority assigned.
4. The CE's request for the next ready actor is fulfilled with the token with highest priority. The graph-level priority techniques described in [8] are used here.

This cache management policy guarantees that the target microprocessor does not encounter cache misses or page faults. It can, however be idle whenever the RQ is empty. The cache coherence issues of the TSU-based machines are much simpler than those of pure control-flow computing because all computational variables are only written once i.e., we adhere to the single assignments semantics. The synchronization values of the graph memory are the ones that behave like conventional variables because they are decrement possibly more than once. However, since the only operation performed on them is decrement and check if zero a few times only, they do not exhibit much locality characteristics and thus not need to be moved around the different caches of the system. So the owner cache will always perform the decrement.

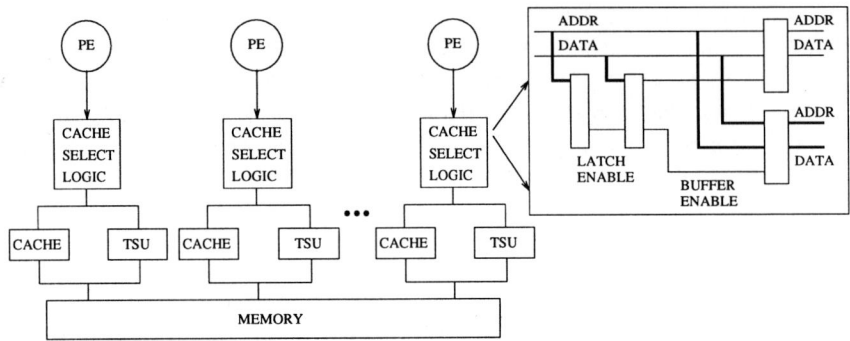

Fig. 5. Schematic of a dual machine with TSU

The overall memory hierarchy of the TSU-based machines has some unique features. The synchronization frames (of the graph memory) of each context block are all initialized with identical values. Therefore, it is enough to just load the initial values into the memory and then each subsequent "page-faults" does not require transfer of data from the disk. It is enough to just copy generic values from the memory or better just transfer the values of the generic frames to the cache. Of course we have to make sure that the cache tags do reflect the correct frames. This can be improved if we modify the cache is such a way as to keep the generic values in the cache and just do the copy within the cache itself.

The situation is similar with the data-frames of the dynamic values. Each data-frame is initialized to the same generic values. So, we can apply the same policies as the ones about the synchronization frames of the graph-cache.

The demands of the TSU-based machines on virtual memory might appear very large. However, if we closely study the use of the virtual space of the machine we will notice that all data-frames have a relatively short life span after which they are not used again. So the maximum size of the virtual space needed might be large but only a fraction need be "active" at any time.

2.2 Implementation Issues

In order to achieve rapid prototyping we are designing an FPGA-based TSU as an off-chip module for an existing design. This augmented design will be able to operate under both their native mode and also as multi-threading machine. We call such designs Dual-machines. Figure 5 shows such a configuration. The TSU is placed in parallel with the off-chip cache. The operating system can switch between the two modes by "writing" to a predetermined address. When this address is access the comparator of the selection unit decodes the address and the data associated with the address is latched. The data word contains the enable signal for one of the two units (the original cache or the TSU).

Both the RQ and the AQ are memory-mapped. Thus, the target processor can access them using load, jump and store instructions. The TSU has a simple snooping hardware that watches the cache traffic for these two addresses. If it *sees* a write to AQ or a jump relative to the RQ it intercepts the operation and passes them to the DFGE for processing. The rest of the instructions execute in the normal mode of the machine. Thus, the implementation of the `Ret` instruction of Figure 4 is simply a combination of a `jump (Rrq)` and `store RAQ, Thread_ID`. This will cause the first instruction of the thread on top of the RQ to be fetched and also will notify the DFGE which thread has completed. Delayed-branching and other similar techniques are utilized to hide the effect of the brancbing instructions for the switching between threads.

3 Performance Analysis

The D^3-simulation facility comprises an X-window based GUI. It operates under two major modes: In the **graph mode**, there is no partitioning of the D^3-graph. At each time step, all ready operations (both graph and computational) are executed (assuming an infinite number of processors). This mode allows the user to determine the maximum parallelism achievable for a specific D^3-graph. In the **Graph mode with Clipping** The user can specify the number of PEs and GEs which are present in the machine. The following simulation parameters are set by the user:

- T_L: Thread length; Number of actors in a thread.
- N_{DFGE} (N_{CE}): Number of DFGE's (CE) available.
- L_{CE} (L_{DFGE}): Time elapsed from the time an actor finishes executing in a CE (DFGE) and the time it enters the AQ (RQ).
- T_{CE} (T_{DFGE}:): Time required for a computation (graph) operation to complete execution.

The **vector inner product** was used for the first set of experiments. The results of these experiments are tabulated in Table 1 (experiments 1.0-1.2). In each of these runs, the time to execute a computation instruction is set to 1 cycle. By varying the DFGE's processing time per instruction (T_{DFGE}), we have simulated varying execution times required to process the graph portion of an actor. In the past, dynamic data-flow machines have been suspected of introducing too much overhead. In the D^3-machine, all the overhead introduced by the decoupled dynamic model is taken care of inside the DFGE. Thus, by varying the DFGE execution speed, we are able to study the effect of increased data-flow scheduling overhead. The results of our experiments indicate that a ten-fold increase in the DFGE overhead (T_{DFGE}) results in an overall increase of execution time of no more than 40% for the experiments with TL=10.

The **trapezoidal rule** for the numerical estimation of a definite integral was also coded and used as a programming example in our experiments. A third degree polynomial was chosen as the example integrand. Table 1 (experiments 2.0-2.2) lists some results obtained from various runs with different thread lengths and DFGE processing times (T_{DFGE}).

This data demonstrates the same behavior as the Inner product example did. Also, another phenomena can be noticed in both examples. Consider the first column in the tables where the DFGE processing time ranges from unity, to five and to ten. The longer thread length examples actually took longer execution

	N = 10	N = 50	N = 100	N = 1000	N=5000	N = 10000
Experiment 1.0: $L_{CE} = L_{DFGE} = 0, T_{CE} = T_{DFGE} = 1, N_{CE} = N_{DFGE} = \infty$						
TL=1	117	557	11007	11007	55007	110007
TL=2	71	291	566	5516	27517	55016
TL=10	100	200	325	2575	12575	25075
Experiment 1.1: $L_{CE} = L_{DFGE} = 0, T_{CE} = 1, T_{DFGE} = 5, N_{CE} = N_{DFGE} = \infty$						
TL=1	227	1067	2117	21017	105017	210017
TL=2	131	551	1076	10526	52527	105026
TL=10	120	240	390	3090	15090	30090
Experiment 1.2: $L_{CE} = L_{DFGE} = 0, T_{CE} = 1, T_{DFGE} = 10, N_{CE} = N_{DFGE} = \infty$						
TL=1	337	1577	3127	31027	155027	310027
TL=2	191	811	1586	15536	77535	155034
TL=10	138	280	455	3605	17605	35105
Experiment 2.0: $L_{CE} = L_{DFGE} = 0, T_{CE} = T_{DFGE} = 1, N_{CE} = N_{DFGE} = \infty$						
TL=1	120	560	1110	11000		
TL=2	112	372	697	6547		
TL=10	212	344	509	3479		
Experiment 2.1: $L_{CE} = L_{DFGE} = 0, T_{CE} = 1, T_{DFGE} = 5, N_{CE} = N_{DFGE} = \infty$						
TL=1	236	1076	2126	21026		
TL=2	192	652	1227	11577		
TL=10	252	404	594	4014		
Experiment 2.2: $L_{CE} = L_{DFGE} = 0, T_{CE} = 1, T_{DFGE} = 10, N_{CE} = N_{DFGE} = \infty$						
TL=1	351	1591	3141	31041		
TL=2	272	932	1756	16606		
TL=10	292	464	679	4549		

Table 1. Execution time for the Inner product and Trapezoid Experiments

time than the shorter thread length runs. This is because the advantages of longer thread length as given are outweighed by the decrease in parallelism due to lengthening the thread length with a very small amount of computations. This fact indicates that increasing the thread length of actors only yields run time improvement as a function of the total number of computations to be performed.

Figure 6.a shows the effect of increasing the data-driven synchronization overhead (Time spent in DFGE per thread) to the overall execution time. For a relatively modest thread size (i.e., 10), the effect of the increased overhead is negligible with respect to execution time. These results indicate that increasing the T_{DFGE} results in linear growth in execution time, and more importantly, the rate of this linear growth is inversely proportional to the thread length. The results obtained in experimenting with the communication latencies are shown in Figure 6.b. This data shows how increasing communication latency affects overall execution time. Increasing the communication latencies from 1 cycle to 15 cycles barely doubles the total execution time for the thread length 10 example. These results show that data-driven nature of the D^3-machine does indeed provide tolerance to high communication latencies.

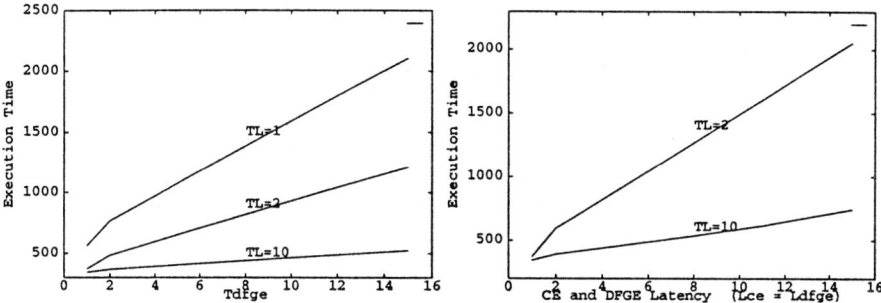

Fig. 6. (a)Execution Time vs Time spend by the DFGE for synchronization processing for the Trapezoidal rule example and (b) Execution time vs communication latency for the Trapezoidal Rule Example

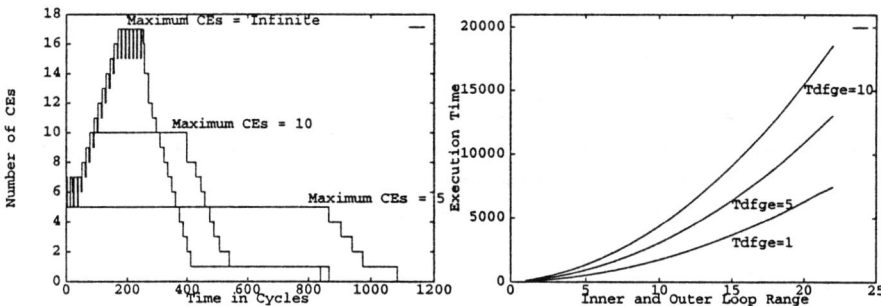

Fig. 7. (a)Number of PEs vs Execution Time for the Trapezoidal example (b) Execution Time vs Loop Range for the Jacobi Example

Figure 7.a illustrates the results obtained from experimenting with a finite number of available PEs. This figure contains the results of three different runs using the Trapezoidal rule example for a problem size of 200. The three experiments consisted of running the example with an unlimited number of PEs, 10 PEs available, and 5 PEs available. The histogram in Figure 7.a indicates that approximately one half of the running time consisted of a high degree of parallelism with the remaining half being used for single processor activity. This unequal load balancing is attributed to the particular implementation of the D^3-graph and is not an artifact of the D^3-architecture. Therefore, by limiting the available PEs to 10, runtime is increased by an insignificant amount and load balancing is increased somewhat but the run is fairly unbalanced overall. However, by restricting the number of available PEs to 5, the runtime is increased by approximately 20% with a decrease in available resources by 70% and the computational workload is much more balanced as is evident by the fact that the majority of the execution time has all 5 PEs executing.

Our final example calculates a single iteration of the **Jacobi** algorithm. For this example we have implemented the hierarchical matching scheme that we have described in [2]. Figure 7.b show the execution time versus problem size for three different DFGE processing speeds ($T_{DFGE} = 1, 5, 10$). These curves show similar behavior with the ones obtained for the other examples; A 500 % and 1000 % increase in the T_{DFGE} results to an 70% and 146% increase of the total execution time respectively.

4 Concluding Remarks

The TSU is a hardware mechanism that enables the design of large scale multiprocessors that exhibit tolerance to high communication and memory latencies. It mode of operation is based on the Decoupled Data-Driven (D^3) model of of execution. The D^3-model of execution provides a very promising approach for building hybrid systems because the decoupling of the synchronization and computation and their asynchronous execution allows us to concentrate in designing only a portion of the needed modules and readily adopt the state-of-the-art technology for the rest. The TSU provides a minimal hardware approach in designing Multiprocessors based on the D^3 model of execution. The performance analysis showed that the TSU-based machines can exhibit tolerance to long latencies and also exploit locality. The results obtained show that the overhead of the data-driven synchronization of the TSU is very small and has very minimal effect in the perfromance of the machine. Furthermore, the TSU-based machine can benefit from the improved cache placement and replacement policies that can be optimal in the short term.

References

1. P. Evripidou and J-L. Gaudiot. The USC Decoupled Multilevel Data-Flow Execution Model. In *Advanced Topics in Data-Flow Computing*. Prentice Hall, 1990.
2. P. Evripidou and J-L. Gaudiot. A Decoupled Graph/Computation Data-Driven Architecture with Variable Resolution Actors. In *Proceedings of the 1990 International Conference on Parallel Processing*, August 1990.
3. R. A. Iannucci. Toward a Dataflow/von Neumann hybrid architecture. In *Proceedings of the 15th Annual Symposium on Computer Architecture*, May 1988.
4. L. Bic. A process-oriented model for efficient execution of Dataflow programs. In *The 7th International Conference on Distributed Computing*, Berlin, FRG, 1987.
5. R.S. Nikhil and Arvind. Can dataflow subsume von neumann computing. In *Proceedings of the 16th Annual Symposium on Computer Architecture*, May 1989.
6. P. Evripidou and R. Barry. Mapping FORTRAN Programs to Single Assignment Semantics for Efficient Parallelization. *Parallel Processing Letters*, In Press, 1997.
7. Arvind and K.P. Gostelow. The U-Interpreter. *IEEE Computer*, pages 42–49, February 1982.
8. P. Evripidou and J-L Gaudiot. Block Scheduling of Iterative Algorithms and Graph Priority in a Simulated Data-flow Multiprocessor. *IEEE Transactions on Parallel and Distributed Systems*, 4(4), April 1993.

Data Dependence Path Reduction with Tunneling Load Instructions

Toshinori Sato

Microelectronics Engineering Lab.,
Toshiba Corp., Kawasaki 210, Japan
toshinori.sato@toshiba.co.jp

Abstract. The technique for reducing the length of the data dependence path is presented. This technique, named *tunneling-load*, utilizes the register specifier buffer in order to hide the load latency, and thus reduces the length of the data dependence path. True data dependences can not be removed by any techniques such as register renaming, and are the unavoidable obstacle limiting the instruction level parallelism. The length of the data dependence path including the load instructions is longer than those of other instructions, because the latency of the load instruction is longer than those of other instructions. In order to reduce the dependence path length including the load instructions, we propose the tunneling-load technique. We have evaluated the effects of the tunneling-load, and found that in an in-order-issue superscalar platform the instruction level parallelism is increased by over 10%.

1 Introduction

True data dependences can not be removed by any techniques such as register renaming[12], and are the unavoidable obstacles limiting the instruction level parallelism (ILP). The latency of the load instruction is longer than those of other instructions, and hence the temporal distance of the data dependency caused by the load instruction is also longer.

Among the factors increasing the load latency, the penalty of the data cache miss has been dealt with by a variety of techniques such as non-blocking cache[9], software prefetching[14], hardware prefetching[7], and dynamic code scheduling[6]. Even if the penalty of cache miss is solved, the load latency still remains. This load latency is the load-use hazard explained as follows. Fig. 1 shows the traditional 5-stage pipeline, which consists of the instruction fetch (IF), the instruction decode and register fetch (ID), the execution and effective address generation (EX), the memory access (MEM), and the write back (WB) stages.

The load instruction is processed in the following fashion. First, the effective address is generated during the EX stage. Second, the data cache is accessed with this effective address at the beginning of the MEM stage. And finally, the result is obtained at the end of the MEM stage. Therefore, the result is unavailable to the instruction which immediately follows the load instruction. In other words, because the load instruction executes two operations including

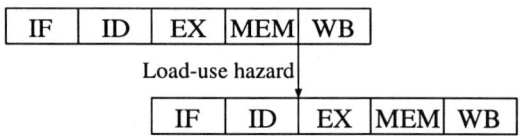

Fig. 1. Load-Use Hazard

the effective address generation and the memory access, the latency of the load instruction has to be longer than those of other instructions.

Recently, the impact of the load-use hazard has become serious. This is because the load latency becomes a major factor obstructing the processor performance, as the instruction issue rate increases. For example, increasing the load latency from 1 to 2 degrades the performance of four-issue processors by approximately 30%[5]. This means that the processor performance could be improved by 30%, if the load latency existing in the 5-stage pipeline were hidden.

In order to hide the load latency, several address prediction techniques have been proposed[1, 2, 8, 11, 15]. Using these techniques, the load address is calculated a few cycles earlier before it would normally be computed in the pipeline, and the load latency can be hidden if the prediction is correct. These techniques carry out the speculative data cache fetching, which causes the explosion of the memory traffic and the pollution of the data cache.

This paper proposes the technique solving the problems described above. The proposed technique, named *tunneling-load*, does not predict the effective address, but generate it. The correctness of the effective address generated at earlier pipeline stage is confirmed before starting the data cache access, and thus the useless speculative cache fetching is removed. Using the tunneling-load, the data dependences caused by the load instructions can be hidden, and therefore it becomes possible to reduce the length of the data dependence path including the load instructions. As the results, the more ILP can be exploited.

The organization of the rest of this paper is as follows. Section 2 surveys previously proposed related works. The tunneling-load mechanism is explained in Section 3. In Section 4, the evaluation methodology is presented and the effect of the tunneling-load is evaluated in Section 5. Section 6 discusses the future study. And finally, our conclusions are presented in Section 7.

2 Previous Work

In order to hide the data cache access latency, several load address prediction techniques have been proposed [1, 2, 8, 11].

Golden et al.[11] and Eickemeyer et al.[8] proposed the load target buffer (LTB) and the load delta buffer (LDB), respectively. These mechanisms are very similar. The LTB (LDT) is indexed by the instruction address and accessed at the IF stage. The load address is predicted by the actual address previously used and the stride value, and the predicted and actual addresses must be compared when the actual one is generated at the end of the EX stage. If these two addresses match, the load latency is hidden by 2 cycles on the traditional 5-stage pipeline. The accuracy of prediction by the LTB (LDT) is relatively low, because these techniques cannot use the value of register file after modification. Hence, there is a tradeoff between the prediction accuracy and the hardware cost. In order to achieve sufficient accuracy, a lot of entries are necessary. In [8], the reduction of the entries is examined, but the reduction degrades the prediction accuracy. Furthermore, these techniques have a serious drawback. They execute both the actual effective address generation and the comparison between the predicted and actual addresses in one cycle. This can easily become a critical path limiting the cycle time of the pipeline. Thus, in practical implementation, there might be misprediction penalties in spite of their conclusion that the performance is not degraded by any miss penalties.

Austin et al.[1] examined the data cache organization, and proposed the technique whereby the effective address was generated and the data cache was accessed during the EX stage. Because the tag field of the effective address is required after the data cache has been read by the index field, only the set index must be calculated early in the stage. By analyzing the reference type and the offset distribution of several benchmark programs, the fast prediction mechanism of the set index is explored, which is constructed with one stage of logical OR gates. However, the simplest prediction circuit has a significant impact on the cycle time of the processor, because it is on the critical path through the cache memory array. That is, it is difficult to implement the prediction mechanism in the situation that the access time of the data cache dominates the cycle time of the processor, as has recently become commonplace. Another proposal by Austin et al.[2] is the base register and index cache (BRIC), which holds the base and index register values. Since the BRIC is accessed earlier than the register fetch stage and the effective address is generated using the fast address calculation[1], the load instructions can be executed up to 2 cycles earlier than normal execution. This technique relies on the fast address calculation, and thus it is also difficult to implement the prediction mechanism due to the same reason explained above.

All of previous works are promising only if the prediction is correct. However, when the prediction fails, there are several drawbacks related to the speculative cache access. Useless cache access results in the explosion of memory traffic and the pollution of data cache causing the performance degradation. Furthermore, the exception signaling for the speculative cache access is not desirable, because the exception might not occur when the data cache is accessed with the actual address. Thus, the special exception handling mechanism for the speculative cache access must be implemented.

3 Tunneling-Load

In this section, we propose the tunneling-load technique in order to reduce the length of data dependence path. Firstly, we present the mechanism for the tunneling-load technique. And next, we explain why speculative data cache fetching can be eliminated.

3.1 Tunneling-Load Mechanism

We have already proposed the address prediction scheme using the register specifier buffer (erstwhile address prediction buffer)[15], which has the same problems explained at previous section. In order to solve the problem, we propose the tunneling-load technique. The tunneling-load technique utilizes the register specifier buffer (RSB), which is similar to the branch target buffer (BTB), but it has three differentiate features. First, the RSB does not contain the target addresses, but the register specifiers which store the base and index addresses. Second, the effective address can be generated, even if there are not any entries where the register specifiers are contained. In such a case, the RSB supplies the stack pointer and the zero register specifiers. And finally, the RSB is not accessed by the program counter (PC), but the target program counter (TPC) which walks ahead of the PC. Fig. 2 shows an entry of the RSB. It mainly consists of three fields, namely the tag instruction address field, and the base and index register specifier fields. In addition, each entry contains the valid bit (**v**) and the information field for the least recently used replacement (**LRU**).

Fig. 3 indicates the tunneling-load mechanism. Let us see the pipeline attached with the RSB. As shown in Fig. 3, the traditional 5-stage pipeline can be reconstructed into a 6-stage pipeline including the RSB access (RSB), the instruction fetch (IF), the instruction decode and register fetch (ID), the execution and effective address generation (EX), the memory access (MEM), and the write back (WB) stages. The PC and TPC indicate the IF and RSB stages, respectively. The reconstruction of the pipeline into 6-stage does not increase the branch misprediction penalty because the PC indicates the IF stage. During the recovery process, the TPC is also restored with help of the branch prediction after the PC is corrected.

The technique requires additional hardware support consisting of the RSB, the TPC, a scoreboard, an address calculating adder, and two 5-bit comparators. The scoreboard consists of n 1-bit entries, where n is the number of registers. If a bit is set, this indicates that the corresponding register is updated at immediately previous cycle. The comparators compare the register specifiers obtained from the RSB and those decoded from the instruction code. In addition, the register file must provide two more read ports, and the data cache has to be dual-ported.

The process of the technique is as follows. It consists of five steps. **Step 1**: The RSB is accessed by the TPC and supplies the base and index register

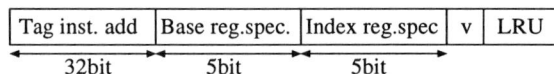

Fig. 2. RSB Entry Field

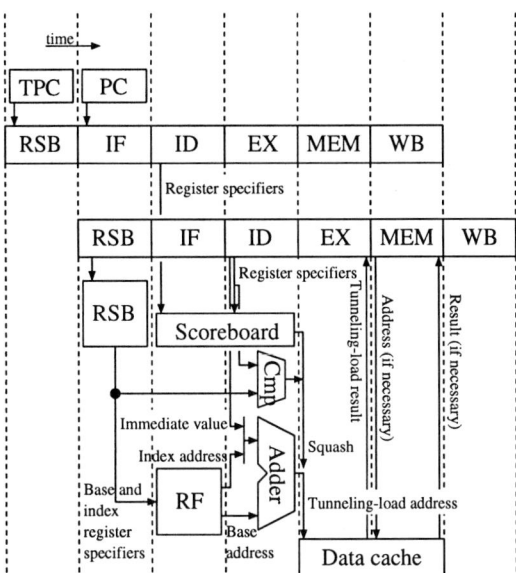

Fig. 3. Tunneling-Load Mechanism

specifiers to the pipeline.

Step 2: At the next cycle, the PC indicates the target instruction indicated by the TPC at the previous cycle, and the target instruction is fetched. The base and index values are obtained from the register file with the base and index register specifiers supplied by the RSB.

Step 3: At the next cycle, the effective address is generated. If the addressing mode is `register + register`, it is generated with the base and index values. Otherwise, it is generated with the base value and the immediate offset which is decoded at this cycle. Simultaneously, the target instruction is decoded. The scoreboard bits for the base and index registers are checked. If any scoreboard bits are set, the generated address is squashed because it causes the address-generation miss[1]. The comparisons between the register specifiers obtained from the RSB and those decoded from the instruction code are also executed. If they do not match, the generated address is squashed because it causes the primary or bad TPC misses. When the target instruction is not a load instruction, the generated address is discarded. Otherwise, if the address of target instruction is not contained in the RSB, the address and the base and index register specifiers are stored in the RSB. And lastly, the scoreboard bits for the destination registers of all instructions at this stage are set.

Step 4: Next, when the tunneling-load succeeds, the datum is fetched from the data cache by the generated address and the load latency is successfully hidden. Otherwise, there is not any operation, thus useless speculative cache access is eliminated.

---

[1] The address-generation, primary, and bad TPC misses are explained in Section 3.2.

Step 5: Finally, if the tunneling-load fails, the datum is fetched by the normally generated address at the MEM stage normally. Thus, there are no miss penalty cycles.

As described above, no speculative data cache fetching belongs to the tunneling-load. Hence, this technique is free from the problems of any load address prediction techniques: the memory traffic explosion, the data cache pollution, and the misprediction penalty. Note that even though the speculative cache access is not executed, the load latency is successfully hidden by the tunneling-load because the speculative cache access is useless.

3.2 Eliminating Useless Cache Fetching

In this section, we explain how the speculative data cache fetching can be eliminated. The address generation misses are classified into three categories: the address-generation, the primary, and the bad TPC misses[15].

The address-generation miss is caused when the base and index register values are modified by the instructions immediately before the target load instruction. This register modification can be found before the load address is generated. Let us see Fig. 3. The destination register specifiers of the preceding instructions which modify the base and index values for the succeeding load instruction are supplied at the ID stage. Similarly, the base and index register specifiers of the succeeding load instruction are obtained from the instruction operand at the ID stage. Therefore, it is possible to detect the register modification in question by scoreboarding between the destination registers of the preceding instructions and the source registers of the succeeding load instruction at the ID stage.

The primary miss is meant that the address of the target load instruction is not cached in the RSB. The primary miss does not always cause the misprediction, because the stack pointer specifier supplied by the RSB as the default may be proper. The misprediction occurs when the stack pointer is not the base register for the load instruction in question, or when the stack pointer is the base register but its value is modified. Therefore, the misprediction in this category is verified by two mechanisms. One is the scoreboarding described above, which checks the modification of the stack pointer value. The other is a comparator which compares the stack pointer specifier and the register specifiers obtained from the instruction code.

The bad TPC miss is caused by the branch mispredictions including two cases explained as follows. One is that the instruction in question is not a load instruction. In this case, even if an incorrect address is generated, the address is discarded at Step 3 and the misprediction never occurs. The other is the case that the instruction is a load instruction. In this case, the specifiers supplied by the RSB are not guaranteed as correct ones, so the same mechanism as to the primary miss is enough for verification: the scoreboarding and the comparing.

From these investigations above, we obtain the following. The correctness of any load address predictions can be verified at Step 3. Therefore, it becomes possible to eliminate useless data cache accesses.

Table 1. Baseline Processor Configuration

Fetch Width	1, 2, 4, 8 instruction(s)
Branch Predictor	1024 entry direct-mapped BTB, gshare scheme[13], 12-bit BHR, 4096 entry PHT, 2 cycle miss penalty
Issue Width	1, 2, 4, 8 instruction(s)
FU's	1, 2, 4, 8 universal functional unit(s) except Ld/St
FU Latency	iALU 1, iMul 3, iDiv 6, Ld/St 2, fAdd 2, fMul 3, fDiv/Sqrt 6
RF's	32 32-bit integer registers, 32 32-bit floating point registers
I-Cache	16K 2way set-associative, 32 byte blocks, 6 cycle miss penalty
D-Cache	16K 2way set-associative, 32 byte blocks, 2-port,write-back, non-blocking load, hit under miss, 6 cycle miss penalty
L2 Cache	ideal

4 Evaluation Methodology

4.1 Processor Model

We evaluated the effect of the proposed mechanism by using the SimpleScalar tool set (version 1.0.2)[3]. The SimpleScalar architecture is based on the MIPS architecture, and the cycle-by-cycle simulator is executed on a SPARCstation.

The baseline model is an in-order-issue superscalar processor. The superscalar degree is between 1 and 8. The pipeline is a 6-stage pipeline shown in Fig. 3. The configuration of the baseline processor is summarized on Table 1. The functional units are universal and there are no restriction on the combination of instructions issued together, except that the number of the load instructions which can be issued per cycle is one. The instruction fetch width and issue width are same. The RSB is constructed full-associatively, and the number of its entry is 64.

4.2 Workload

The SPEC92 benchmark suite is used for this study. The reference input files which are provided by SPEC are used with slight modifications. The left side of Table 2 shows the summary. The Fortran programs were converted to C programs using AT&T F2C (version 1994.11.03), and then all programs were compiled by GNU GCC (version 2.6.3) with the optimization option, -O3. Each program was executed to completion or for the first 1 billion instructions. The right side of Table 2 shows the dynamic instruction count executed. Note that the statistics in the table do not include the cycles executed by the operating system.

5 Evaluation

In this section, we present experimental results. Firstly, we evaluate the performance improvement when the RSB is attached to the baseline model. Secondly, we make an evaluation with an alternative memory system for the baseline

Table 2. Benchmark Programs

Benchmark	Input	Modification	# of inst(mil.)
008.espresso	tial.in		1000.0
022.li	li-input.lsp		1000.0
023.eqntott	int_pri_3.eqn		1000.0
026.compress	in		664.4
072.sc	loada2		1000.0
085.gcc	insn-recog.i		164.2
013.spice2g6	greycode.in		1000.0
015.doduc	doducin		1000.0
034.mdljdp2	mdlj2.dat	MAX_STEPS=250	1000.0
039.wave5			1000.0
047.tomcatv		N=129	480.8
048.ora	params	ITER=15200	116.8
052.alvinn		NUM_EPOCHS=50	1000.0
056.ear	args.short		420.5
077.mdljsp2	mdlj2.dat	MAX_STEPS=250	1000.0
078.swm256	swm256.in	ITMAX=120	1000.0
089.su2cor	su2cor.in		766.2
090.hydro2d	hydro2d.in		11.2
093.nasa7			1000.0
094.fpppp	natoms		1000.0

Table 3. Examined Models

Name	B1L1	B2L1	B4L1	B8L1	B1L2	B2L2	B4L2	B8L2
Issue	1	2	4	8	1	2	4	8
RSB	No							
Ld/St	1				2			
D cache	1-port				2-port			
Name	R1d	R2d	R4d	R8d	R1s	R2s	R4s	R8s
Issue	1	2	4	8	1	2	4	8
RSB	Yes							
Ld/St	2				1			
D cache	2-port				1-port			

model. The baseline model is attached with two load-store units and its data cache is modified to be dual-ported. And lastly, we investigate the performance degradation when the data cache of the evaluated model is single-ported. The designs we examine are listed in Table 3. First row shows the names for evaluated models. Second row indicates the instruction fetch width, and the next row shows if there are the RSB attached to the model. Fourth row indicates the number of load store unit. And last row shows the number of the data cache port.

We use the Instruction Per Cycle (IPC) as a performance metric.

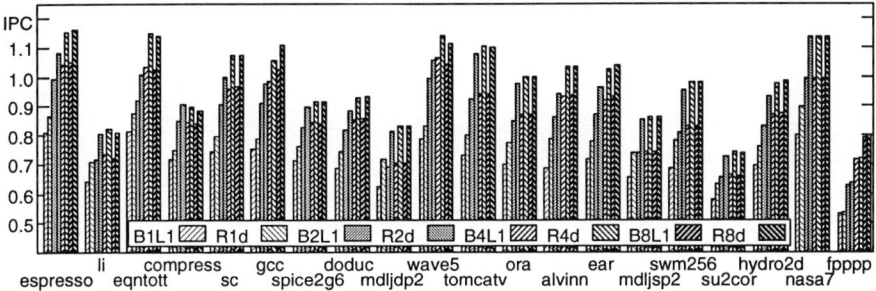

Fig. 4. Simulation Results (1)

5.1 Effect of Tunneling-Load

Fig. 4 shows the performance improvement when the tunneling-load mechanism is attached to the baseline model, which has only one load-store unit and whose data cache is single-ported. Note that the data cache of the evaluation model is dual-ported.

From Fig. 4, in the case of the single-issue model, the performance for the integer programs is increased by 6.7% on average with the maximum of 9.7%. For the floating-point programs it is increased by 9.7% on average with the maximum of 15.2%. The results are different from those explained in [15], which says the performance of the single-issue processor is improved by 1.5% on average for the floating-point programs. This is because the latencies of floating-point units are different between the evaluate model in this paper and that in [15].

In the case of the dual-issue model, the performance is increased by 9.1% on average with maximum of 11.5%, and 11.7% on average with maximum of 17.7%, for the integer and floating-point programs, respectively. As can be seen from Fig. 4, when the issue width is increased from one to two, the performance is jumped up significantly in both the baseline and evaluated models.

In the case of the four-issue model, the performance is increased by 9.9% on average with maximum of 12.2%, and 12.0% on average with maximum of 18.1%, for the integer and floating-point programs, respectively. In the case of the eight-issue model, the performance is increased by 9.8% on average with maximum of 11.9%, and 12.1% on average with maximum of 16.8%, for the integer and floating-point programs, respectively. It should be noted that in some programs the performance of the eight-issue model is lower than that of the four-issue model. One of the reasons is the miss rate of the instruction cache. In the case of wave5, for example, the instruction cache miss rate of the eight-issue model is 10.8% worse than that of the four-issue model.

5.2 Hardware Cost Consideration

In comparison with the baseline model, the evaluated model described above has an additional load-store unit and its data cache is dual-ported. In order to

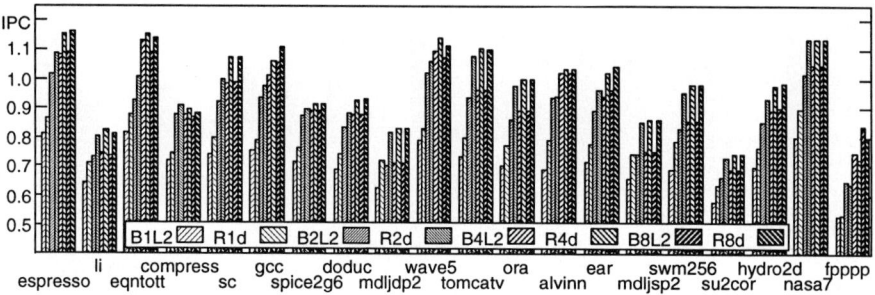

Fig. 5. Simulation Results (2)

make a fair comparison in terms of the hardware cost, we take up an alternative memory system for the baseline model, including two load-store units and the dual-ported data cache. The results are shown in Fig. 5.

In the case of the single-issue model the performance for the integer programs increased by 6.7% on average with the maximum of 9.7%. For the floating-point programs it is increased by 9.7% on average with the maximum of 15.2%. In the case of the dual-issue model, the performance is increased by 6.9% on average with maximum of 9.8%, and 9.0% on average with maximum of 16.6%, for the integer and floating-point programs, respectively. In the case of the four-issue model, the performance is increased by 5.7% on average with maximum of 10.2%, and 8.4% on average with maximum of 16.8%, for the integer and floating-point programs, respectively. In the case of the eight-issue model, the performance is increased by 6.2% on average with maximum of 10.1%, and 8.3% on average with maximum of 16.8%, for the integer and floating-point programs, respectively.

As can be seen, the performance improvement is slightly reduced in general, if the memory system of the baseline model is extended.

5.3 Single-Ported Cache Case

This section presents the results for the evaluated model whose data cache is single-ported. Because the die area of the dual-port cache is twice larger than that of the single-ported cache, the single-port cache is desirable. The results are shown in Fig. 6. The performance is comparable to the model with dual-port cache. Its degradation is less than 1%, except spice2g6. Thus, it is found that the tunneling-load is effective without additional hardware cost into the memory system.

6 Current Status and Future Study

In this section we explain the current status and future direction of this study.

As mentioned in [15], it is possible to hide the load latency that is grater than two if the distance between the TPC and the PC is enlarged. Recent study shows

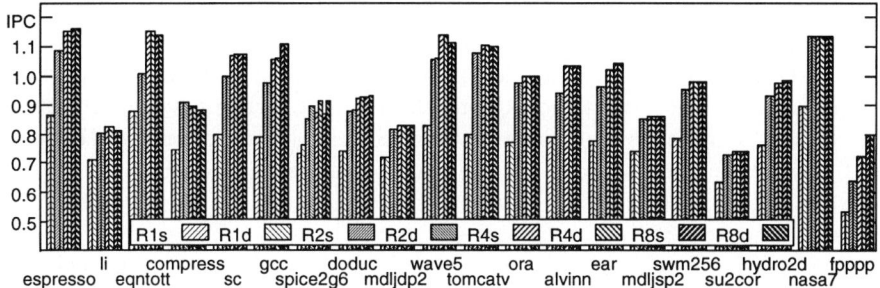

Fig. 6. Simulation Results (3)

that the tunneling-load is effective when the TPC points ahead even farther. Due to lack of space, detailed and extensive studies will be presented in future publications.

Future study dealing with the tunneling-load will investigate the register port consideration and the attachment to the out-of-order execution engine.

The drawback of the tunneling-load is the impact on the access speed of the register file when the multiple loads per cycle are supported. The number of ports on the register file would increase by four rather than two per load pipe. The additional ports may slow the register file, and this has a serious impact. Since the register port requirement is a common problem among the high-issue-rate processors, several techniques which attack to the register port requirement[4, 10, 17] are proposed. We are now investigating techniques which relieve the demand on the register ports.

Clearly, out-of-order-issue processors will not suffer as much from the pipeline interlocks caused by the load-use hazard. However, there are still existed the possibilities to remove true data dependences between an unresolved load instruction and succeeding instructions which are dependent upon the load instruction. In such a case the speculatively resolving the load instruction is effective[16]. The tunneling-load might be applicable to this purpose, and an algorithm is being examined. Note that the speculative load resolution is different in purpose from the speculative data cache fetching.

7 Concluding Remarks

We have proposed the new technique for reducing the length of the data dependence path and exploiting the ILP, and evaluated its effect by the cycle-by-cycle simulation. This technique, named tunneling-load, utilizes the RSB. Using the tunneling-load, useless speculative cache access is eliminated and the problems included in any load address prediction techniques are solved.

We started with the investigation of the load address prediction techniques, and explained the problems belonging to them. We studied the address prediction miss and presented that the speculative cache access is unnecessary for reducing the load latency if the new verification mechanism is introduced. And then we proposed the tunneling-load technique, which does not predict the load address but generate it. By eliminating the speculative data cache access, the problems such like the memory traffic explosion and the data cache pollution are solved.

The experimental results show that the tunneling-load exploits the ILP, especially for floating point workloads. We have found that on the four-issue machine the tunneling-load with the RSB which has 64 entries improves the performance by approximately 10% on average.

We are encouraged that the tunneling-load is very effective for improving the pipeline performance.

References

1. T.M.Austin et al., "Streamlining data cache access with fast address calculation", Proc. of ISCA22, pp.369-380, 1995.
2. T.M.Austin et al., "Zero-cycle loads: microarchitecture support for reducing load latency", Proc. of MICRO28, pp.82-92, 1995.
3. D.Burger et al., "Evaluating future microprocessors: the SimpleScalar tool set", Technical Report CS-TR-96-1308, University of Wisconsin Madison, July 1996.
4. A.Capitanio et al., "Partitioned register files for VLIWs: a preliminary analysis of tradeoffs", Proc. of MICRO25, pp.292-300, 1992.
5. P.P.Chang et al., "IMPACT: an architectural framework for multiple-instruction-issue processors", Proc. of ISCA18, pp.266-275, 1991.
6. P.P.Chang et al., "Comparing static and dynamic code scheduling for multiple-instruction-issue processors", Proc. of MICRO24, pp.25-33, 1991.
7. T-F.Chen et al., "Effective hardware-based data prefetching for high-performance processors", IEEE Trans. Computers, vol.44, no.5, pp.609-623, May 1995.
8. R.J.Eickemeyer et al., "A load-instruction unit for pipelined processors", IBM J. Res. Develop., Vol.37, No.4, pp.547-564, July 1993.
9. K.I.Farkas et al., "Complexity/performance tradeoffs with non-blocking loads", Proc. of ISCA21, pp.211-222, 1994.
10. M.Franklin et al., "Register traffic analysis for streamlining inter-operation communication in fine-grain parallel processors", Proc. of MICRO25, pp.236-245, 1992.
11. M.Golden et al., "Hardware support for hiding cache latency", Technical Report CSE-TR-152-93, University of Michigan, Feb. 1993.
12. R.M.Keller, "Look-ahead processors", ACM Computing Surveys, vol.7, No.4, pp.177-195, Dec. 1975.
13. S.McFarling, "Combining branch predictors", WRL Technical Note TN-36, Digital Western Research Laboratory, 1993.
14. T.C.Mowry et al., "Design and evaluation of a compiler algorithm for prefetching", Proc. of ASPLOS V, pp.62-73, 1992.
15. T.Sato et al., "Hiding data cache latency with load address prediction", IEICE Trans. Inf. & Syst., vol.E79-D, no.11, pp.1523-1532, Nov. 1996.
16. T.Sato, "Data dependence speculation combining memory disambiguation with address prediction", Proc. of SWoPP'97 (IPSJ SIG Notes), Aug. 1997.
17. S.Wallace et al., "A scalable register file architecture for dynamically scheduled processors", Proc. of PACT'96, pp.179-184, 1996.

Performance Estimation of Embedded Software with Pipeline and Cache Hazard Modeling

Norbert Imlig, Akihiro Tsutsui

NTT Optical Network Systems Laboratories
1-1, Hikari-no-oka, Yokosuka-shi
Kanagawa, 239 Japan
{imlig, akihiro}@exa.onlab.ntt.co.jp

Abstract. A major challenge in telecommunication design is introducing flexibility while still meeting real-time performance goals. Keeping both flexibility and performance while minimizing cost, leads to mixed hardware-software systems. In the absence of a generic partitioning algorithm, accurate cost and performance modeling become crucial when exploring architectural alternatives. This paper presents a case study in which we apply an efficient software performance estimation method to an ATM (Asynchronous Transfer Mode) network application. Since the execution efficiency of pipelined RISC machines heavily depends on the characteristics of the application and the underlying memory hierarchy, effects from pipeline- and cache stalls must be taken into account. The aim of our methodology is to increase the predictability of software execution time in order to minimize expensive hardware implementation.

Key-words: Software estimation, RISC, Cache, ATM, Co-desing

1 Introduction

Apart from flexibility and performance, the cost of ATM systems [1], [2] plays a major role. In recent years, two trends could be observed in ATM system design. The first one maximizes performance and paves the way to "Gigabit Networking". In order to accelerate system speed, software oriented functionality is transferred to hardware. The second trend tries to minimize overall system cost by moving low layer tasks to software. In either case, finding the most appropriate cut-off point between hardware and software is a key problem in embedded system design. While hardware performs rather primitive operations on fixed bit-width data structures in a short cycle period, software is strong at applying sophisticated algorithms on complex structures. An important advantage of software is scaleability. While additional hardware takes extra resources and cost, additional software stored in memory has only little resource penalty. Due to the *locality of code* principle, execution performance increases only slightly if we make use of an appropriate memory hierarchy. However, cache and pipeline performance varies heavily from application to application which complicates accurate estimation.

The layered nature of telecommunication protocols leads to an architecture as shown in Fig. 1. The system can be divided into three basic partitions:

1. Pre-processing (mostly done with application specific hardware circuits)
2. Co-processing (embedded hardware-software systems)
3. Post-processing (host system with general purpose CPU)

Application specific integrated circuits (ASIC) connect the system to the physical layer (PHY). Stream oriented tasks have high performance constraints but only low functionality. They are standardized early on and thus can be implemented as fixed hardware circuits. Middle layer protocols interface the low level hardware block with the host system at the end of the chain. In the case of ATM functionality is rather complex. Apart from tasks like operation and management functions some signaling and reservation protocols may also be implemented here since both flexibility and performance targets must be met [3]. Dedicated hardware and software engines run these tasks in an embedded system manner [4]. Since this functionality is often not completely standardized, an approach that allows updates not only during product design but also in the field later on is preferred.

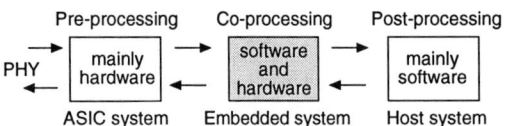

Fig. 1. Telecommunication systems

In this paper we analyze ATM layer functionality and investigate implementation issues for a mixed hardware-software system. Our methodology starts with a C description. After compiling, we simulate the code on the target processor and track real execution time. Postsimulation of the application's memory trace allows us to evaluate stalls caused by cache architecture. Cost and performance of hardware functions are obtained by synthesizing their RTL description.

The rest of the paper is organized as follows. After a short overview on related work in Sect. 2, we describe our system model and the target architecture. In Sect. 4 we introduce the measurement method and the used tools. Then the results of an evaluation using ATM transmission circuits are presented. We conclude our paper with some final thoughts and proposals on future work.

2 Related Work

The task of implementing a certain amount of functionality in a mixed hardware-software environment is referred as co-design problem [5], [6]. Given a specification and a target architecture, co-design can be divided into the following tasks, as shown in Fig. 2.

1. Behavioral modeling (functional verification of system specification)
2. Partitioning (evaluating architectural alternatives, cost-performance evaluation, software-hardware tradeoff)
3. Synthesis of hardware, interfaces and software modules
4. Validation (testing whether implementation meets overall system requirements)

These subtasks are not isolated but interconnected by feedback loops. The goal of partitioning could be described as mapping the system's functionality to software and hardware while minimizing overall cost. Flexibility and cost speak for software oriented implementation while performance favors specific hardware. Recently, a new type of dynamically reconfigurable hardware such as FPGA's (field-programmable gate arrays) have been introduced. With this, the boundaries between hardware and software have even become more fuzzy. In the absence of generic tools, partitioning relies heavily on experience and heuristic methods. Therefore accurate estimation of cost and performance of a chosen partition is a major task in co-designing.

Partitioning can occur at different levels of granularity (i.e., statement, basic block[1], function, task). However, fine-grain partitioning is not needed for modeling simple, control oriented telecommunication circuits. Partitioning at basic block level makes it also difficult to estimate the communication overhead between hardware and software. Thus we partition an algorithm at the function level. Hardware and software can communicate by just passing arguments. This allows us to localize computation which minimizes interfacing overhead.

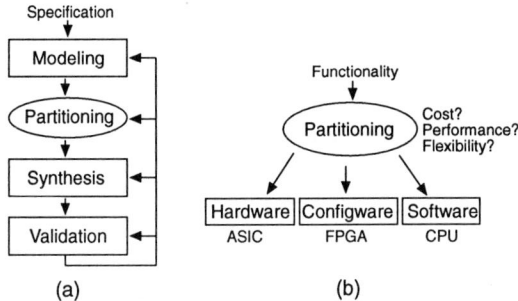

Fig. 2. Hardware-software co-design: (a) design flow, (b) partitioning

Software performance estimation methods can be divided into two major categories. *Dynamic profiling* keeps track of the sequences of all executed instructions and data references. This type of measurement requires a simulator close to the target architecture [7]. The second approach is *static estimation*. Extended path analysis is performed to determine minimum and maximum execution times [8]. However, static analysis makes it difficult to accurately model effects of pipeline- and cache organization. Further, careful user annotations are necessary to set upper bounds for loops and dynamic data structures.

Because we intend to measure the performance of real telecom input workloads, we choose dynamic profiling for our methodology. Rather than in worst case, we are interested in the average performance (i.e., throughput) of our system. Variations of execution time can be compensated by appropriate input buffering.

3 Target Architecture

A systematic approach in system design starts with the definition of the target architecture. We employ a typical embedded system which consists of a general purpose RISC processor surrounded by application specific, co-processing hardware engines. The "system-on-a-chip" has a direct I/O interface like UTOPIA (Universal Test and Operations Physical Interface for ATM) [9]. As a target processor we chose DLX [10] because of its similarity with other popular RISC architectures like MIPS [10]. The memory system consists of a main memory module connected to the processors' data and instruction caches by buses.

Hardware-software communication takes place in a shared memory fashion. Since we assume not to run an operation system, only one software task can be executed on the processor at any time. Concurrent execution is possible between hardware engines and the main CPU. Synchronization between hardware and software tasks is achieved in two ways. If the latency of a hardware engine is small, we stall the processor and poll for the result. The second type of communication uses a timer. After starting,

---

[1] A basic block is a sequence of instructions that always execute in succession.

software and hardware execute in parallel. When the hardware task finishes, the expired timer fetches the result with an interrupt. Both types of handshaking can be modeled as structural CPU stalls. Further, this communication scheme allows accurate overall estimation and supports flexible scheduling.

An similar system was implemented in a dedicated device [11]. A 32 bit DLX processor controls various ASIC engines and an embedded 8 bit CPU. It is fully programmable and can support a wide range of ATM networking applications.

4 Methodology

Measurement takes place in two phases, as shown in Fig. 3. First, every function is implemented as software. By simulating the whole system we verify its functionality and correct errors.

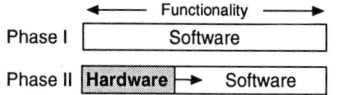

Fig. 3. Design methodology: "stay soft-go hard"

After evaluating instruction mix and time, we trade complexity off against performance. In order to minimize communication overhead, data and control dependencies are analyzed. Selected functions are transferred to hardware. This incremental approach, called "stay soft-go hard", is repeated until overall performance and cost requirements are fulfilled. Since hardware engines are still simulated on a functional level in software, efficient exploration of the whole design space can be achieved in a short time.

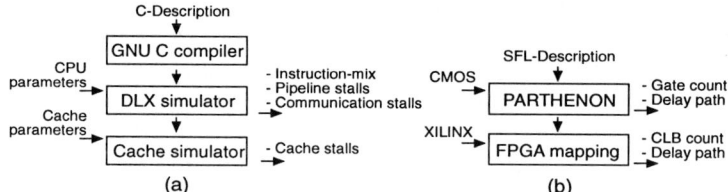

Fig. 4. Estimation: (a) software, (b) hardware

4.1 Software Estimation

The left side of Fig. 4 summarizes the different stages of software estimation. After coding the system in C, we compile it with a 32 bit compiler based on *GCC* version 2.6.3. We then run the application on the DLX software simulator *dlxsim+*. Our tool is an extended version of *dlxsim* [12] and tracks stalls and instruction mix per function.

In order to estimate cache performance, we feed a memory trace of the application and cache parameters into the simulator *dinero* [13]. A trace is a finite sequence of memory references obtained by running the system on dlxsim+. Cache parameters like block size, associativity can be freely configured. In the next section we describe how time is measured.

4.2 Measurement of Time

Execution time of a CPU can be decomposed into 3 separate components:

$$\text{CPU}_{\text{time}} = \text{CPI}_{\text{total}}\left[\frac{\text{cycles}}{\text{instr}}\right] \times \text{IC}\left[\frac{\text{instr}}{\text{program}}\right] \times t_{\text{clock}}\left[\frac{\text{sec}}{\text{cycle}}\right] \quad (1)$$

- Clock cycle time (t_clock) is the inverse of the CPU clock rate. It depends on overall organization and implementation technology (VLSI process).
- Instruction count (IC) is the total number of commands a program executes to produce desired results. It is influenced by the instruction set architecture of the machine and compiler technology. We measure it with the simulator dlxsim+.
- Cycles per instruction (CPI) is the most challenging metric. It tells how many cycles of the CPU are bound to the execution of a specific instruction. The CPI_base of a machine can be calculated by summing the product of each individual component (CPU_i) by the fraction of occurrences (Mix_i) of that instruction in the program.

$$\text{CPI}_{\text{base}} = \sum_{i=1}^{n} \text{CPI}_i \times \frac{\text{IC}_i}{\text{IC}} = \sum_{i=1}^{n} \text{CPI}_i \times \text{Mix}_i \quad (2)$$

Due to pipeline, cache, and interface stalls (communication between hardware and software), overall CPI further increases:

$$\text{CPI}_{\text{total}} = \text{CPI}_{\text{base}} + \text{CPI}_{\text{pipeline}} + \text{CPI}_{\text{memory}} + \text{CPI}_{\text{interface}} \quad (3)$$

In order to measure overall CPI we proceed in 3 steps:
1. In step 1 we assume a perfect memory system and evaluate the CPI caused by pipeline stalls on a per function basis. While DLX has a CPI_base of one, there are two types of pipeline stalls recorded by the simulator.

$$\text{CPI}_{\text{pipeline}} = \text{CPI}_{\text{data\_stall}} + \text{CPI}_{\text{control\_stall}} \quad (4)$$

Data_stall (also load or *read-after-write* stalls) arise when one instruction depends on the result of the previous one. Since DLX has a clock latency of one after load instructions, this leads to a one cycle pipeline interlock. Control_stall arise from the pipelining of branches and other instructions that change the program counter. Since DLX uses a simple delayed branch technique, the penalty depends on the degree of static optimization achieved by the compiler. Moreover, load stalls are scheduled by the compiler to minimize the conflicts caused by data dependency.

2. In step 2 we measure the penalty caused by the memory hierarchy. After feeding the cache simulator with the memory trace of the application and the cache parameters, we insert the output of the simulation into the following formula:

$$\text{CPI}_{\text{memory}} = \frac{\text{accesses}}{\text{IC}} \times \text{miss rate} \times \text{penalty}_{\text{cache}} \quad (5)$$

Accesses per instruction and the miss rate are recorded by the simulator. Cache penalty depends on the bus architecture and the latency of main memory access.

3. In step 3 we estimate the overhead caused by the software-hardware interface. The handshaking described in section 3 allows us to model communication hazards as structural stalls. In the case of co-processing, the CPU stalls until the result is available.

$$\text{CPI}_{\text{interface}} = \frac{\text{calls}}{\text{IC}} \times \text{penalty}_{\text{interface}} \quad (6)$$

We track the numbers of calls to the hardware function and multiply it by the latency penalty. In the case of concurrent execution, we take the penalty caused by timer interrupts.

4.3 Hardware Estimation

The right part of Fig. 4 summarizes the different stages of hardware estimation. Coding is done in the RTL (Register Transfer Level) language called SFL (Structured Function description Language) [14], [15]. It enables the designer to model single phase clock- synchronized circuits with low redundancy. Since the language is small and precise, it makes efficient use of the designer's time. Synthesis is done with PARTHENON [16]. It transfers the SLF description to a netlist and outputs gate count and estimated circuit delay. Because this estimation is based on a demonstration library (1.5μm CMOS), we feed the netlist of the design into the Xilinx tools [17] for a second verification. Cost is associated with CLB (Configurable Logic Block) count and time is measured using the *Xdelay* tool.

5 Results

Our case study is based on ATM circuits described in [18]. They synchronize an input byte stream and perform an ATM layer protocol. A similar system can also be found in [19]. We assume the following parameters of the embedded processor architecture.
- Clock cycle time t_clock = 25ns (40 MHz)
- Simple, direct mapped cache (write through, read allocated), blocksize is 8 byte, latency penalty of memory hierarchy is 2 CPU cycles (per 8 byte from main memory), size of instruction- and data cache is 4kB

As a subsystem the embedded CPU shares the chip resources with various ASIC cores as in [11]. On-chip caches of embedded systems generally differ from traditional caches. Because chip area is restricted, they are small and simple and must share the resources with other performance enhancements (i.e., pipelining, buffering).

Figure 5 shows the instruction mix of all software system as compiled with *gcc* option -O2 and run through *dlxsim+*. The frequency of transfer, arithmetic and control instruction varies from function to function. Load-store overhead is dominant. Pipeline stalls related to load interlocks (d_stall) and control hazards (c_stalls) are in the range from 10 to 20 percent.

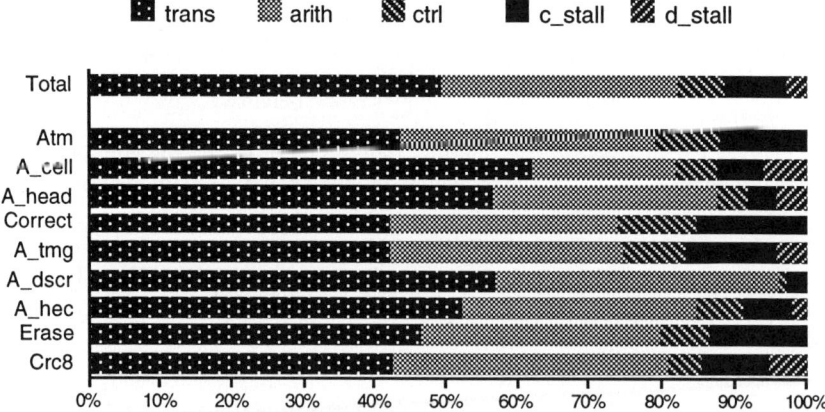

Fig. 5. Instruction mix of ATM functionality: transfer- (trans), arithmetic- (arith) and control- (ctrl) instructions; control stalls (c_stall) and data stalls (d_stall)

Figure 6 shows the complexity-performance graph of all software system. Branch frequency of a task can be correlated to the amount of complexity involved. The higher the frequency of conditional statements, the more complex the task becomes. Complex systems coded in hardware often suffer from state explosion and have high resource penalties. Thus this graph supports the designer in the difficult task of allocating a function to software or hardware.

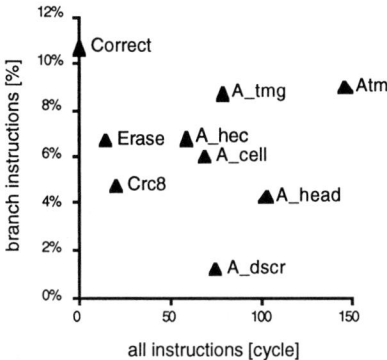

Fig. 6. Complexity-performance graph

Complex tasks such as Atm (control oriented, finite state machine) are located in the upper half of the graph. On the other hand functions like A_dscr (descrambler) are low in complexity, but still take many cycles to compute.

Figure 7 shows the cost-performance graph of the same system, but now implemented as all hardware. The left chart shows gates versus time estimated by the PARTHENON synthesis and on the right is the CLB count (100% equals 400 CLBs) of the XILINX 4010PQ208-5 FPGA versus time. This clearly shows the non-linear scaling penalty of hardware, especially in the case of the fixed cell FPGA architecture with place and route constraints. The cost for implementing the descrambler A_dscr in hardware is relatively low (i.e., about 500 gates). On the other hand, the corresponding software function takes a lot of CPU resources (about 75 cycles). That is why A_dscr is a candidat for hardware implementation.

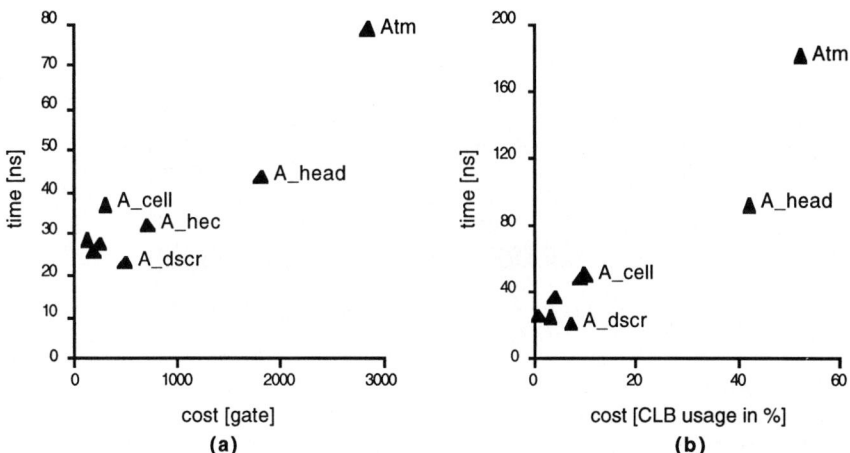

Fig. 7. Cost-performance graph: (a) PARTHENON estimation, (b) XILINX 4010 estimation

The cache performance reflects the scaling penalty of software. Figure 8 shows various cache sizes and resulting miss rates. Cache performance is heavily correlated to the instruction and data code size of the application and the amount of temporal and spatial locality. The code of our software implementation needs about 7kB instruction and 2 kB data memory. Since low layer ATM functionality consists of only a few loops, locality of code is low. This sensitivity is represented by the strong increase in miss rate at 8kB and 2kB.

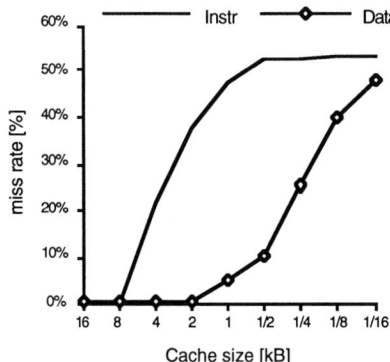

Fig. 8. Cache performance: miss rate of instruction and data memory accesses

The left of Fig. 9 shows software time of each function and the accumulated time of the entire software system. The penalty resulting from pipeline stalls (d_stall, c_stall) is low compared to that of memory hierarchy (m_stall). With a 4 kB instruction cache, the total miss rate is about 15 percent. Thus, cache performance dominates overall performance. The right chart shows cost of the hardware circuit as estimated by PARTHENON.

We now shift functions of low complexity but high performance constraints to hardware. Figure 10 shows the situation after partitioning. Tasks Crc8, Erase, A_hec, A_dscr and A_head are implemented as hardware circuits.

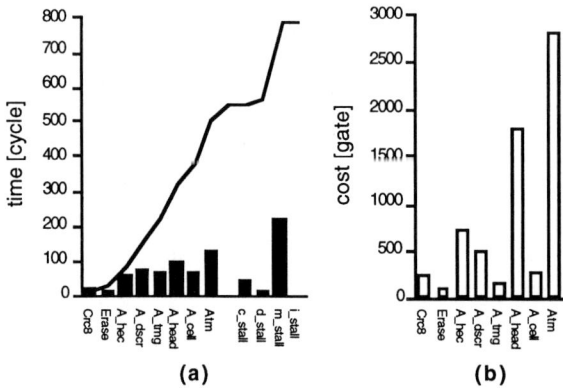

Fig. 9. All software system: (a) execution time of the software tasks, pipeline stalls (c_stall, d_stall), memory stalls (m_stall) and interface stalls (i_stall). (b) cost of the corresponding hardware task

Since code size has decreased, cache performance is better. It is now in the same dimension as the penalty caused by pipelining. On the other hand, interface overhead (i_stall) must be added. Since latency of the hardware tasks is low, we employ co-processing and stall the CPU until the results are available. We calculate the interface penalty of a task as mentioned in Equation (6) by dividing the latency of the corresponding hardware function with the clock cycle time of the CPU.

The left of Fig. 11 shows the cost-performance design space of our application. The performance gap between software and hardware implementation is significant. Apart from the instruction count and clock cycle time, performance depends heavily on the "execution efficiency" measured as clock cycles per instructions (CPI). The contribution of the CPI varies strongly from partition to partition, as shown on the right of Fig. 11.

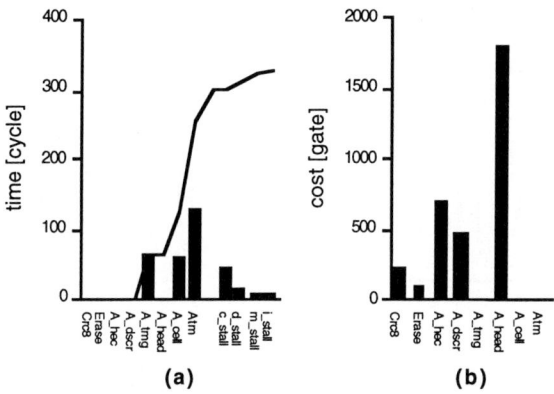

Fig. 10. Mixed hardware-software system: tasks with low complexity (Crc8, Erase, A_hec, A_dscr, A_head) are implemented as hardware circuits. (a) software time, (b) hardware cost of the chosen partition

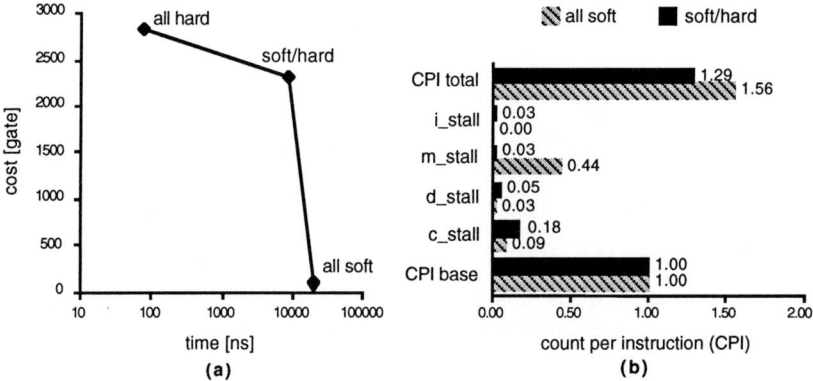

Fig. 11. Hardware-software gap: (a) cost-performance design space, (b) count per instruction (CPI) which reflects "execution efficiency" of the CPU: pipeline stalls (c_stall, d_stall), memory stalls (m_stall) and interface stalls (i_stall)

The cache penalty seems to be the most unpredictable factor when evaluating software performance. We tried to improve the cache predictability by carefully inlining functions and thus increase the locality of the code. Figure 12 shows the trace of the instruction references of the all software system before and after inlining. The very symmetric execution of the inlined program leads to an cache miss rate improvement as shown in Fig. 13. With additional prefetching (load-forward) the miss rate can be further decreased without a significant increase of the bus traffic.

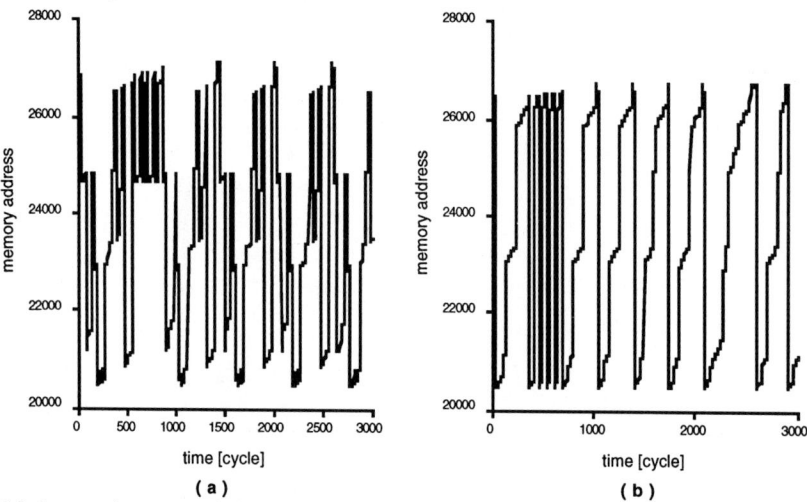

Fig. 12. Instruction trace of all software implementation: (a) accesses to the program memory without code inlining, (b) with inlining

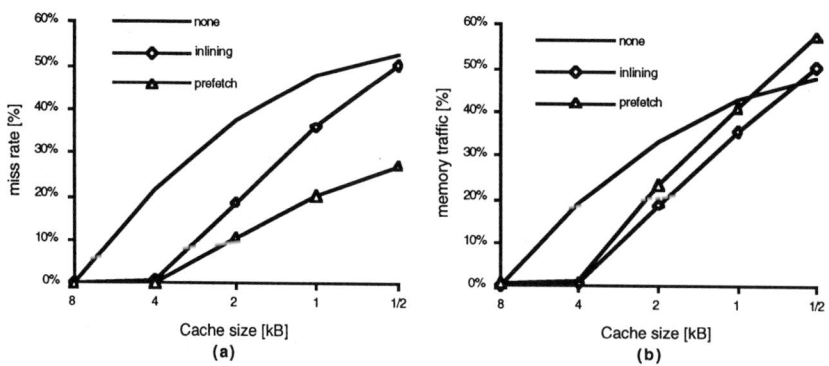

Fig. 13. Cache performance: (a) miss rate of traces in Fig. 12, (b) bus traffic between main memory and instruction cache (words form memory divided by demand fetches)

6 Conclusion

In a case study we have analyzed ATM networking functionality and investigated various cost and performance issues of software and hardware implementation. Our methodology starts with a C description and models performance in an incremental approach. After evaluating the CPI of pipeline stalls, we add the CPI resulting from cache architecture and overhead caused by hardware-software interfacing. The example considered reveals that the execution efficiency of a CPU modeled with corresponding CPI's heavily depends on the characteristics of the application, and especially the underlying memory hierarchy. We have shown, that the predictability of cache performance can be increased by inlining code segments and prefetching. The limiting factor in a mixed hardware-software system is generally the execution time of software tasks. By increasing the accuracy of CPU performance estimation, more system functionality can be allocated to software while reducing expensive hardware implementation. Since our approach accurately estimates this time, we are able to find the best partition in an effective way.

Further work must be done when modeling the performance of parallel hardware and software execution. Communication overhead increases if CPU and hardware engines access the shared memory at the same time. Due to port and bus restrictions these accesses lead to structural hazards which are difficult to predict.

Acknowledgments

The authors would like to thank H. Nakada, K. Shirakawa, M. Inamori, K. Ishii and T. Miyazaki from NTT Optical Network Systems Lab. for supporting this work. Thanks are also due to the reviewers for their helpful suggestions.

References

1. A. Alles: ATM Internetworking. Cisco Systems Inc., 1995.
   http://cell-relay.indiana.edu/cell-relay/docs/cisco.html
2. E. Hoffman, A. Mankin and M. Perez: VINCE: Vendor Independent Network Control Entity. Naval Research Laboratory, 1993.
   ftp://hsdndev.harvard.edu/pub/mankin/
3. The ATM Forum: Traffic Management Specification Version 4. atmf95-0013R6, 1995.
4. K. Buchenrieder: Hardware/Software Codesign- An Annotated Bibliography. IT Press, Hartenstein, Chicago, 1995.
5. R. Gupta: Hardware-Software Co-design, Tools for Architecting Systems-On-A-Chip. Proc. of the Asia and South Pacific Design Automation Conf. (ASP-DAC), pp. 285-289, 1997.
6. W. Wolf: Hardware-Software Co-Design of Embedded Systems. Proc. of IEEE, Vol. 82, No.7, pp. 967-989, 1994.
7. W. Ye, R. Ernst, T. Brenner, and J. Henkel: Fast Timing Analysis for Hardware-Software Co-Synthesis. Proc. of Int. Conf. Computer Design, IEEE CS Press, pp. 452-457, 1993.
8. Y-T. S. Li, S. Malik, and A. Wolfe: Performance Estimation of Embedded Software with Instruction Cache Modeling. Proc. of Int. Conf. on Computer-Aided Design (ICCAD), pp. 380-387, 1995.
9. The ATM Forum: An ATM-PHY Interface Specification Level 1. Version 2.01, 1997.
10. J. L. Hennessy and D. A. Patterson: Computer Architecture, A Quantitative Approach. Morgan Kaufmann Publishers Inc., Second Edition, 1996.

11. M. Inamori, K. Ishii, A. Tsutsui, K. Shirakawa, and T. Miyazaki: A New Processor Architecture for Digital Signal Transport Systems, to appear in Proc. of ICCD, 1997.
12. L. B. Hostetler, B. Mirtich: DLXsim- A Simulator for DLX. Reference Manual, 1996.
    ftp://max.stanford.edu/pub/hennessy-patterson. software
13. M. D. Hill: Cache simulator dineroIII. Reference Manual, 1989.
    ftp://max.stanford.edu/pub/hennessy-patterson.software
14. Y. Nakamura, K. Oguri, A. Nagoya, M. Yukishita, R. Nomura: High-Level Synthesis Design at NTT Systems Labs. IEICE Trans. on Information and Systems, Vol. E76-D, No 9, pp. 1047-1054, 1993.
15. J. Suzuki and S. Ono: Entropy CODEC from Behavioral Description Based LSI-CAD for Fully Programmable Image Coding System. Proc. of Design Automation for Embedded Systems, pp. 231-255, 1996.
16. Y. Nakamura, K. Oguri, A. Nagoya, and R. Nomura: A hierarchical behavioral description based CAD System. Proc. of EURO ASIC, pp. 282-287, 1990.
17. Xilinx: The programmable Logic Data Book. Xilinx Inc., 1994.
18. N. Ohta, H. Nakada, K. Yamada, A. Tsutsui, and T. Miyazaki: PROTEUS: Programmable Hardware for Telecommuncication Systems. Proc. of ICCD, 1994.
19. Y. Takabatake, M. Hashimoto, T. Tsujita, J. Takeda, and Y. Shobatake: A Software-Based ATM Interface Card and its Evaluation. IEICE Trans. on Communications, Vol. E-80-B, No. 1, pp. 127-134, 1997.

An Implementation and Evaluation of a Distributed Shared-Memory System on Workstation Clusters Using Fast Serial Links

Hironori Nakajo, Akihiro Ichikawa and Yukio Kaneda

Department of Computer and Systems Engineering, Faculty of Engineering, Kobe University, Rokko-dai Nada Kobe 657, Japan

Abstract. We summarize an implementation of a distributed shared-memory system on a workstation cluster. In this paper, we introduce fast serial links called Serial Transparent Asynchronous First-in First-out Link (STAFF-Link). By using these links we construct a parallel processing system based on the workstation cluster. In the workstation cluster, a distributed shared-memory mechanism is utilized for inter-process communication with software controlled cache. We evaluate the performance of the system for several applications.

1 Introduction

Recent improvements of integrated circuits and network technologies allow a number of workstations to be connected to obtain high performance of parallel/vector computing systems.

In general, multiprocessor architectures can be classified based on processor communication mechanisms. The communication can either be through explicit messages sent directly from one processor to another, or through accesses to shared-memory. In message-passing systems, programmers observe a set of separate computers that communicate only by sending explicit messages. In contrast, in shared-memory systems, memory is accessible by all processors, and communication is done through shared variables or messages stored in shared-memory buffers.

In shared-memory systems, there is no requirement for programmers to manage data movement. While on message-passing systems, programmers must explicitly keep all data passed between processors. Therefore, the latter is quite difficult to describe parallel programs if communication patterns are irregular. Tightly coupled multiprocessor systems have achieved commercial successes, such as SUN SPARCcenter1000 or SGI Origin2000.

Though parallel programming libraries such as PVM[1] or MPI[2] are now available in workstations clusters[3], the need of inter-process communication through a shared-memory style in a parallel computing environment for multiple workstations increases. From the above situation, we are interested in the implementation of the shared-memory style in workstation clusters.

The shared-memory system on workstation clusters has two major disadvantages compared with message-passing systems or tightly-coupled shared-memory machines. First, there is no physical shared-memory in the system. Second, the local area network which connects workstations usually does not hold higher bandwidth.

Since there is no centralized shared-memory in workstation clusters, virtual shared-memory mechanisms using message passing have been proposed, such as IVY[4], TredMark[5], Quarks[6] and Tempest[7]. In such systems, caching techniques are indispensable against low bandwidth of LANs.

In our previous study[8], we have implemented a distributed shared-memory (DSM) system based on a 10-base Ethernet local area network. In this study, because of the low bandwidth and frequent contentions of our LAN, we could not achieve sufficient performance for all applications.

From the above study, we have learned that flexible and sophisticated network facilities are required for implementing efficient shared-memory parallel computing environment. Therefore, we have implemented a workstation cluster which consists of workstations connected via fast serial links in a point-to-point manner. The serial link is called Serial Transparent Asynchronous First-in First-out link (STAFF-Link).

In the remaining sections of this paper, we introduce the hardware configuration of a STAFF-Link and our parallel computing environment of a workstation cluster connected via STAFF-Links. We measure the performance of our system with several application programs for validation, and discuss the result.

2 STAFF-Link

STAFF-Link is adopted to connect multiple I/O units and processing elements in order to configure a scalable I/O subsystem of JUMP-1(Japan University Massively Parallel computer)[9].

The interface of the STAFF-Link is designed for a general purpose, thus we aim the use of the fast serial link at constructing a high performance workstation cluster.

2.1 Transfer speed and distance of Local and Wide Area Networks

In a parallel computing system, each processing element is connected via a fast bus on a backplane board or with flat cables. When conducting data communication at high-speed using the cables with many signal lines, it is difficult to extend the cables because of the influence of noise arising from cable induction properties, and the cables become thick. Furthermore, if the transfer bit width is enlarged, the area occupied by the connector on the substrate increases in proportion to the number of router ports.

Considering the tradeoff between physical and spatial restrictions and transfer speeds, we have implemented a communication link which is able to be accessed as if FIFO memory is located between the ends of the link. This link is referred to STAFF-Link *(Serial Transparent Asynchronous First In First Out Link)*.

In the field of wide area networks such as B-ISDN and ATM, high-speed serial communication LSIs are now cheaply available. The reliability of communication using these LSIs has improved, and this, combined with improvement in technology for mounting of these LSIs on printed circuit boards, holds much promise for serial communication among workstations.

In order to show the position of STAFF-Link's transferring speed and distance, we describe transferring speed and distance in current computer interfaces and LAN and WAN in Fig.1. It can be supposed that a bundle of multiple STAFF-Links makes transfer speed to range from tens of Mbps to a few Gbps with valid distances of a few meters to a few hundred meters.

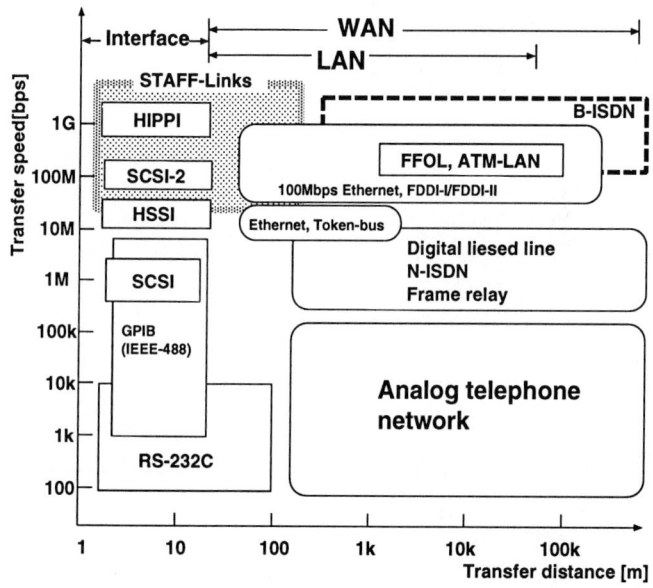

Fig. 1. Transfer speed and distance in computer interface / LAN / WAN

2.2 Configuration of a STAFF-Link

Conventional serial communication interfaces involve delays due to conversion of parallel data to serial one, with consequent limits on transfer speeds. Serial communication may be considered in terms of the following five phases.

1. Data write.
2. Parallel–serial conversion.
3. Data transfer.

4. Serial–parallel conversion.
5. Data read.

With using a STAFF-Link, phases 2 – 4 are handled by a high-speed serial communication LSI. Buffers are provided at both the send and receive ends of the link, and the five phases overlapped to raise throughput.

Fig.2 shows the configuration of a STAFF-Link. The communication block consists of a high-speed bidirectional serial communication LSI (TAXI chip[10]), and two FIFO memories (send and receive), and an asynchronous communication controller (X flow control) for the purpose of handshaking to prevent FIFO overflow. A current single STAFF-Link is able to handle transfer rates of up to 140 Mbps.

Connection of two communication blocks by Category 5 twisted-pair cables results in a virtual bidirectional FIFO at each end, and in turn effectively hides the physical communication distance between nodes to ensure a transparent communication path. The communication controller provides asynchronous communication control simultaneously with Xon/Xoff control. Namely, when the receiving FIFO at the destination is less than half full of data, the receiving side sends an Xoff message requesting an interruption of transmission. When the receiving FIFO is able to receive again, the Xon message is sent requesting resumption of transmission. This control is performed automatically between the two communication controllers. At the transmission side, data is written to the sending FIFO until it is full, and at the receiving side the data is able to be read from the FIFO until it gets empty. The use of multiple STAFF-Links allows a workstation cluster system with a wide communication bandwidth.

3 DSM system configuration of a workstation cluster

3.1 Design concept

The design concept of our system is shown in Fig.3.

In stead of a single specific server node, each node manages shared-memory space. Thus the load of management of accesses to shared-space can be distributed in the whole system.

As mentioned before, caching technique is needed for workstation clusters with a shared-memory style, thus our system supports a cache mechanism with a write invalidate protocol. Shared-memory of our system consists of pages which are cached into each local memory.

3.2 Software configuration in each workstation

The software configuration in each node is shown in Fig.4.

Manager process manages **DSM** in main memory by page and updates a directory according to accesses or invalidation messages from other nodes.

Cache memory also takes place in main memory through which an application program accesses a shared-memory space.

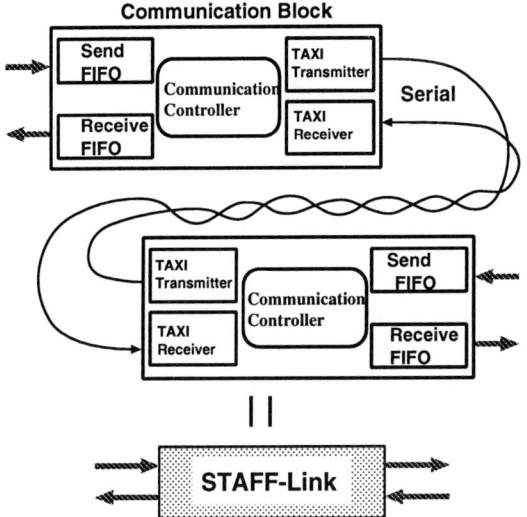

Fig. 2. Serial Transparent Asynchronous First-in First-out Link (STAFF-Link)

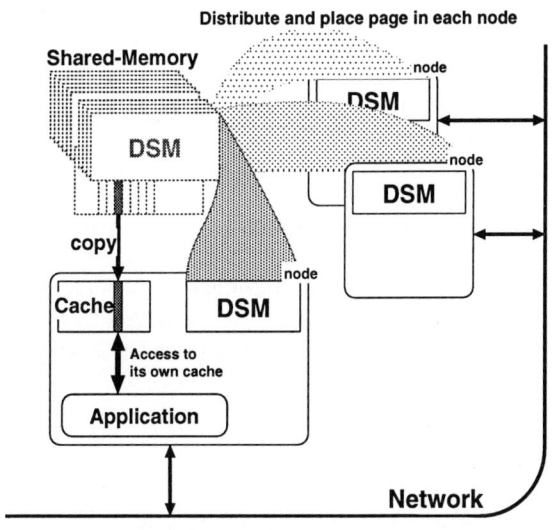

Fig. 3. The concept of our DSM system

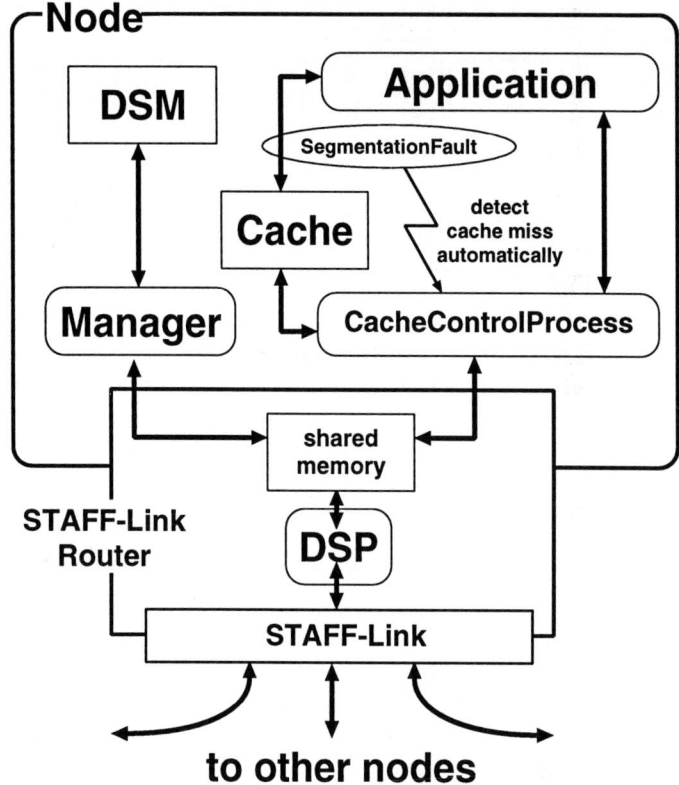

Fig. 4. Structure of one node

Cache Control process is linked to application programs and executed as a thread. After the process detects miss accesses in the application programs, it requests a required page to the **Manager process** and also locks a page for the requested page. Moreover **Cache Control process** conducts an invalidation message according to a request from other nodes in order to keep cache memory coherent.

3.3 Detecting cache misses

In a tightly coupled bus-based multiprocessor system, data accesses from processing elements are always snooped by hardware. However, in workstation cluster environments, operating systems cannot keep watch over all accesses.

In our system, miss accesses to DSM cache are detected by UNIX *Segmentation fault*(SIGSEGV) which is commonly used in software controlled cache systems. Consequently, there is no need to check the state of DSM cache in each access, thus cache hit accesses do not bring any overheads.

3.4 Efficient invalidation by chained multi-cast

In a write access to a page in DSM, clean pages which are a copy of the page have to be invalidated to keep coherence.

In a bus network, it is easy to broadcast invalidation messages at the same time, however, the system with the bus network lacks scalability because of the upper bound limit of its bandwidth. In contrast, it is supposed that a point-to-point network environment makes the system robust in the case of a large number of workstations.

Moreover, in invalidation based cache coherence protocols, it is reported that the number of copies of shared pages do not increase (at most two or three) in most application programs[11], thus the invalidation using a broadcast message does not take the advantage of the simultaneous request.

Therefore, our system adopts the chained multi-cast message as an invalidation message which is a single message and travels around just the required nodes. The chained multi-cast invalidation message includes the address of the node which issues the invalidation request, thus the message is finally sent to the original node as an acknowledge message. The basic concept of chained multicast mechanism is introduced in [12]. This mechanism is very similar to myrinet[13].

3.5 Hardware configuration of Staff-Link router board

Since software controlled routing in a workstation may cause significant overheads which degrades system performance by wasting CPU power, a hardware routing independent of operating systems in the workstation is needed.

From this reason, we have designed and implemented a STAFF-Link router board as shown in Fig.5.

The specification of the I/O network router board is shown below.

- Board : SBus double-height
- Routing controller : DSP (TMS320C40)
- Program memory : 512KWord (2MB)
- Data memory : 512KWord (2MB)

Performance of data transfer using the router board is shown in Table 1, which shows communication time in detail.

From the result shown in the table, it is found that a bottleneck exists in accessing main memory in the workstation, thus the data transfer speed of STAFF-Link is limited up to 64Mbps.

Fig. 5. I/O network router board

4 Preliminary performance evaluation of the DSM system

4.1 Purpose

We have configured an experiment system using STAFF-Link network router boards as shown in Fig.6.

The purpose of this experiment system is to investigate current achievements and problems as well as the future improvements by measuring transfer speed and access speed via STAFF-Links of each workstation.

The workstation cluster consists of sixteen workstations (Sun Microsystems: SPARCstation5 (SS5) with microSPARC-II).

Before evaluating the performance of the workstation cluster using STAFF-Links, we have evaluated software overheads of our system using the software configuration shown in Fig.7.

In the configuration, we have implemented **Cache Control process** and **Invalidate Handler** in Fig.4 included into application programs as a single thread. Ethernet is used as an interconnection network in stead of STAFF-Links.

Fig. 6. Current configuration of the workstation cluster

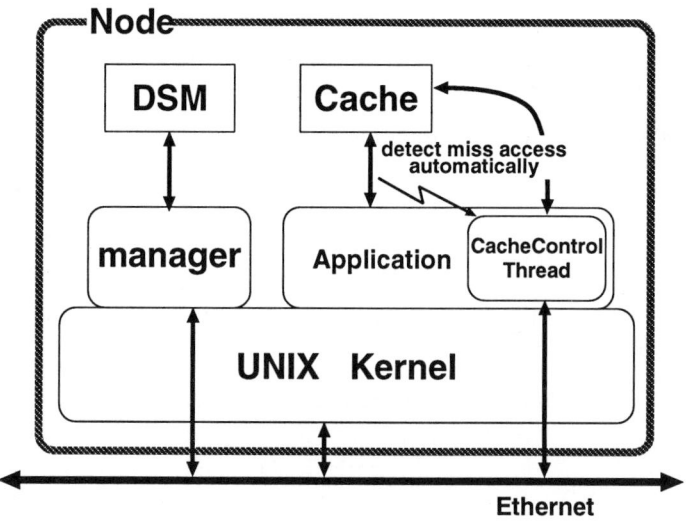

Fig. 7. Configuration of software controlled system

Table 1. Communication time of a STAFF-Link router board

process	time or through put
access start up on communication buffer	$45\mu S$
write to communication buffer	$128Mbps$
communication buffer $\rightarrow$ FIFO	$200Mbps$
communication start up	$320nS$
send FIFO $\rightarrow$ receive FIFO	$140Mbps$
FIFO $\rightarrow$ communication buffer	$200Mbps$
read from communication buffer	$64Mbps$
routing(per 1 node)	$1\mu S$

The specification of workstations used in the experiment is shown in Table 2.

Table 2. Specification of the workstation node

Workstation	Sun SPARCStation5
CPU	microSPARC II (110MHz)
OS	Solaris2.4 (SunOS5.4)
Memory	32MB
Network	Ethernet(10Mbps)

4.2 Performance of cache accesses

In the case of cache hits, it takes $64nS$ and $84nS$ for a read access and a write access respectively. These values are almost the same to the conventional cache memory access.

Table 3. Access time of cache hit(nS)

read hit	write hit
64.121	83.509

In the case of cache misses, we have to study further investigation as follows.

With no copy to be invalidated. If the accessed page is not in any cache memory, the requested page has to be sent from DSM. Table 4 shows each access time in the case of cache misses, however, it is assumed that these accesses do not cause any invalidations nor replacements of the page.

Table 4. Access speed of cache miss(mS)

Manager	read miss	write miss	clean write
local	5.065	5.103	3.237
remote	8.894	8.872	3.757

In a read miss or a write miss which causes page transfer between cache memory and DSM, it takes $5.7mS$ in the case that the application process and the **Manager process** which is responsible to the requested page are in the same node. On the contrary, about $9.3mS$ is taken when they are in different nodes. From the above result, transfer time in a unit of page sized 4Kbyte is about $3.6mS$.

In the case of a clean-write, which is defined as a write access to a clean page, about $3.2mS$ is taken when the **Manager process** is in the same node, otherwise about $3.8mS$.

Fig.8 shows communication penalties in cache miss accesses.

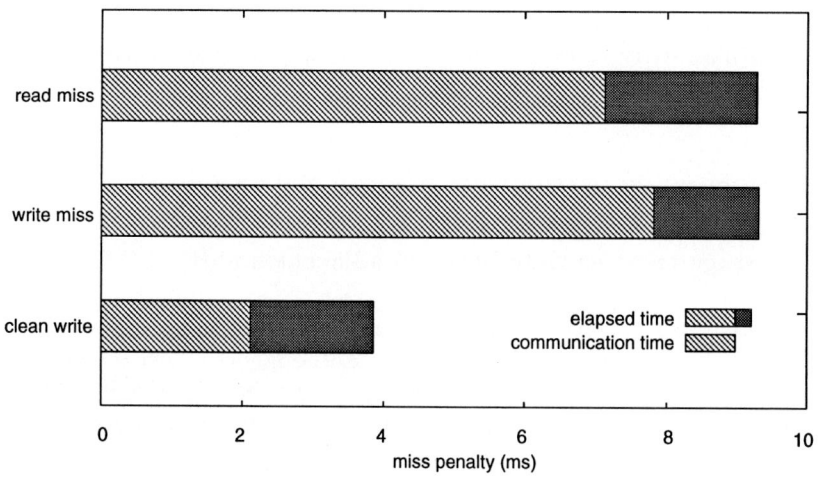

Fig. 8. Miss penalty in cache access time

The time of the communication penalties calculated from the Table 4 does not include communication protocol overheads. On the other hands, communication time in Fig.8 includes the time in communication protocol as well as data transfer time.

Thus the startup time in a read-write miss and a clean-write is about $3.5-4mS$ and about $2mS$ respectively. It occupies about $75-85\%$ in a read or write miss and about 50% in a clean-write.

Consequently, the coherence control overhead is about $1.5-2mS$ in a read miss or a write miss and the time in a clean-write is about $1.7mS$.

In the case of invalidations. Table 5 shows the time spent in a write miss which causes invalidation of copies in other nodes.

To calculate from the former case which does not include invalidations, it takes about $3mS$ for an invalidation.

Table 5. Cache miss with invalidations (mS)

read miss	write miss
10.8	11.0

A write miss which causes an invalidation varies as shown in Fig.9 depending on the number of nodes which have copies to be invalidated.

In our system, chained multicast mechanism is adopted, thus each invalidation requires about $1mS$ per copy of the target page.

Since a clean-write does not cause page transfer, the time is reduced about $5mS$ comparing to a write miss.

5 Performance estimation using a STAFF-Link network

5.1 Message transfer time between adjacent nodes

In our system using STAFF-Links, since most requests and acknowledge messages consist of 3 - 6 bytes, their overhead can be ignored even though header information is supplemented to each message. Since the page size is 4,096 byte, the time to transfer a message between adjacent nodes is estimated as about $602\mu S$ from the specification of the router board.

Using the transfer time, we have estimated the performance of the DSM using STAFF-Links considering the performance results of the software controlled DSM.

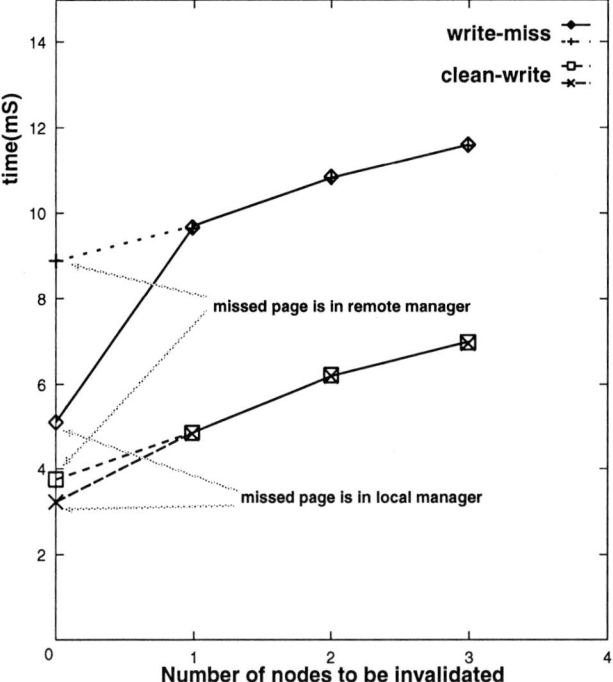

Fig. 9. Write access with invalidation

5.2 Preliminary performance of cache accesses with STAFF-Links

The STAFF-Link network has a configuration of a bi-directional ring structure with bypasses as shown in Fig.10. In this case, an average diameter is about 2.1 hops, thus the average time of routing is supposed to be $2.1\mu S$. This value can be ignored since the message transfer time between nodes is large enough.

By replacing the communication time of the Ethernet environment with the STAFF-Link network environment, the time of a cache miss can be obtained. We have estimated average access time in the case of cache misses as shown in Table 6.

In the Ethernet environment, the time of a cache miss access is about $3.8mS$ in a clean-write, and $8.9mS$ in a read or write miss from Table 4.

In the STAFF-Link network environment with 12 workstation nodes, it takes about $0.7mS$ and $2.1mS$ in a clean-write and a read/write miss respectively. Consequently the read/write miss penalty using STAFF-Links is reduced to $18\% - 24\%$ of the Ethernet environment.

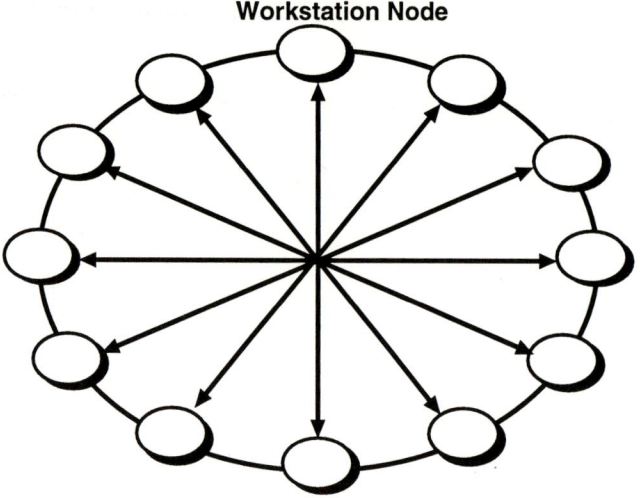

Fig. 10. Ring structure with bypasses of clusters

Table 6. Average access time of cache miss (mS)

read miss	write miss	clean-write
3.09	3.05	0.669

5.3 Access time with invalidations using STAFF-Links

In the STAFF-Link network environment, since a chained multicast message is also used for invalidations, the time of a write miss which causes invalidations are obtained from Fig.9.

Thus, the time of a write miss with invalidations is estimated as shown in Fig.11.

It takes about $3mS$ and $0.6mS$ in a write miss and a clean-write respectively. Access time is not heavily affected by the number of nodes to be invalidated.

6 Summary

In this paper, we have described a workstation cluster system with distributed shared-memory using a fast serial link called STAFF-Link as an interconnection network. We have configured an experimental system with software controlled cache memory, and measured its preliminary performance. From the result, we have estimated the preliminary performance of the DSM with STAFF-Links.

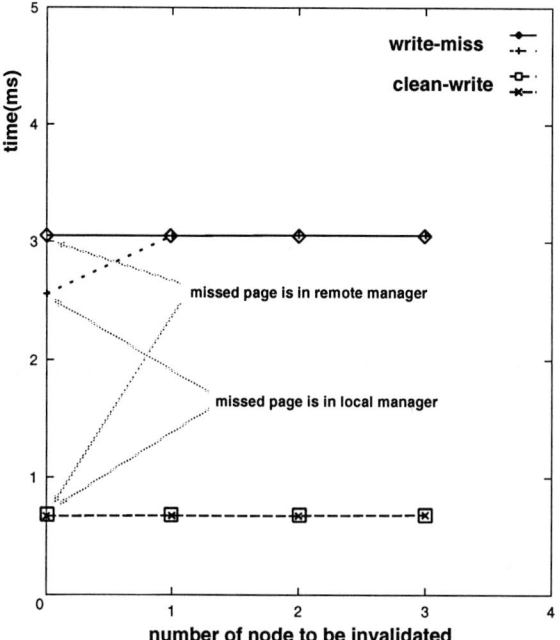

Fig. 11. Prediction of write access with invalidation in the STAFF-Link network environment

We have confirmed the communication performance of a STAFF-Link to be up to 140 Mbps within a bare link.

By using STAFF-Link router boards, we have estimated that we can reduce miss penalty to about $2/3 - 1/4$ of that in the Ethernet environment.

Our future work for the purpose of higher performance is as follows.

- Developing a fast synchronization mechanism
- Implementing weaker consistency models
- Developing a parallelizing compiler based on the above models

Furthermore, we are going to evaluate our system using other applications such as SPLASH[14] or SPLASH2[15].

Acknowledgements

We would like to thank Shinji Tomita of Kyoto University for his support of implementing a STAFF-Link router board. This work was supported in part by the Japan Ministry of Education, Science and Culture under Grant in Aid for Scientific Research 042235103 entitled "Research on Massively Parallel Hardware Architecture" and 06508001 entitled "Development and Testbed of Massively Parallel Computer Prototype".

References

1. V. S. Sunderam, "PVM: A Framework for Parallel Distributed Computing," *Concurrency: Practice and Experience*, Vol. 2, Number 4, pp. 315–339, (1990).
2. Message Passing Interface Forum, "MPI: A Message-Passing Interface Standard," *International Journal of Supercomputer Applications and High Performance Computing*, Vol. 8, Number 3/4, pp. 159–416, (1994).
3. Craig C. Douglas, Timothy G. Mattson and Martin H. Schultz, "Parallel Programming Systems for Workstation Clusters," *Technical Report TR-975, Yale University Department of Computer Science Research*, (1993).
4. Li, K and Hudak, P. : "Memory Coherency in Shared Virtual Memory Systems", ACM Trans. Comput. Syst., Vol.7, No.4, pp.321-359 (1989).
5. Pete Keleher, Sandhya Dwarkadas, Alan L.Cox, and Willy Zwaenepoel: "TreadMarks: Distributed Shared Memory on Standard Workstations and Operating Systems", Rice COMP TR93-214 (1993).
6. Dilip Khandekar:"Quarks : Portable Distributed Shared-Memory on Unix", *quarks/doc/tech-report.ps including ftp://jaguar.cs.utah.edu/pub/dsm/ Quarks.tar.Z*, (1995).
7. Mark D. Hill, James R. Larus and David A. Wook: "Tempest: A substrate for Portable Parallel Programs.", Proc. of COMPCON'95, pp.327–332 (1995).
8. H. Nakajo, K. Kuramae, Y. Kaneda and S. Maekawa: The Implementation and Evaluation of Software Distributed Shared-Memory (DSM) for Workstation Clusters (in Japanese), Trans. IPS Japan, Vol.36, No.7, pp.1719–1728 (1995).
9. H. Nakajo, S. Ohtani, T. Matsumoto, M. Kohata, K. Hiraki and Y. Kaneda: An I/O Network Architecture of the Distributed Shared-Memory Massively Parallel Computer JUMP-1, Proc. of 11th Int. Conf. on Supercomputing (ICS97) (1997) (to appear).
10. Advanced Micro Devices, Inc, *Am7968 / Am7969-175 TAXI-175 Transmitter / Receiver Data Sheet and Technical Manual* (1992).
11. Daniel E. Lenoski and Wolf-Dietrich Weber: "Scalable Shared-Memory Multiprocessing", Morgan Kaufmann Publishers, (1995).
12. Hironori Nakajo, Takeshi Yoshinaga, Koichi Wada and Yukio Kaneda, "Ring-Connected Parallel Computer KORP – Coherence Protocol for Distributed Shared-Memory – , Proc. of Int. Conf. on Parallel and Distributed Systems ICPADS'92, pp.504–511 (1992).
13. N. J. Boden, D. Cohen, R. E. Felderman, A. E. Kulawik, C. L. Seitz, J. N. Seizovic and W. K. Su: "Myrinet: A Gigabit-per Second Local Area Network.", IEEE Micro, 15(1): pp.29–36 (1995).
14. Singh, J. P. et al. : "SPLASH: Stanford Parallel Applications for Shared-Memory", Computer Systems Laboratory, Stanford University, CA 94305.
15. Steven Cameron Woo, Moriyoshi Ohara, Evan Torrie, Jaswinder Pal Singh, and Anoop Gupta: "SPLASH-2Programs: Characterization and Methodological Considerations," *Proceedings of the 22nd International Symposium on Computer Architecture*, pp.24-36, (1995).

Designing and Optimizing 3-connectivity Communication Networks Using a Distributed Genetic Algorithm

Jianhua Ma, Runhe Huang and Eiju Tsuboi

Computer Software Department, The University of Aizu, 965-80, Japan

Abstract. In this paper, a distributed genetic algorithm (DGA) for 3-connectivity communication network design is proposed and implemented on a transputer based parallel machine, ParsyTec Gcel-1/64. It is emphasized that how parallelism can be used with the genetic algorithm. Performance of the (sequential) genetic algorithm (GA) is compared to Dijkstra algorithm (DA) in terms of computation time and total link costs versus various network graph sizes. The efficiencies of the distributed genetic algorithm over the genetic algorithm and Dijkstra algorithm are reported and discussed.

1 Introduction

With the rapid development and use of fiber optic technology, survivable network design against any single link or single node failure for traffic with cost effective objective becomes an extremely important issue for communication network. In particular, a network topology, which has at least three diverse paths called 3-connectivity between any required source-destination node pair, is considered in this paper in order to meet higher level of network survivability, such as military communication networks. Although problems of communication network design have been intensively studied and many heuristic methods proposed by Monma and Shallcross [10], they are all 2-connectivity network oriented approaches. As it is well known that conventional and heuristic approaches are lack of adaptability: it is difficult to extend those methods for 2-connectivity networks to 3-connectivity networks. Recently, few researchers have focused on using genetic algorithms for network design instead of heuristic solution procedures [1] [2]. Davis described how a genetic algorithm can be used to design the backbone of high-performance network under routing and survivability constraints. Esbensen proposed a genetic algorithm for the Steiner problem in a graph. However, due to the solution representation in Davis's genetic algorithm, it is necessary to check routing and survivability constraints for each of solution generated. This leads to a very high computation cost and necessitates a repair mechanism invoked during solution evaluation.

In this paper, a genetic algorithm for 3-connectivity communication network design is presented. Our genetic algorithm uses a good solution representation, in which the routing, diameter and survivability constraints are encoded so that both checking of constraints and a repair mechanism can be avoided. To speed up the computation phase, a transputer based parallel machine is employed to implement the genetic algorithm for 3-connectivity networks. However, it seems clear that the process of a genetic algorithm is inherently sequential, the challenging question is how can parallelism be used with the genetic algorithm? To this question, a distributed genetic algorithm is proposed and implemented on a transputer based parallel machine, ParsyTec Gcel-1/64. This work is the extension of our previous work on designing 2-connectivity communication networks by using a genetic algorithm[6] [7].

2 The Algorithm for 3-connectivity Network Designs

2.1 The Problem Definition

The problem of 3-connectivity network design can be defined as follows: starting with a complete network graph $G = (V, L)$, for a given traffic demand matrix R, cost matrix of placing a link C_f, link capacity matrix C_l, and the cost per unit matrix C_c, the objective of designing and optimizing a communication network is to find a minimal cost network graph $G' = (V', L')$ over link placement and capacity which can meet the given traffic demands, under the constraints of 3-connectivity, and diameter of 3-connected network (≤ 7), i.e.:

$$\min_{G'}(\sum_{s,d}(\sum_{(i,j)\epsilon P^*_{s,d}(G')} R(s,d)C_c(i,j)) + \sum_{(i,j)\epsilon G'} C_f(i,j)), \quad (1)$$

$$\sum_{s,d} R_{s,d}|((i,j)\epsilon P^*_{s,d}(G')) \leq C_l(i,j) \quad (2)$$

where, $R(s,d)$ is denoted the traffic demand between source node s and destination node d and $C_f(i,j)$, $C_c(i,j)$ and $C_l(i,j)$ are denoted the cost of placing a link $l_{i,j}$ between node v_i and node v_j, cost per unit capacity over the link $l_{i,j}$, and capacity of the link $l_{i,j}$, respectively. $P^*_{s,d}(G')$ is the paths from s to d in G'.

2.2 The algorithm description

The algorithm starts from a complete network graph $G = (V, L)$ with given matrices R, C_f, C_c and C_l. Then, a genetic algorithm is employed to search for three optimal disjoint paths for each demand as follows: for the 1st traffic demand R_{s_1,d_1}, the genetic algorithm is used to search for three optimal disjoint paths $P^*_{s_1,d_1}(G_1)$, the resulting subgraph $G_1 = (V_1, L_1)$ has minimal total link costs, i.e.,

$$\min_{G_1}(\sum_{(i,j)\epsilon P^*_{s_1,d_1}(G_1)} R(s_1,d_1)C_c(i,j) + \sum_{(i,j)\epsilon G_1} C_f(i,j)) \quad (3)$$

where, subgraph $G_1 = (V_1, L_1) \subseteq G$, a subset $V_1 \subseteq V$ and $L_1 \subseteq L$.

Similarly, for the 2nd traffic demand $R(s_2, d_2)$, the genetic algorithm is again used to search for three optimal disjoint paths $P^*_{s_2,d_2}(G_2)$, the resulting subgraph $G_2 = (V_2, L_2)$ has minimal total link costs, i.e.,

$$\min_{G_2}(\sum_{s,d}(\sum_{(i,j)\epsilon P^*_{s,d}(G_2)} R(s,d)C_c(i,j)) + \sum_{(i,j)\epsilon G_2} C_f(i,j)) \quad (4)$$

where, $G_1 \subset G_2 \subset G$, and $\sum_{s,d}$ is the sum of all link costs in G_2 which meets the demands $R(s_1, d_1)$ and $R(s_2, d_2)$.

The procedure repeats until the genetic algorithm has run for all n demands. The final resulting network graph $G'=G_n$ has

$$\min_{G_n}(\sum_{s,d}(\sum_{(i,j)\epsilon P^*_{s,d}(G_n)} R(s,d)C_c(i,j)) + \sum_{(i,j)\epsilon G_n} C_f(i,j)) \quad (5)$$

and $G_1 \subset G_2 \subset ... G_i ... \subset G_n=G' \subset G$.
The algorithm is working as shown in Fig.1.

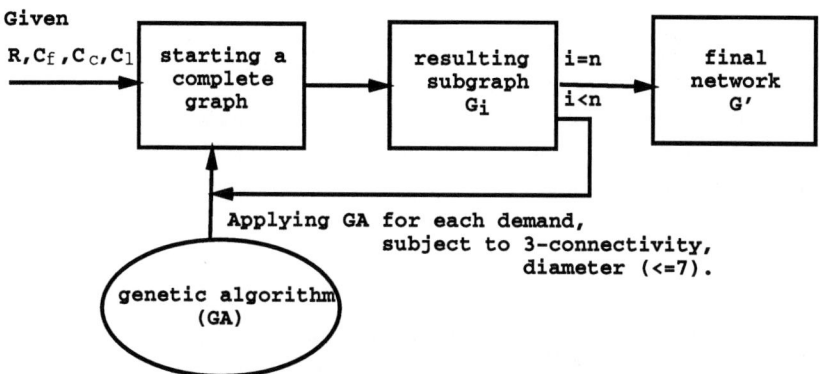

Fig.1. The algorithm in diagram

3 The Genetic Algorithm

In nature, the individuals constituting a population adapt to the environment in which they live. The fittest individuals have the highest probability of survival and tend to increase in numbers, while the less fit individuals tend to die out [4] [3] [5]. Genetic algorithms are based on natural evolution. The genetic algorithm in this context, is to search for three optimal disjoint paths in terms of cost corresponding to each demand $R(s, d)$ so that a resulting graph consisted of all the necessary links having certain capacity satisfies all demands and has minimal total link costs. Three disjoint paths mean that they have no any common shared links or nodes.

3.1 Feasible Solution Representation

The most critical step in applying a genetic algorithm to a survivable communication network design is to choose a way to represent a solution to the problem in a chromosome. A chromosome is the corresponding representation or genetic encoding of the individual (solution). Although the template of genetic algorithms is same, different solution representations lead to complete different algorithms with different efficiency. In our genetic algorithm, to avoid inefficient checking of violation of the constraints and a repair mechanism, a chromosome is a feasible solution, that is, it does not break 3-connectivity and diameter constraints. In order to achieve this, special care in choosing chromosome representation and genetic operators should be taken.

A chromosome in our genetic algorithm consists of three block strings with the total length of $(d_c - 1) * 3 + 2$, where d_c is the diameter of 3-connected network. Each block string represents a path from source node s to destination node d. The value of each bit in a string takes *0* or a node index from *1* to *n*. The number n is the total node number in an initial complete graph. A node index is only allowed to appear at most once in a string in order to guarantee three disjoint paths and to avoid unnecessary circles in a path. For example,

10 0 5 3 0 0 18 0 0 12 17 0 13 0 0 1 0 0 19 15

represents a feasible solution, that is, three disjoint paths from between node 10 and node 15, form the traffic demand $R(10, 15)$.

This representation has an outstanding advantage: diameter and 3-connectivity constraints are encoded in a chromosome to avoid checking of constraints and a repair mechanism invoked because of violation of the constraints during chromosome evaluation.

3.2 Feasible Solution Oriented Crossover Operator

For selected two parent chromosomes p_1 and p_2, the crossover operator generates two offspring, ch_1 and ch_2. One of the criteria for the crossover operator is obviously to generate feasible offspring, that is, solutions generated by the crossover operator should be feasible and do not break the constraints. In order to guarantee this, a modified crossover operator with two-point crossover and the operation of swapping duplicated nodes is employed. By randomly selecting two crossover points, each of two parent chromosomes is then partitioned into three parts. Two resulting chromosomes r_1 and r_2 are generated by exchanging middle part of two parent chromosomes. And such exchange may result in invalid chromosomes since duplicated nodes may occur after the exchange. If there are duplicated nodes in the resulting chromosomes, the operation of swapping duplicated nodes between the two resulting chromosomes is, thus, necessarily employed. For example, there are two parent chromosomes,

$p_1 =$ 10 0 5 3 16 0 18 9 0 12 17 0 13 0 0 1 0 0 19 15
$p_2 =$ 10 0 5 2 17 11 0 1 14 0 0 0 16 9 0 0 19 0 0 15

If two crossover points are randomly selected as 4 and 10, then resulting chromosomes are obtained by exchanging the middle part of two selected chromosomes as below:

$r_1 =$ 10 0 5 2 **17** 11 0 **1** 14 0 **17** 0 13 0 0 **1** 0 0 19 15
$r_2 =$ 10 0 5 3 **16** 0 18 **9** 0 12 0 0 **16** **9** 0 0 19 0 0 15

As we can see, nodes 17 and 1, 16 and 9 are duplicated in the resulting chromosomes r_1 and r_2, respectively. Applying the operation of swapping duplicated nodes to r_1 and r_2, then we can obtain feasible offspring like

$ch_1 =$ 10 0 5 2 17 11 0 1 14 0 **16** 0 13 0 0 **9** 0 0 19 15
$ch_2 =$ 10 0 5 3 16 0 18 9 0 12 0 0 **17** **1** 0 0 19 0 0 15

As it can be seen that two offspring are feasible solutions.

3.3 Fitness Evaluation

As each solution generated initially or by genetic operators is feasible, evaluation of each solution becomes simply calculation of the total link costs over three disjoint resulting paths. By given i^{th} demand $R(s_i, d_i)$ and k^{th} solution in a population of solutions, the cost for each link in the three disjoint paths $P^*_{s_i,d_i}(G_i)$ consists of a cost for placing the link and a required capacity cost. If the link is already existed in the previous resulting graph G_{i-1} and available capacity of the link can meet the demand, then the existent link can be used for satisfying the demand $R(s_i, d_i)$, thus, the cost of placing the link should not be counted again, thus, is set to zero. The total link costs of the kth solution, are the sum of each link l_j cost in the three disjoint paths $P^*_{s_i,d_i}(G_i)$ as follows:

$$C_k = \sum_{l_j \epsilon P^*_{s_i,d_i}(G_i)} (C_f(l_j) + R(s_i, d_i) * C_c(l_j)) \qquad (6)$$

And the fitness of a solution is defined as a function of the inverse of its cost and scaled into the range of [0, 1].

$$f_k = \frac{C_{\max} - C_k}{C_{\max} - C_{\min}} \qquad (7)$$

where, $C_{\max}$ and $C_{\min}$ are the maximum and minimum cost of solutions in the population, respectively.

A solution is evaluated with respect to its fitness and a selection of parents for mating is a fitness based randomized scheme conducted from the whole population of chromosomes. Therefore, the fitter solutions are more likely to make contributions to next generation and the less fit solutions are more likely to be replaced by offspring in next generation. The population of solutions, thus, are evolved toward higher and higher quality over many generations until an optimal solution is reached. It is an optimization process via natural evolution.

4 Dijkstra Algorithm vs. Genetic Algorithm

As it is well known that given a graph, Dijkstra algorithm [9] can be used to search for a shortest path from a given node s to all other nodes. The problem of searching for three shortest disjoint paths in terms of cost for a given source node s and destination node d can be achieved by applying Dijkstra algorithm at least three times to a given graph with special cares in not violating three disjoint paths and diameter constraints. The special cares include (1) checking whether the total number of links in a resulting shortest path is greater than the diameter constraint or not. If so, Dijkstra algorithm has to search for another shortest path for replacement, (2) deleting links and nodes which are in the shortest path from the given graph before Dijkstra algorithm is applied again to search for the second or third shortest path to guarantee three shortest paths are disjoint.

Generally speaking, for simple shortest path searching without other combinational conditions, Dijkstra algorithm can give slight better performance than the genetic algorithm. However, Dijkstra algorithm has time complexity of $O(r_n * c_n * m^2)$ and it is necessary to checking whether diameter constraint is violated or not, and genetic algorithm does not need to conduct any checking procedure and has time complexity of $O(r_n * p * g)$. Where, m is total nodes number in a given graph, r_n is the number of required demands, c_n is connectivity, and p and g are population size and total generation number in the genetic algorithm, respectively.

As it can be seen that computation time of Dijkstra algorithm is increased significantly with increase of the graph size because it is proportional to square of the graph size, m^2, while computation time of the genetic algorithm is not directly relevant to the graph size. It is proportional to the multiple of population size p and generation number g in the genetic algorithm. However, population size and generation number may have to increase slightly when the graph size is increased significantly because increase of the graph size makes the searching space larger. Therefore, Dijkstra algorithm is suitable to small size network designs while the genetic algorithm is better for large scale network designs. For future super-highway systems, genetic algorithms are certainly better choices to large scale communication network designs in a super-highway system.

5 A Distributed Genetic Algorithm

Initially, it seems clear that the process of a genetic algorithm (GA) is inherently sequential. Each generation must be produced before it can be used as the basis for the following generation, and it is antithetical to the evolutionary scheme to jump forward. However, how can parallelism be used with the genetic algorithm approach? It is believed that the most natural way to distribute a genetic algorithm over the processors is to partition the population of solutions and assign one subset of the population to the local memory of each processor and each processor runs the genetic algorithm on its own subpopulation, periodically selecting good solutions from its subpopulation and sending copies of them to one of its neighbors. It will also received copies of this neighbor's good solutions, with which it will replace bad solutions in its own subpopulation. However, the efficiency of a distributed genetic algorithm must consider the effects of mutation, crossover, and selection on the entire population in a GA. It has proved

[13] that converting a GA to a DGA does not change the effect of the mutation, crossover operators but the survival rate of schemata in a DGA differs from the survival rate of schemata in a GA because the selection is conducted on the entire population in a GA while the selection is conducted on the subpopulation in a DGA. To remedy the problem, the exchange of good solutions between neighboring processors mentioned above is indispensable. The frequency of exchange and the number of solutions exchanged are two adjustable parameters. A DGA can be expressed as follows:

BEGIN {Distributed Genetic Algorithm}
  PAR for each nodal GA
    randomly generate initial subpopulation of solutions;
    While (*terminate()==FALSE*)
      evaluate solutions;
      IF (*generation* mod *interval_of_exchange* == *0*) THEN
        select *number_of_exchange* good solutions from the subpopulation;
        send copies of them to the neighboring processors;
        receive copies of good solutions from the neighboring processors;
        replace *number_of_exchange* bad solutions with the received ones;
      endIF
      select parent solutions from the subpopulation for mating;
      generate offspring by applying genetic operators;
      advance the subpopulation by the fitter solutions.
  endPAR
END {Distributed Genetic Algorithm}

The parallel machine to implement the distributed genetic algorithm is ParsyTec Gcel-1/64 [12]. ParsyTec Gcel-1/64 is a MIMD machine with 64 T805 transputers. Each transputer, with 4 MB own local memory, has four serial links which are used for communications between transputers. Each transputer has an identification and their physical network structure is 2D grid, in which 16 transputers (16 nodes) as a cluster and there are 4 clusters in total for Gcel-1/64 machine. Although all transputers in the network is physically linked in the topology of 2D grid, they actually can communicate to each other via *Virtual Links* so that any kind of communication topologies called *Virtual Topologies* can be created by sets of those virtual links. To achieve optimal throughput and convenient application modeling, the communication networks need to be mapped optimally onto the physical processor and link network.

The virtual topology of the network used to implement the distributed genetic algorithm for communication network design is torus. The virtual torus topology of the network is believed the most suitable topology for this application. In each processor, the genetic algorithm runs on the subpopulation just like a sequential genetic algorithm except for the occasional migration of the best solutions from and to other processors. That is after certain generations, these processors will swap their best individuals to achieve global searching rather than local searching based on the subpopulation.

Comparing with other topologies, the advantage of the virtual torus topology for this application has threefold: first, it is efficient for neighboring processors to carry out communications, i.e. exchange of good solutions both in the horizontal and vertical directions. Second, it is easy to obtain optimal mapping of virtual torus topology onto the physical processor of the parallel machine so as to obtain optimal throughout. Third, all processors can swap their best solutions efficiently with their neighbors to reach global searching after certain generations since each processor has four neighboring processors which is the maximum number it can has in a transputer based network.

The way of its broadcasting and gathering a data is shown in Fig.2(a). Each processor runs the genetic algorithm on its own subpopulation and can swap their best solutions with its four nearest neighboring processors as shown in Fig.2(b). When each processor exchanges good solutions with its 4 neighboring processors, it has to be synchronized with its 4 neighboring processors. Exchanging their best solutions is synchronized as 4 sequential steps as marked *1, 2, 3, 4*, while sending information to one direction (right, left, up, or down) is carried out in parallel. Where, *1, 2, 3* and *4* are to send information to its right, left, up and down neighbors, respectively.

The interval of exchange and the number of solutions exchanged are two ad-

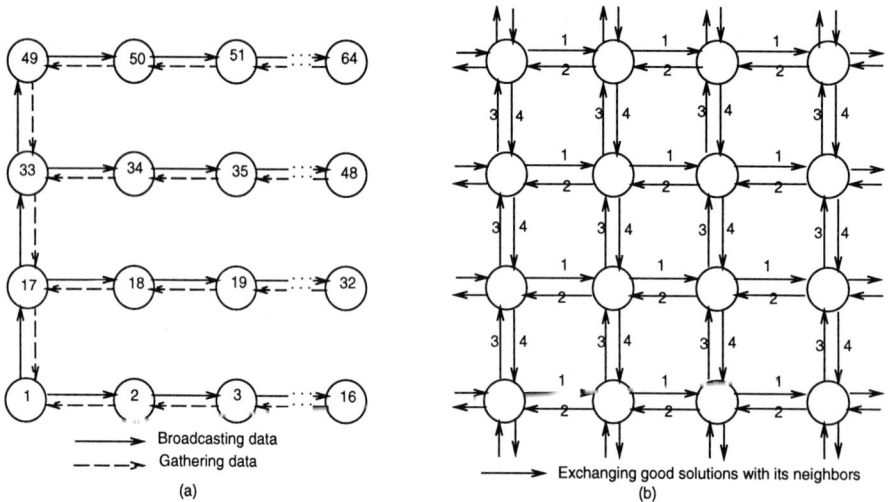

Fig.2. The virtual torus topology

justable parameters which can balance the performance of the genetic algorithm and the communication overheads. The interval of exchange and the number of solution exchanged depend upon problem size and subpopulation size in each processor. With carefully selecting these two parameters, it is possible to obtain better performance with reasonable communication overheads.

6 Implementations and Results

For the genetic algorithm run on each processor, subpopulation size is 10, generation number is 100, two points crossover probability is 0.96 and mutation probability is 0.03. For the network considered, the cost for placing a link $C_f(i,j)$ is randomly set in the range of [100, 1000]. The capacity cost for a link is the multiple of the total demands over the link and the cost per unit capacity, i.e., $(\sum R_{sd})*C_c$.

6.1 Performance of the DGA in Terms of Link Cost

In this implementation, the network has 30 nodes and has to meet 20 demands. Each link has a constant capacity of 300. The cost per unit capacity C_c is assumed to be 100. The demand R_{sd} is simply set to 1 unit capacity. In each processor, the DGA runs on the subpopulation with migration of one best solution from and to its four neighboring processors in every generation, the resulting network has the total link cost of 33838. DA is run for the same design problem, the resulting network has the total link cost of 32370. The results show that DA gives a slightly better performance in terms of cost than the DGA. However, if a network considered is more complicated, the DGA is expected to produce a better solution than DA because genetic algorithms are more effective and efficient to combinational searching problems.

Let us consider another network which has 20 nodes and has to meet 16 demands. Each link has two possible types of links with a capacity of 300 or 160. The link with 300 capacity costs 100 and the link with capacity 160 cost 60. The cost per unit capacity C_c is 80, and each demand is one of the three possible types, i.e. 1, 2 or 3 units. Obviously, if the total capacity required on a link is less than or equal to 160, then it is economic to use a link with capacity of 160 rather than to use a link with capacity of 300. If the total capacity required is more than 160 and less than or equal to 300, it is better to use a link with capacity of 300 rather than to use two links, each of them has capacity of 160.

To this network design, the DGA generates a resulting network with the total link cost of 15394 but a resulting network by using Dijkstra algorithm has the total link cost of 18722. The resulting network generated by the DGA is shown in Fig.3. As we can see that to a complicated network design problem, the distributed genetic algorithm gives a better solution than Dijkstra algorithm. In fact, most practical applications of communication network design are combinational design and optimization problems. The genetic algorithm is, thus, of practical value over Dijkstra algorithm.

6.2 Performance of the GA in Terms of Time Cost

As we analyzed in Sect. 4 that Dijkstra algorithm has the time complexity of $O(r_n * c_n * m^2)$ and the genetic algorithm has the time complexity of $O(r_n * p * g)$. When a network graph size is reasonable small, DA is expected to give an optimal solution with less computation time than the GA. But when the network graph

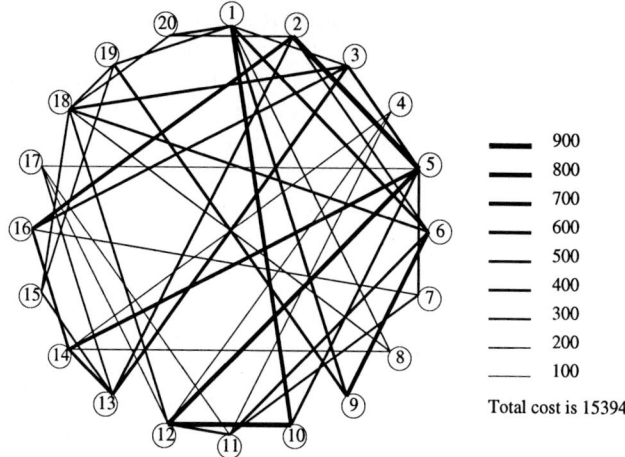

Fig.3. Performance of the GA in terms of cost

size reaches a certain size M, the GA is expected to produce an optimal solution faster than DA.

Let us assume that a network considered is 30-node network, both DA and the GA for one demand give the same optimal solution with the cost of 1880. However, the GA takes 1.20387 seconds and DA takes 0.124398 seconds. Obviously, to this network design, DA is quicker. What we are interested in is to find out a graph size M, when $m \geq M$, the GA takes less time than DA for obtaining an almost same optimal solution. When $m = 30$, we have

$$1 * 50 * 390 * k_1 = 1.20387 \qquad (8)$$

$$1 * 3 * 30^2 * k_2 = 0.124398 \qquad (9)$$

where, 50, 390 are the population size and the generation number used in the genetic algorithm, respectively. 1, 3, 30 are the demand number, connectivity, and graph size, respectively. k_1 and k_2 are the coefficients used in order to obtain computation time from the time complexity. From equations (8) and (9), we obtain

$$\frac{k_1}{k_2} = 1.339970 \qquad (10)$$

When $m = M$, we have

$$\frac{1 * 50 * 390 * k_1}{1 * 3 * M^2 * k_2} = 1 \qquad (11)$$

From equations (10) and (11), we get

$$M = 183.028 \qquad (12)$$

This result indicates that the genetic algorithm is more efficient than Dijkstra algorithm when designing a network with more than 183 nodes because the time complexity of the genetic algorithm is not directly relevant to the network size and the time complexity of Dijkstra algorithm is proportional to the square of the network size.

6.3 Computation Efficiency of the DGA

To speed up computation time, the DGA is developed based on the GA. Because of necessary migration of the best individuals from and to 4 neighboring processors, communication overheads of the parallel network is unavoidable and significant comparing to its computation time. Table 1 is implementation results of the DGA versus the GA and DA in terms of time cost for various graph sizes.

Table 1. The DGA vs. the GA and DA

Graph Size	30	100	300
DA	0.124398	1.281072	11.44219
The GA	1.203870	2.560077	4.426864
The DGA	0.387054	0.467282	0.573867
Efficiency of DGA			
over DA	0.321397	2.741540	19.93875
over the GA	3.110341	5.478655	7.714094

where, efficiency of DGA over DA is defined as $\frac{time\_cost\_of\_DA}{time\_cost\_of\_DGA}$ and efficiency of DGA over GA is defined as $\frac{time\_cost\_of\_GA}{time\_cost\_of\_DGA}$. It can be seen from the results that efficiency of the DGA is increased with increase of network graph size. Therefore, both of the DGA and the GA are effective and efficient to large scale communication network design.

7 Conclusions

A genetic algorithm is successfully developed for 3-connectivity communication network designs. The success of our genetic algorithm is able to encode 3-connectivity and diameter constraints in a chromosome representation so that checking of the 3-connectivity and diameter constraints can be avoidable. As a result, the computation time of our genetic algorithm is significantly reduced compared to Davis' genetic algorithm [1] which uses a different chromosome representation.

In order to further speed up computation phase, a distributed genetic algorithm for 3-connectivity communication network designs is proposed and implemented in a transputer based parallel machine. It is emphasized how a parallelism can be used with the genetic algorithm and it is pointed out that *interval_of_exchange* and *number_of_exchange* are two critical parameters to the performances in terms of time cost and link cost. It is necessary to tune these two parameters to achieve global optimization with minimal communication overheads. Both the genetic algorithm and the distributed genetic algorithm are efficient for large scale of communication network designs. In our future work, we would like to focus on how to further increase of efficiency of the distributed genetic algorithm.

8 Acknowledgments

Authors would like to thank Professor D. Frank Hsu for his valuable discussions.

References

1. Davis L., Orvosh T., Cox A. and Qiu Y.: A genetic algorithm for survivable network design. The Proceedings of the Fifth International Conference on Genetic Algorithm, (1993) 408–415.
2. Esbensen H: Computing near-optimal solutions to the Steiner problem in a graph using a genetic algorithm. Networks, **26**, (1995) 175–185.
3. Goldberg D. E.: Genetic algorithms in search, optimization and machine learning. ISBN 0-201-15767-5, (1989), Addison Wesley.
4. Holland J. H.: Adaptation in natural and artificial systems. Ann Arbor: The University of Michigan Press, (1975).
5. Huang R., Fogarty T.C.: Learning prototype control rules for combustion control with genetic algorithm. Periodicals of Modeling, Measurement and Control, C, **38**, No. 4, (1993) 55–64.
6. Huang R., Ma J., Tsuboi E.: Communication network design via a genetic algorithm based learning algorithm. The Proceeding of the IASTED International Conference on Artificial Intelligence, Expert Systems and Neural Networks, (1996) 15–18.
7. Huang R., Ma J., Kunii T. L., Tsuboi E.: Parallel genetic algorithms for communication network design. The Proceeding of the Second Aizu International Symposium on Parallel AlgorithmsArchitectures Synthesis, (1997) 370–377.
8. Jog Prasanna, Gucht Dirk Van: Parallelization of probabilistic sequential search algorithms. The Proceedings of the 2nd International Conference on Genetic Algorithms, (1987) 170–176.
9. Kingston, J. H.: Algorithms and data structures. Addison-Wesley Publishing Company, ISBN 0 201 41705 7, (1990).
10. Monma C. L., Shallcross D. F.: Methods for designing communications networks with certain two-connected survivability constraints. Operations Research, **37**, No. 4, (1989) 531–541.
11. Palmer C. C., Kershenbaum A.: An approach to a problem in network design using genetic algorithms. Networks, **26**, (1995) 151–163.
12. ParsyTec: Software documentation, (1993).
13. Pettey Chrisia C., Leuze Michael R.: A theoretical investigation of a parallel genetic algorithm. The Proceedings of the 3rd International Conference on Genetic Algorithms, (1989) 398–405.

Adaptive Routing on the Recursive Diagonal Torus

A. Funahashi[1] and T.Hanawa[1] and T.Kudoh[2] and H. Amano[1]

[1] Keio University, 3-14-1 Hiyoshi, Kohoku-ku, Yokohama, 223 Japan
[2] Real World Computing Partnership, 1-6-1 Takezono, Tsukuba, Ibaraki, 305 Japan

Abstract. Recursive Diagonal Torus, or RDT consisting of recursively structured tori is an interconnection network for massively parallel computers. By adding remote links to the diagonal directions of the torus network recursively, the diameter can be reduced within $log_2 N$ with smaller number of links than that of hypercube.
For an interconnection network for massively parallel computers, a routing algorithm which can bypass a faulty or congested node are essential. Although the conventional vector routing is a simple and near-optimal method, it can only use a deterministic path. In this paper, adaptive routing algorithms on RDT are proposed and discussed. The first algorithm is based on Duato's necessary and sufficient condition. With this method virtual channels are effectively used while paths with redundant routing steps are prohibited. Another algorithm based on the turn model is proposed. By prohibiting certain turns on RDT, it permits paths with additional hops. Both algorithms are proved to be deadlock free.

1 Introduction

Communication network is one of the critical components of a highly parallel multicomputer. Recently, multicomputers providing more than a thousand computation nodes are commercially available, and efforts have been exerted to implement Massively Parallel Computers (MPCs) with tens of thousands nodes. In these systems, the connection topology often dominates the system performance.

Instead of hypercube used in first-generation multicomputers, most recent machines take the 2-D or 3-D mesh (torus) network[1][2][3]. Although the diameter of a mesh network is large ($O(\sqrt{M})$ or $O(\sqrt[3]{M})$ for M nodes), it only requires four or six links per node unlike the hypercube which requires $log_2 M$ links per node.

However, in an MPC with more than ten thousands nodes, the large diameter of the mesh network is intolerable. To address this problem, we proposed a novel extension of mesh network called Recursive Diagonal Torus (RDT) [4], which consists of recursively structured mesh (torus) connection. It supports a smaller diameter and degree than that of the hypercube if the number of nodes is 1000-10000. Through the computer simulation, the bandwidth and latency are much improved compared with 2-D/3-D tori [4]. The router chip providing the vector routing algorithm with multicasting was implemented for a massively parallel machine JUMP-1[5].

In this paper, deadlock-free adaptive routing algorithms on RDT are proposed. In Section 2, the structure of RDT and the vector routing algorithm are briefly introduced. An adaptive routing using minimal paths based on Duato's method is proposed in Section 3. More flexible routing algorithm based on the turn model is also proposed in Section 4.

2 Interconnection Network: RDT

Recursive Diagonal Torus (RDT) is a novel class of networks which consists of recursively structured mesh (torus) connections of meshes with different sizes in the diagonal directions[4][6].

When four links are added between node (x, y) and nodes $(x \pm n, y \pm n)$ (n: *cardinal number*) respectively, additional links result in a new torus-like network. New torus-like network is formed at an angle of 45 degrees to the original torus, and the grid size is $\sqrt{2}n$ times that of the original torus. We call this new torus-like network the rank-1 torus. On the rank-1 torus, we can form another torus-like network (rank-2 torus) by providing additional links in the same manner. Figure 1 shows rank-1 and rank-2 tori when n is 2. The RDT consists of such recursively formed tori.

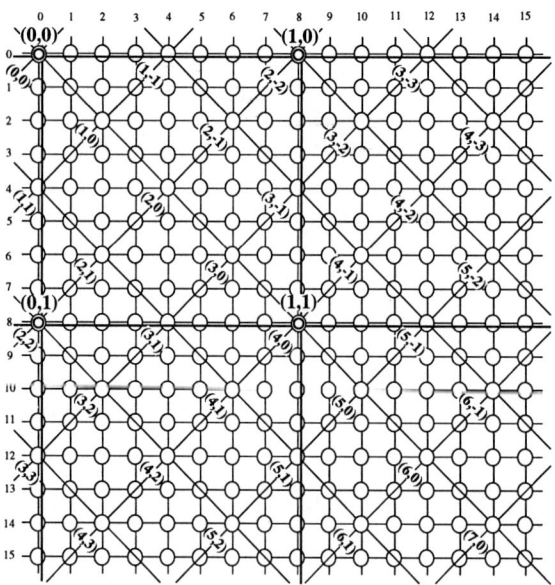

Fig. 1. Upper rank tori

RDT(n,R,m) can be defined as a class of networks in which each node has links to form base (rank-0) torus and m upper tori (the maximum rank is R) with cardinal number n. Note that, each node can select different rank of upper tori from others.

The RDT in which every node has links to form all possible upper tori is called the perfect RDT (PRDT(n,R)) where n is the cardinal number (usually, 2) and R is the maximum rank. Although PRDT is unrealistic due to its large degree (4(R+1)), it is important as the basis for establishing routing algorithm, broadcasting/multicasting, and other message transfer algorithms.

For a system with thousand of nodes, the RDT whose degree is 8 and the maximum rank of upper tori is 4, that is, RDT(2,4,1) is suitable. In the RDT, each node can select different rank tori from others. Thus, the structure of the RDT(2,4,1) also varies with the rank of tori which are assigned to each node. This assignment is called the *torus assignment*. Various torus assignment strategies can be selected considering the traffic of the network.

2.1 The vector routing

The vector routing is an assignment independent of routing algorithm which represents the route of a message with a combination of unit vectors each of which corresponds to each rank of tori.

On the torus structure, a vector from a source node to the destination node is represented with a vector $\mathbf{A} = a\mathbf{x_0} + b\mathbf{y_0}$ where $\mathbf{x_0}$ and $\mathbf{y_0}$ are unit vectors of the base (rank-0) torus. The goal of the routing algorithm is to represent the vector $\mathbf{A}$ with a combination of vectors each of which corresponds to a unit vector of each rank of torus.

First, the direction of the unit vector corresponding to each rank torus must be defined. Here, the direction of the unit vector for each rank torus is changed clockwise at an angle of 45 degrees. That is, the unit vectors of rank-(i+1) torus $\mathbf{x_{i+1}}, \mathbf{y_{i+1}}$ can be represented with the unit vectors of rank-i $\mathbf{x_i}, \mathbf{y_i}$ as follows:

$$\mathbf{x_{i+1}} = n\mathbf{x_i} + n\mathbf{y_i} \quad (1)$$

$$\mathbf{y_{i+1}} = -n\mathbf{x_i} + n\mathbf{y_i} \quad (2)$$

First, the target vector $a\mathbf{x_0} + b\mathbf{y_0}$ is represented with a combination of $\mathbf{x_1}, \mathbf{y_1}, \mathbf{x_0}$ and $\mathbf{y_0}$ as follows:

$$a\mathbf{x_0} + b\mathbf{y_0} = g\mathbf{x_1} + f\mathbf{y_1} + j\mathbf{x_0} + k\mathbf{y_0} \quad (3)$$

Here, we select maximum g and f in order to use the upper torus as possible. From equations (1) and (2), maximum integers for g and f are represented as follows:

$$g = \frac{a+b}{2n}, f = -\frac{a-b}{2n}$$

In order to minimize j and k corresponding to the remaining unit vectors of the rank-0 torus (thus, the required message transfers using the rank-0 torus), the integer divisor used here is rounded to the nearest whole number (If the remainder is greater than n, increment the divisor).

Thus, j and k are represented with g and f:

$$a\mathbf{x_0} + b\mathbf{y_0} = g(n\mathbf{x_0} + n\mathbf{y_0}) + f(-n\mathbf{x_0} + n\mathbf{y_0}) + j\mathbf{x_0} + k\mathbf{y_0}$$

$$a = ng - nf + j, b = ng + nf + k$$

$$j = a - ng + nf, k = b - ng - nf$$

Then, $g\mathbf{x_1} + f\mathbf{y_1}$ are represented with a combination of vector $\mathbf{x_2}$, $\mathbf{y_2}$, $\mathbf{x_1}$, and $\mathbf{y_1}$ in the same manner. By iterating this process to the maximum rank, vectors for message routing are obtained.

The routing vectors for each rank are obtained in the array $vector[rank]$.

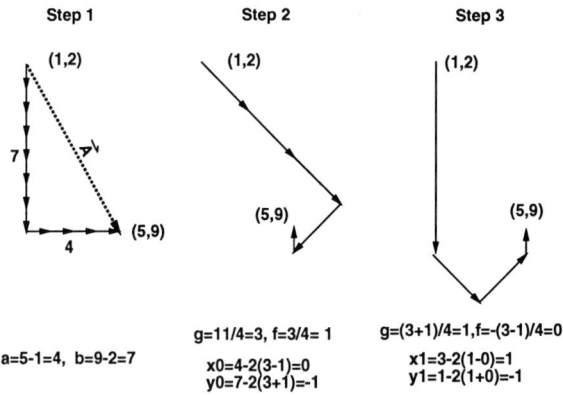

Fig. 2. An example of the vector conversion

Figure 2 shows an example of a vector from (1,2) to (5,9) converted into a combination of unit vectors of rank-0, rank-1, and rank-2.

3 Adaptive routing with minimal paths

Adaptive routing is a technique to select the route of packet dynamically. When a packet encounters a faulty or congested node, the packet can select another bypassing route. The vector routing is useful for a basis of an adaptive routing, since as alternative routes can be easily obtained by changing the order of vectors. However, we must not forget that an adaptive routing has a possibility of deadlock. There are a lot of researches on deadlock free adaptive routing techniques[7]. These techniques are classified into two methods: using only minimal paths, and using alternative paths with additional routing steps. The former method does not require extra routings while the latter can use alternative routes more flexibly. First, deadlock free adaptive routings with minimal routes are proposed for the RDT based on Duato's protocol. Then, another algorithm which permits redundant routing steps is proposed based on the turn model.

3.1 Duato's protocol in the k-ary n-cube

Duato states a general theorem defining a criterion for deadlock freedom and then uses the theorem to propose a fully adaptive, profitable, progressive protocol[8], called Duato's protocol (DP). The theorem states that by separating virtual channels on a link into restricted and unrestricted partitions, a fully adaptive routing can be performed and yet be deadlock-free. This is not restricted to a particular topology or routing algorithm. Cyclic dependencies between channels are allowed, provided that there exists a connected channel subset free of cyclic dependencies.

Simple description of Duato's protocol is as follows.

a. Provide that every packet can always find a path toward its destination whose channels are not involved in cyclic dependencies(escape path).
b. Guarantee that every packet can send to any destination node using escape path and the other path which cyclic dependency is broken by escape path.

By selecting these two routes a. and b. adaptively, deadlock can be prevented. Duato applied this method to the k-ary n-cube[9].

3.2 Applying Duato's protocol on PRDT

Here, we apply this routing algorithm for PRDT.

Definition 1. : Duato's protocol on PRDT

1. Provide an escape path C_1 on a torus of PRDT as well as the case for the k-ary n-cube.
2. Next, the order of rank usage is restricted. Let X_i and Y_i be channel of each dimension in the rank i torus. Use the channel in the X first and descending order of the rank. That is, for PRDT(2,4), the channel is used in the following order
$X_3 \rightarrow Y_3 \rightarrow X_2 \rightarrow Y_2 \rightarrow X_1 \rightarrow Y_1$
We refer this escape path C_1'.
3. Add a new virtual channel $C_F(Fully\ adaptive)$ which is used for the fully adaptive routing. There are two algorithms: D-A and D-B.
Algorithm: D-A
Provide the virtual channel C_F directly for the escape channel C_1'. In C_F, each direction of $+X$ and $+Y$ in odd rank and even rank must be the same direction. In the vector routing, the unit vector for each rank torus

is changed clockwise at an angle of 45 degree as represented in function(1) and function(2), the unit vector for odd rank torus must be same direction with the unit vector for rank 0 torus $(\mathbf{x}_0, \mathbf{y}_0)$ and the unit vector for even rank torus must be the same with the one for rank 1 torus$(\mathbf{x}_1, \mathbf{y}_1)$.

Algorithm: D-B

Provide the virtual channel C_{Fn} not for C'_1 but for C_1 in each rank. C_{Fn} channels can cross dimensions in any order following a minimal path, but must cross ranks in descending order.

□

Figure 3 illustrates the fully adaptive virtual channel C_F in Algorithm D-A. Since C_F is directly assigned to the escape path C'_1, the C_F itself must be a minimal routing. This means that a packet must not use the opposite direction which used in the past.

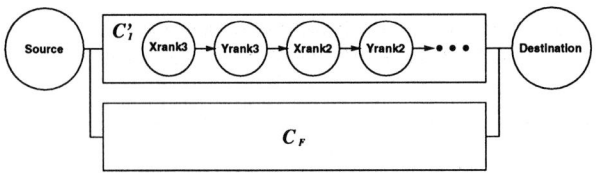

Fig. 3. Channels using Algorithm D-A

On the other hand, the fully adaptive path is assigned to the escape path C_1 of each rank in Algorithm D-B(Figure4). Therefore, there is no restriction for using unit vector, while the order of ranks is restricted.

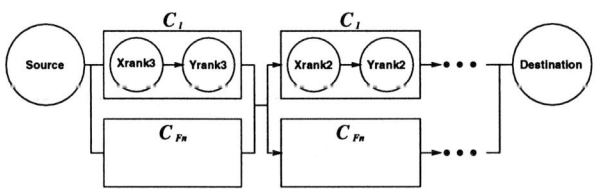

Fig. 4. Channels using algorithm D-B

Figure5 illustrates the possible path and impossible path for algorithm D-A and D-B. The path (b) which uses rank 2 before rank 3 is allowed in the algorithm D-A while it is prohibited in the algorithm D-B, since the rank is not be used in the descending order. On the contrary, path (c) in which the unit vectors of rank 1 and rank 3 are directed opposite to each other is prohibited in the algorithm D-A but allowed in the algorithm D-B.

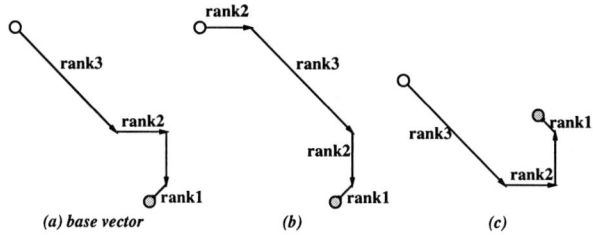

Fig. 5. Examples of vectors in algorithm D-A and D-B

Theorem 2. *Algorithm D-A is deadlock-free.* □

Proof Since the order of the rank is the same as that of the e-cube routing[10], the escape path $C'1$ is deadlock free. In C_F, the opposite direction which used in the past is prohibited, and so C_F is a minimal path. From Duato's theorem[9], Algorithm D-A is deadlock-free. □

Theorem 3. *Algorithm D-B is deadlock-free.* □

Proof C_1 is the same escape path used in Duato's protocol, and is deadlock free. C_{Fn} is a minimal path in each torus. From Duato's theorem[9], Algorithm D-B is deadlock-free in each rank of torus. Since the order of used rank is the same as the e-cube routing, C_1 nor C_{Fn} in any rank does not cause a cycle each other. Therefore, Algorithm D-B is deadlock-free.□

4 Adaptive routing based on the turn model

Although Duato's protocol is powerful approach for bypassing the congestion, only minimal paths can be used. For selecting paths with additional steps, another adaptive routing based on the turn model[11] is proposed here.

4.1 Turn model for Two-Dimensional Meshes

Deadlock in the wormhole routing is caused by message packets waiting for each other in a cycle. The turn model proposed by Glass is a method which prevents deadlock by prohibiting certain turns.

For two-dimensional meshes, Figure6(a) shows the possible turns and simple cycles. Deadlock can be prevented by prohibiting only one turn from each cycle,

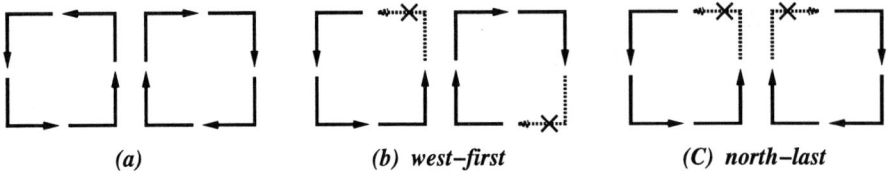

Fig. 6. The turn model for two-dimensional meshes.

as shown in Figure6(b),(c). These routing algorithms are called the west-first routing algorithm and north-last routing algorithm, respectively. Although this model is for a simple mesh network without cyclic links, it is easily used in the torus by introducing extra channels like the e-cube routing.

4.2 The turn model for RDT

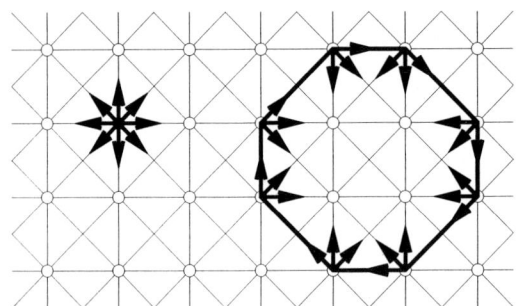

Fig. 7. The possible turns and simple cycles in RDT.

Definition 4. :   **North-last routing for PRDT**

Here, we extend the turn model for two-dimensional meshes of the RDT. The possible turns in RDT are expressed in Figure7. As shown in Figure7, there are eight different directions in the RDT, so there exists sixteen 45-degree turns, sixteen 90-degree turns and sixteen 135-degree turns.

Here, like the north-last routing algorithm for two dimensional mesh, the right top turns and left top turns of cycles are prohibited as shown in Figure8(a).

However, these restrictions are not sufficient for RDT. Cycles without left top turns or right top turns are still possible as shown Figure8(b). In order to break such triangle cycles, dotted turns shown in Figure8(c) must be prohibited. As a result, the following turns are prohibited in RDT.

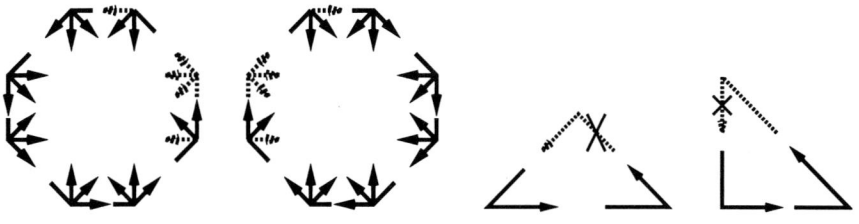

(a) The first step to the north-last routing on RDT.

(b) Particular types of cycles.

(c) Completed form of the north-last routing on RDT.

Fig. 8. The north-last routing on RDT.

North-last routing for PRDT is a routing in which fourteen turns shown in Figure4.2 are prohibited. A packet transfer through cyclic links is also prohibited. □

As well as the turn model for two dimensional torus, cyclic links can be used by introducing an extra channel for the e-cube routing. Also, this routing can be directly applied for any type of RDT including RDT(2,4,1)/α.

For showing that the proposed north-last routing algorithm for RDT is deadlock free, the channel numbering method by Dally and Seitz[10] is applied. In this method, channels in the direct network is numbered so that every packet is transferred along channels with strictly increasing (or decreasing) numbers. If such a numbering is possible, it shows that there is no cyclic path between buffers in channels.

Theorem 5. *The north-last routing for RDT is deadlock-free.* □

Proof Assuming that the size of the base torus of RDT is $m \times n$. Assign two dimensional number of channel from a node (x, y) according to its direction as shown in Figure9, and let the unique number of the channel be $c_x \times m + c_y$.

Since the size of the base torus of RDT is $m \times n$, the range of the possible channel number (c_x, c_y) is represented by the following equations.

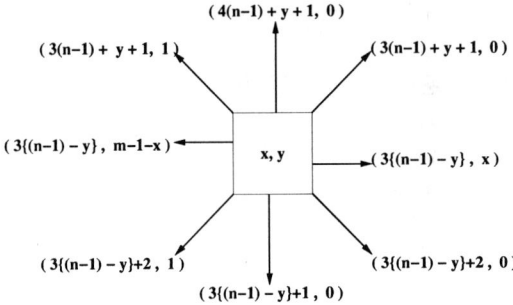

Fig. 9. Numbering of the channels leaving each node (x, y) for the north-last routing algorithm for RDT.

$$0 \leq c_x \leq 5(n-1)$$
$$0 \leq c_y \leq m-2$$

In RDT, there are eight possible input directions. As shown in Figure 10, all possible output channel numbers are larger than the number of input channel. In other words, the packet transfer to an output channel whose number is less than input channel is prohibited by the Definition 2 within the range shown in the above equations.

Therefore, channels are used in the increasing order on RDT. □

Figure11 shows an example of routing on the 4×4 RDT. The blocked channels are bypassed with a path consisting of channels in increasing order. This figure also shows that the number of permitted output channel is lager than that of input channel.

5 Conclusion

Two adaptive routing algorithms on RDT are proposed and proved to be deadlock-free. A simulation study which demonstrates the effect of the proposed routing algorithm is required.

References

1. *Paragon XP/S Product Overview*. Intel Corp., 1991.
2. W. J. Dally A. Chien S. Fiske W. Horwat J. Kenn M. Larivee R. Lethin P. Nuth and S. Wills. The J-machine: A Fine-Grain Concurrent Computer. In *IFIP 11th Computer Congress*, pages 1147–1153, August 1989.

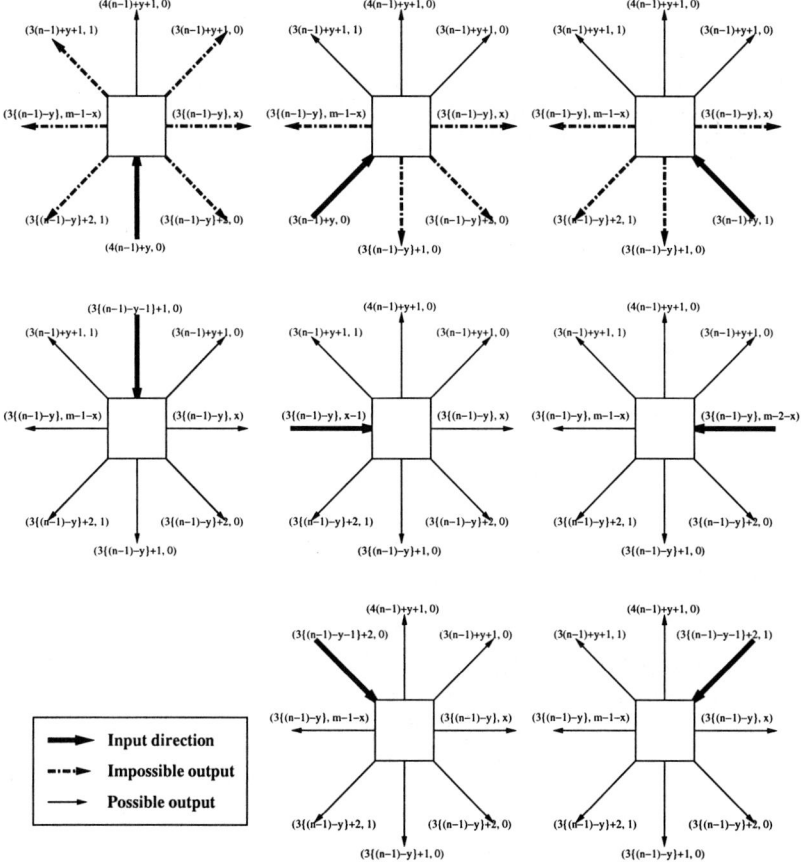

Fig. 10. The possible output channels for each input channel.

3. H. Ishihata T. Horie S. Inano T. Shimizu S. Kato and M. Ikesaka. Third Generation Message Passing Computer AP1000. In *International Symposium on Supercomputing*, pages 46–55, November 1991.
4. Y. Yang, H. Amano, H. Shibamura, and T. Sueyoshi. Recursive diagonal torus: An interconnection network for massively parallel computers. *Proceedings of IEEE SPDP*, 1993.
5. H. Nishi, K. Nishimura, K. Anjo, H. Amano, and T. Kudoh. The JUMP-1 router chip: The versatile router for supporting distributed shared memory. *Proceedings of International Phoenix conference on computers and communications*, 1996.
6. Y. Yang and H. Amano. Message Transfer Algorithms on the RDT. *IEICE Transaction on Information and Systems*, 79(2), 1996.
7. L. M. Ni and P. K. McKinley. A Survey of Wormhole Routing Techniques in Direct Networks. *IEEE Transactions on Computers*, February 1993.

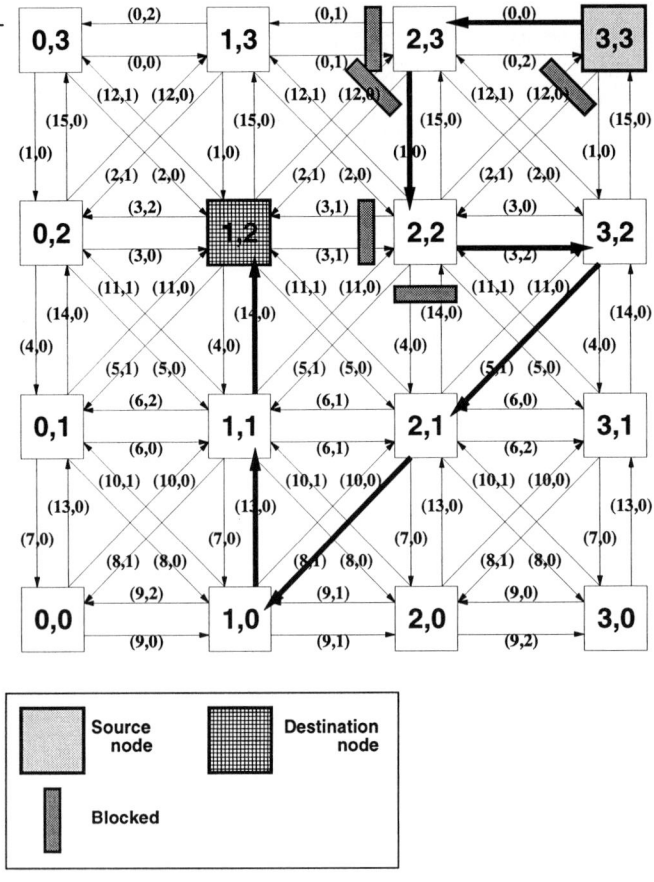

Fig. 11. Example of north-last routing for RDT($m = 4, n = 4$).

8. J. Duato. A Necessary And Sufficient Condition For Deadlock-Free Adaptive Routing In Wormhole Networks. *Proceedings of the International Conference on Parallel Processing*, 1:142–149, 1994.
9. J. Duato. A Necessary And Sufficient Condition For Deadlock-Free Adaptive Routing In Wormhole Networks. *IEEE Transaction on Parallel and Distributed Systems*, 6(10), 1995.
10. W. J. Dally and C. L. Seitz. Deadlock-Free Message Routing in Multiprocessor Interconnection Networks. *IEEE Transactions on Computers*, 36(5):547–553, May 1987.
11. C. J. Glass and L. M. Ni. Maximally Fully Adaptive Routing in 2D Meshes. *Proceedings of ISCA92*, pages 278–287, 1992.

Achieving Multi-level Parallelization

Carrie J. Brownhill[1], Alexandru Nicolau[1], Steve Novack[2] and Constantine D. Polychronopoulos[2]

[1] Department of Information and Computer Science
University of California,
Irvine, CA 92697-3425
[2] Center for Supercomputing Research and Development
University of Illinois at Urbana-Champaign
1308 W. Main St.
Urbana, IL 61801

Abstract. Many modern machine architectures feature parallel processing at both the fine-grain and coarse-grain level. In order to efficiently utilize these multiple levels, a parallelizing compiler must orchestrate the interactions of fine-grain and coarse-grain transformations. The goal of the PROMIS compiler project is to develop a multi-source, multi-target parallelizing compiler in which the front-end and back-end are integrated via a single unified intermediate representation. In this paper, we examine the appropriateness of the Hierarchical Task Graph as that representation.

1 Introduction

The design of the internal representation (IR) of a parallelizing compiler is driven, in a large part, by the compiler's target granularity. For example, a compiler which uses source language transformations, such as converting sequential loops to DOALL loops, will need to store information about source level statements and expressions, as well as information about control flow structures. If transformations use information which is readily available only in the original source code, then the IR may need to store that information. For example, it might store expressions representing the upper, lower, and intermediate values of the loop iteration variable. When working with source level transformations, having the IR reflect the high level structure of the program makes the design of the compiler simpler and more efficient. On the other hand, a parallelizing compiler which works at the instruction level must have an IR which reflects this level of detail. Traditionally, these types of compilers work an on IR which closely mirrors the final generated assembly code, without having access to the original high level language source code. This means that information about control flow structures such as loops and branches must be reconstructed, and often may not produce all of the information that was available in the source program. The PROMIS compiler is intended to fill both roles of source statement level parallelizer and instruction level parallelizer. In order to do this efficiently, it must have an IR which represents the program at both levels, and perhaps at other levels as well. A *hierarchical* representation is necessary. One such hierarchical representation is the Hierarchical Task Graph (HTG). This paper explores the suitability of the HTG as the IR for a multi-level parallelizing compiler.

2 The Hierarchical Task Graph

Hierarchical Task Graph's (HTG's) were originally presented for use at the coarse-grain level in [1, 2]. The HTG is defined in [1] as a directed acyclic graph, HTG = (HV, HE) where HV is a set of nodes and HE is a set of edges that represent control flow through the nodes. HV contains five types of nodes:

1. A unique *START* node that has no incoming edges dominates all other nodes in HV.
2. A unique *STOP* node that has no outgoing edges post-dominates all other nodes in HV.
3. *Simple* nodes at the coarse-grain level are intended to represent tasks that have no sub-tasks.
4. *Compound* nodes represent sub-HTG's (HTG's contained within other HTG's). The compound node X represents the HTG, $HTG(X) = (HV(X), HE(X))$ where $START(X), STOP(X) \in HV(X)$ are the START and STOP nodes of HTG(X). Compound nodes are typically used to represent subroutines and basic blocks (recall that loops are represented by loop nodes).
5. *Loop* nodes represent loops whose loop bodies are sub-HTG's, as defined for compound nodes above.

Figure 1 contains the HTG of a sample code fragment and illustrates the hierarchical nature of the various types of nodes. (This example is taken from [1].) Nodes A and B are *loop* nodes at the top level of the hierarchy. Node C is a *loop* node in the next level down. Node D is a *compound* node, again at the top level of the hierarchy. At the third and lowest level, the HTG contains sixteen nodes corresponding to basic blocks.

A key characteristic of HTGs is the ability of a node to summarize the dependence information of all the nodes it contains. The node can then be treated as one atomic unit, without the need for each of the contained nodes to be visited.

3 Multi-level Parallelization

The ultimate goal of the PROMIS project is to develop a multi-source, multi-target parallelizing compiler in which the front-end and back-end are integrated via a single unified intermediate representation, thus providing several important opportunities for high-level/low-level interactions and trade-offs that are either very difficult, or impossible to do effectively in conventional compilers. (See Figure 2.) Some of the advantages of a unified and integrated approach include the following:

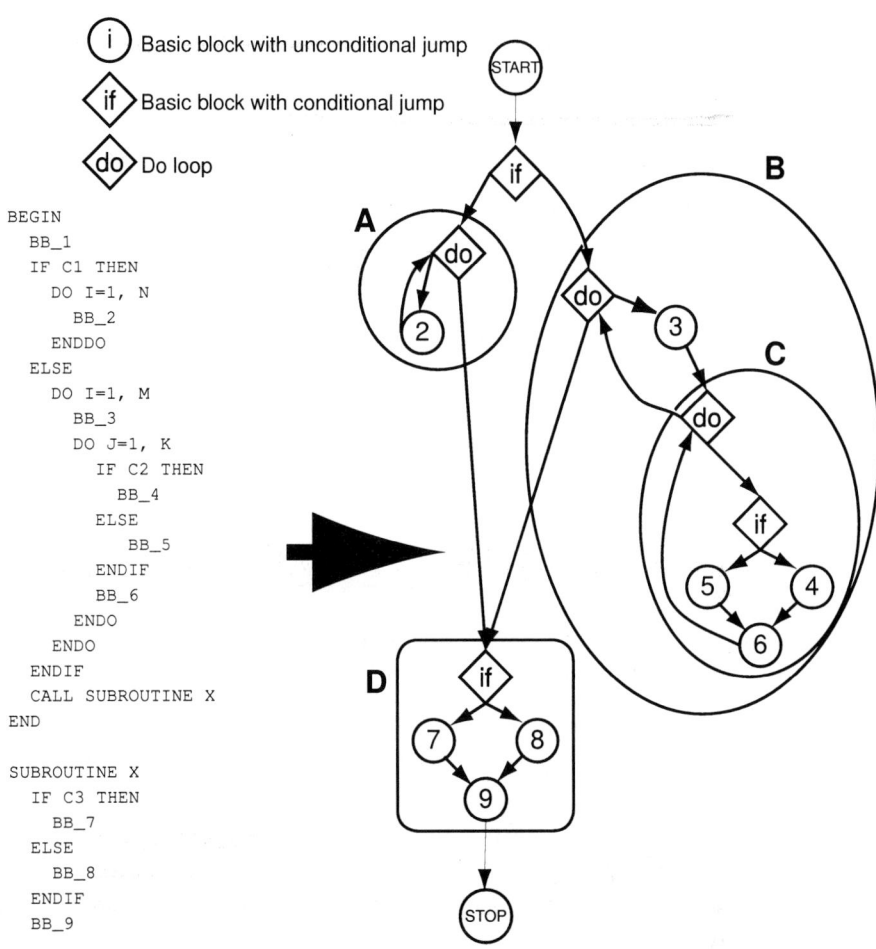

Fig. 1. Hierarchical Task Graph Example from [1]

1. Instruction level parallelization (back-end) can be based on information about the semantics of the source language and algorithm (when available) that is normally not available to compiler back-ends, thus eliminating spurious dependences in the back-end that either cannot be removed, or are too expensive to remove via low-level analysis alone.
2. Instruction level parallelization (back-end) can make use of high-level transformations (e.g., loop interchange, loop fusion, etc.) that have the effect of increasing the availability of instruction level parallelism.
3. Context sensitive trade-offs can be made between coarse-grain and fine-grain parallelism when necessary. Targets having both coarse-grain (multi-processor) and fine-grain (superscalar/VLIW) parallelism are rapidly becoming the standard for workstations and desktop PC's. Some transformations,

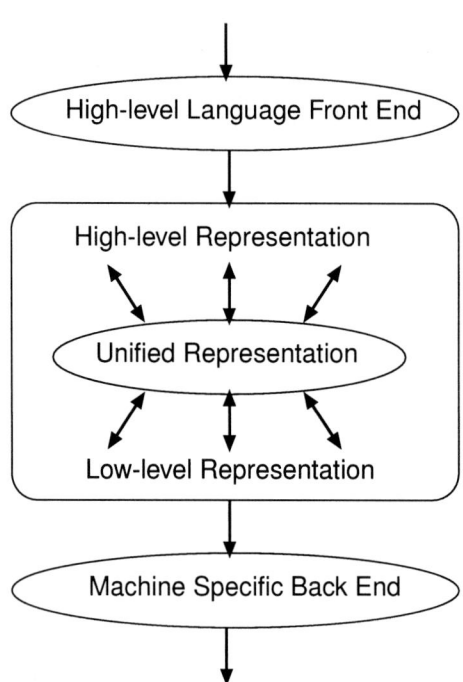

Fig. 2. The PROMIS Compiler

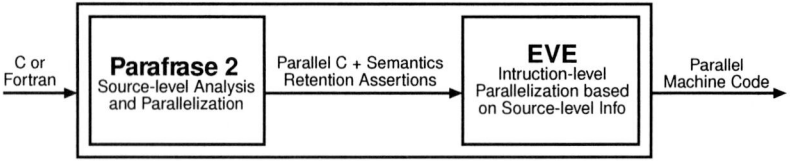

Fig. 3. Structure of the PROMIS Prototype

such as loop interchange, can have the effect of increasing one type of parallelism at the expense of the other (e.g., outer-loop vs. inner-loop parallelism). Integration facilitates context sensitive trade-offs by providing a framework in which different granularities of parallelism can be tried and compared to each other with respect to how each is able to utilize the resources of the specific target given the dependence structure and resource requirements of the particular region of code.

4. Efficiency is improved, in terms of both development cost and execution time of the compiler, since redundancies that traditionally exist between the compiler front-ends and back-ends are eliminated (e.g., we only need one intermediate representation, and analyzes, like symbolic and pointer analysis, need not be duplicated in the back-end).

The above gives only a glimpse of the numerous benefits of a unified and integrated front/backend. One could list many more including development cost, maintainability, extensibility, performance and complexity reasons. However, such a tightly integrated approach to compiler development also comes at the cost of increased design and development complexity. The latter can explain in part the lack of similar efforts in the research community, and particularly in the commercial world where development time can be crucial to the success of a product. For instance, even though there is little doubt that program analysis and parallelization are more effective at a high-level, there is also a profound lack of experience and techniques for passing high-level information to the backend in a way that makes backend analysis redundant or obsolete. At the very minimum, passing data/control flow and dependence information to the backend would necessitate a common representation and internal data structures shared between the parallelizer and the code generator. These are some of the challenges encountered in our project which made the development of the PROMIS Prototype (and the early experimental evidence) all the more important.

As a first step in the PROMIS project, we have developed a proof-of-concept compiler, the PROMIS Prototype, as a test-bed to explore some of the interactions between high-level and low-level transformations, in preparation for full integration in the PROMIS compiler itself. A structural overview of the PROMIS Prototype is shown in Figure 3. The prototype uses Parafrase-2[11] as a front-end for doing parsing, symbolic and dependence analysis[4] and source-level parallelization, and uses the EVE Mutation Scheduling compiler[6, 9, 5, 7, 10, 8] as a back-end for performing instruction level parallelization and code generation.

Because both Parafrase-2 and EVE are capable of generating and using an IR based on a HTG, they are well suited to be used for this research on this subject. The effectiveness of the HTG representation can be seen by comparing the results when it is used and when it is not used.

The prototype compiler generates code for a simulated shared-memory multiprocessor wherein each processor is a pipelined VLIW (see Section 5.1). The input to the prototype is either Fortran or C, which is analyzed and parallelized at the coarse-grain level by the Parafrase-2 front-end. Parafrase-2 then generates C code with parallel directives and Semantics Retention assertions for use by EVE, which continues the parallelization process at the instruction level for the VLIW targets.

Clearly this prototype, with its independent front-end and back-end, does not realize our ultimate goal of fully integrating the two. However, the Semantics Retention mechanism [7] in EVE does provide partial integration by allowing higher level semantics information derived by Parafrase-2 to be retained for use by EVE at the instruction level.

Furthermore, EVE is able to exploit source-level transformations performed by Parafrase-2 that increase the availability of instruction level parallelism. With this prototype we can directly measure some of the benefits of integration, such as its effect on instruction level parallelization, and provides a solid platform for experimentation with high-level/low-level trade-offs by manually tuning Parafrase-2 and EVE.

Parafrase-2 and EVE were both designed as stand-alone compilers. Therefore, the PROMIS prototype uses them as separate passes of the compiler. Since Parafrase-2 performs source level parallelization, it gets called as the first pass of the PROMIS prototype. The source code generated by Parafrase-2 gets passed through the GCC front end of EVE, which generates instruction level code. Then, EVE performs fine-grain parallelization. The design is shown in Figure 3.

4 Use of the Hierarchical Task Graph in a Multi-level Compiler

It seems clear that the most important characteristic for an IR for a multi-level parallelizing compiler is the ability to accurately and efficiently contain any necessary information about the code at the coarse grain level and at the fine grain level. In order to determine if the HTG has this capability, we examine the two passes of the PROMIS prototype compiler, and how they use the HTG structure.

4.1 The HTG in Parafrase-2, a Source Statement Level Parallelizer

For our purposes, the most interesting aspect of the use of an Hierarchical Task Graph in Parafrase-22 is its ability to improve data dependence information. In particular, for loop nodes, variables which are always defined before they are used and which have no outgoing flow dependences are recognized as being local to the loop. Any false loop-carried dependences are recognized and deleted, allowing some previously sequential DO loops to be made into DOALL loops.

```
    DO 100 I = 2,N-1

        R = AA(I,J) * D(I,J - 1)

        D(I,J) = 1.D0 / (DD(I,J) - AA(I,J - 1) * R)

        RX(I,J) = RX(I,J) - RX(I,J - 1) * R

        RY(I,J) = RY(I,J) - RY(I,J - 1) * R

100 CONTINUE
```

Fig. 4. Loop from TOMCATV

For example, in the loop shown in Figure 4, the variable R is defined in the first line of the body of the loop, and then used in the next three lines.

Because R is a scalar, there are loop-carried dependences from the three uses to the define. However, if each loop iteration has its own copy of R, then there is no loop-carried dependences. The loop can be made into a DOALL loop either by moving the declaration of R local to the loop, or by running the loop on a machine with memory local to each processor, and storing R in that local memory.

4.2 The HTG in EVE, an Instruction Level Parallelizer

EVE uses its Hierarchical Task Graph primarily to improve the speed of the compiler and to reduce the size of the generated code. EVE implements Percolation Scheduling transformations, which are normally incremental. By allowing code motion across compound and loop nodes, as data dependences permit, operations can be moved across arbitrarily large regions of code in constant time. In addition, some code duplication that would normally occur as operations moved through multiple control paths can be eliminated.

EVE uses an extended definition of the HTG which allows compound nodes to represent if-then-else blocks, as well as subroutines and basic blocks. This allows EVE to move an operation *across* an if-then-else block (when dependences allow) rather than moving it into the block and causing code duplication. EVE uses heuristics to decide whether to move an operation 'into' a compound node. Because of this, the HTG based program representation can sometimes allow EVE to perform code motions that would otherwise be considered too inefficient. In other cases, EVE may choose not to perform a code motion that may increase code size, and miss some parallelization. In the latter case, the non-HTG based transformations may actually produce a better speedup.

In order to see the effect of the use of the HTG in EVE, the Livermore Loops were parallelized using the PROMIS prototype compiler both with and without the HTG base optimizations enabled in EVE. The resulting speedups are shown in Table 1. Fifteen out of the twenty-four loops show some increase in speedup or stay the same and nine show slight decreases in speedup. Overall, the speedup increased an average of three percent. Recall that these optimizations are primarily intended to improve compiler speed and resulting code size, rather than application speedup. In [5], the use of an HTG based IR in EVE is shown to decrease compile time by a factor of ten.

5 Experiments

The purpose of this phase of our research was to examine the advantages of using the HTG as an IR in both a statement level and an instruction level compiler. To this end, several benchmarks were parallelized and executed on a simulator. Five different versions of the code were generated. Along with the straight sequential code, the following were also produced:

	Without HTG	With HTG
LL1	27.23	27.39
LL2	2.06	2.25
LL3	7.15	7.15
LL4	6.05	6.03
LL5	3.50	4.19
LL6	1.96	1.89
LL7	22.83	24.22
LL8	43.22	45.39
LL9	30.04	30.50
LL10	5.06	4.45
LL11	4.97	4.97
LL12	41.78	42.43
LL13	2.83	2.83
LL14	7.15	7.22
LL15	2.72	2.37
LL16	9.39	9.21
LL17	4.31	4.31
LL18	30.32	30.02
LL19	2.60	2.59
LL20	4.08	3.89
LL21	36.84	42.85
LL22	23.15	24.13
LL23	2.57	2.57
LL24	2.72	1.55
Average	13.52	13.93

Table 1. Speedup for Livermore Loops (original cycles/ new cycles)

- With neither Parafrase-2 nor EVE using an HTG based IR.
- With Parafrase-2 only using HTG based optimizations.
- With EVE only using HTG based optimizations.
- With both Parafrase-2 and EVE using HTG based optimizations.

5.1 Target Specification

The target architecture was a VLIW/MIMD architecture. Six processors with shared memory were simulated. Each processor was a generalized VLIW architecture with a single register file and RISC-like multi-cycle operations with multiple conditionals possible. Each VLIW had three functional units of each of the following types: integer ALU, floating point ALU, shift, floating point multiply, floating point divide, and memory (LOAD/STORE) unit. Each LOAD was considered to take four cycles, and each STORE one cycle.

5.2 Results

Three of the floating point SPEC benchmarks: TOMCATV, SWIM and MGRID were parallelized. The resulting speedups, measured as original dynamic cycle count / final dynamic cycle count, are shown in Table 2. The first column is the speedup when neither Parafrase-2 nor EVE use their respective HTG optimizations. The second column shows the speedup which results when Parafrase-2 uses an HTG to improve its data dependence analysis. Column three shows the speedup when EVE uses its HTG based optimizations (note that these are primarily intended to improve the speed of the compiler, not the application), and the last column is the result when both EVE and Parafrase-2 are using an HTG.

	No HTG	P2 HTG	EVE HTG	Both HTG
TOMCATV	4.68	8.96	5.54	12.46
MGRID	13.27	13.53	13.81	14.03
SWIM	23.08	23.08	23.08	23.08

Table 2. Speedup for three SPEC benchmarks (original cycles/ new cycles)

When examining Table 2, we see that using the HTG based optimization in Parafrase-2 significantly increased the speedup of TOMCATV (from 4.68 to 8.96). It slightly increased the speedup of MGRID, and did not affect SWIM. When the HTG based optimizations of EVE were used, the speedup of TOMCATV was increased from 4.68 to 5.54. The speedup of MGRID increased from 13.21 to 13.81, and again, SWIM was not affected. Finally, the final combined effect was to increase the speedup of TOMCATV to 12.46. MGRID was slightly improved at 14.03, and SWIM was not affected.

6 Other Granularity Levels

The PROMIS prototype compiler only implements two levels of parallelization, statement (or loop) level, and instruction level. Historically, the former has been used for shared and distributed memory multiprocessor architectures; while the latter is only feasible for VLIW or superscalar architectures. As new architectures are developed, other levels of granularity may become feasible. An important issue then arises - the ability to use the HTG structure for other levels of granularity.

For example, on-chip memory made available on VLSI circuit boards may reduce communication costs relative to processor speed. This would imply that the task granularity would drop to a size between statement and instruction level. One way to create tasks this size would be to break long numeric statements into expressions. Arithmetic expressions often have a great deal of implicit parallelism. Note that using an expression level parallelizer is more efficient than an instruction level parallelizer, and would automatically keep tightly-coupled instructions together, reducing interprocessor communication. In Figure 5, a FORTRAN statement is shown along with its HTG. Nodes Q, R, and S are expression nodes.

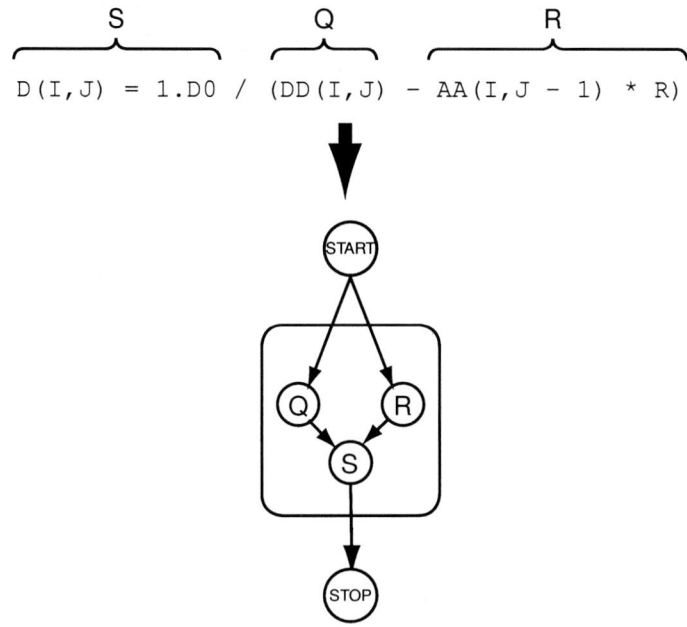

Fig. 5. Hierarchical Task Graph Example with Expression Nodes

7 Communication Between Transformations

Ordering transformations is difficult in parallelizing compilers. Transformations may enhance or inhibit each other in unpredictable ways. Sometimes the same optimization needs to be run several times in order to produce the desired effect, even in conventional compilers. When parallelization on multiple levels is desired, it seems that some backtracking and negotiation between transformations might be beneficial. This implies a structure for holding information which can be incrementally updated at various levels. Information stored might include what transformations had already been run, and the available parallelism inside each node. (See Figure 6). This is similar in flavor to information kept in Region Scheduling [3].

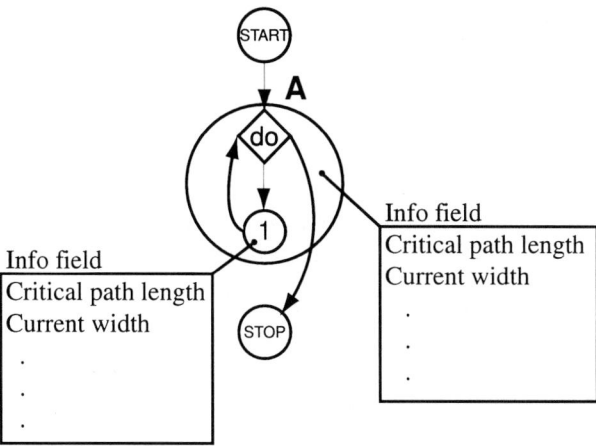

Fig. 6. Hierarchical Task Graph Containing Parallelism Information

8 Conclusion

The PROMIS compiler project has the goal of creating a multi-source, multi-target, multi-level parallelizing compiler. Choosing an appropriate internal representation for a compiler is the first critical design step. In this research, we examine the use of the HTG in a statement level parallelizing compiler, Parafrase-2, and an instruction level parallelizing compiler, EVE, both contained in the PROMIS prototype compiler. It was shown that by allowing data dependence and other information to be summarized for groups of nodes represented as compound and loop nodes, the HTG representation improves the efficiency of the instruction level compiler studied. It also is used to implement key optimizations in the statement level compiler used. The HTG structure allows other levels of granularity to be represented, and can be used as a central database of information and communication for various transformations. While other IR structures are certianly possible, the hierarchical nature of multi-level parallelization seems to match the HTG structure well.

References

1. Milind B. Girkar and Constantine Polychronopoulos. Automatic extraction of functional parallelism from ordinary programs. *IEEE Transactions on Parallel and Distributed Systems*, 3(2), March 1992.
2. Milind Baburao Girkar. *Functional Parallelism: Theoretical Foundations and Implementations*. PhD dissertation, University of Illinois at Urbana-Champaign, December 1991. Available as CSRD Report 1182.
3. R. Gupta and M.L. Soffa. Region scheduling: An approach for detecting and redistributing parallelism. *TOSE*, 16(4), April 1990.

4. Mohammad R. Haghighat and Constantine Polychronopoulos. Symbolic analysis for parallelizing compilers. *ACM Transactions on Programming Languages and Systems*, 18(4).
5. Alexandru Nicolau and Steve Novack. Trailblazing: A hierarchical approach to percolation scheduling. In *Proceedings of the International Conference on Parallel Processing, St. Charles, IL*, pages II120–124. The Pennsylvania State University Press, August 1993.
6. S. H. Novack. *The EVE Mutation Scheduling Compiler: Adaptive Code Generation for Advanced Microprocessors*. PhD thesis, University of California, Irvine, 1997.
7. Steve Novack, Joseph Hummel, and Alexandru Nicolau. A simple mechanism for improving the accuracy and efficiency of instruction-level disambiguation. In *Proceedings of the 8th Annual Workshop on Languages and Compilers for Parallel Computing, Lecture Notes in Computer Science No. 1033*, pages 289–303. Springer-Verlag, August 1995.
8. Steve Novack and Alexandru Nicolau. VISTA: The visual interface for scheduling transformations and analysis. In *Proceedings of the 6th Annual Workshop on Languages and Compilers for Parallel Computing, Lecture Notes in Computer Science No. 768*, pages 449–460. Springer-Verlag, August 1993.
9. Steve Novack and Alexandru Nicolau. Mutation scheduling: A unified approach to compiling for fine- grain parallelism. In *Proceedings of the 7th Annual Workshop on Languages and Compilers for Parallel Computing, Lecture Notes in Computer Science No. 892*, pages 16–30. Springer-Verlag, August 1994.
10. Steve Novack and Alexandru Nicolau. Resource-directed loop pipelining. In *Proceedings of the 9th Annual Workshop on Languages and Compilers for Parallel Computing, Lecture Notes in Computer Science*. Springer-Verlag, August 1996. To appear.
11. Constantine Polychronopoulos, Milind B. Girkar, Mohammad R. Haghighat, Chia L. Lee, Bruce P. Leung, and Dale A. Schouten. Parafrase-2: An environment for parallelizing, partitioning, synchronizing, and scheduling programs on multiprocessors. In *Proceedings of the International Conference on Parallel Processing, St. Charles IL*, pages II39–48, August 1989. also in International Journal of High Speed Computing, Vol. 1, No. 1, 1989.

A Technique to Eliminate Redundant Inter-Processor Communication on Parallelizing Compiler TINPAR

Atsushi KUBOTA[1]*, Shogo TATSUMI[1], Toshihiko TANAKA[1],
Masahiro GOSHIMA[1], Shin-ichiro MORI[1],
Hiroshi NAKASHIMA[2], and Shinji TOMITA[1]

[1] Department of Information Science, Kyoto University
[2] Department of Information and Computer Sciences,
Toyohashi University of Technology

Abstract. Optimizing inter-processor(PE) communication is crucial for parallelizing compilers for message-passing parallel machines to achieve high performance. In this paper, we propose a technique to eliminate redundant inter-PE messages. This technique utilizes a data-flow analysis to find a definition point that corresponds to a use point where the definition and the use are occurred in different PEs. If several read accesses occurred in the same PE use the data defined at the same definition point in another PE, redundant inter-PE messages are eliminated as follows: only one inter-PE communication is performed for the earliest read access and the previously received data are used for the following read accesses. In order to guarantee the consistency of the data, a valid flag and a sent flag are provided for each chunk of received data. The control of these flags is equivalent to the coherence control by the self invalidation on a compiler aided cache coherence scheme.

keywords: parallelizing compiler, message-passing multiprocessor, dependence analysis, data-flow analysis, message coalescing

1 Introduction

In our research project, we are developing a parallelizing compiler TINPAR(TINy PARallelizer)[3] for message-passing multiprocessors. TINPAR accepts scientific programs with data parallelism as many HPF[1] compilers. TINPAR accepts sequential programs with data distribution directives written in an imperative language such as FORTRAN and produce an SPMD code for message-passing multiprocessors. Each statement is scheduled at compile time with the owner computes rule.

For such compilers, optimizing inter-processor(PE) communication is crucial to achieve good performance. There are two major optimization techniques:

- reducing communication costs and

---
* Research Fellow of the Japan Society for the Promotion of Science

- hiding communication latency by overlapping communication and computation.

Reducing communication costs is achieved by

- reducing the number of invocation of communication and
- eliminating unnecessary communication.

The former technique is achieved by packing elements of the same array to be transferred to the same PE into one message (*message vectorization*) and elements from the different arrays to be transferred to the same PE into one message (*message aggregation*). In our earlier work, we have already implemented these techniques for TINPAR.

This paper presents a technique called *message coalescing* that eliminates unnecessary communication and its implementation for TINPAR.

In the remainder of this paper, we show requirements for a program to apply message coalescing in Section 2. In Section 3, we describe a data-flow analysis to obtain the def-use chain of data, which is necessary for message coalescing. In Section 4, we discuss an implementation issue of message coalescing using the data-flow analysis. Finally in Section 5, we show concluding remarks.

2 Message Coalescing

2.1 Message Coalescing

Message coalescing[2] is a technique to reduce the volume of messages; when an element e of an array in the remote memory of a PE s is read by another PE r and the element e is read by the PE r multiple times, the transfer of the value of the element e can be performed just once instead of multiple transfers. The received value is kept in the local memory of the PE r and reads of the value of the remote element e are performed by supplying the value kept in the local memory.

Fig. 1 is an example program of message coalescing shown in [2]. When the arrays A and B are distributed among several PEs, read accesses to an element of the array B in the right hand side of the assignment statement may require inter PE communication. In this case, the element used as B(i+2) in the l'th iteration is used again as B(i+1) in the $l+1$'th iteration. In this way, when the same data is used multiple times, the transfer of the data is performed just once and send/receive library routines are not called on the following uses.

2.2 Conditions to Apply Message Coalescing

Hiranandani, et al[2] did not show conditions for programs required to apply message coalescing clearly. We therefore clarify the required condition for message coalescing.

To apply message coalescing, it must be examined, for each read(*use*) of an element, which statement writes(*defines*) the value of the element. If the

```
do i=1,n
  A(i) = B(i+1) + B(i+2)
enddo
```

⇓ Parallelized with block-wise distribution of arrays A,B

```
/* P:the number of PEs, p:PE ID */
send B(1..2) to p-1
recv tmp_B(1..2) from p+1
for l_i=1,n/P-2
  A(l_i) = B(l_i+1) + B(l_i+2)
endfor
A(n/P-1) = B(n/P) + tmp_B(1)
A(n/P) = tmp_B(1) + tmp_B(2)
```

Fig. 1. Message coalescing

statement is executed multiple times in a loop, the iteration in which the value is defined must be determined. We call these statements and iterations in the program in which the values of data are defined and used as the *definition point* and the *use point*. For data which is defined and used repeatedly in the program, it is required to examine the sequence of definition and use points. Let d be a definition point in this sequence and d' be the next definition point. The value to be read at all use points $r_1, r_2, \ldots r_n$ put between d and d' is the value updated at the definition point d. At the use point r_1, the data is received and stored in a buffer area on the receiver PE then it is read from this buffer area, while at the other use point $r_2, \ldots r_n$, the data is simply read from the buffer area.

3 Data-Flow Analysis

We adopt a tree structure called LWT(Last Write Tree)[4] to represent the precise data-flow of programs. Because many data elements are defined and used in a loop nest, we focus on the analysis of definition and use points in a loop nest.

3.1 LWTs

In this section, we show how to construct LWTs for a loop nest. We use an example code shown in Fig. 2.

Let (i^r, j^r) be the iteration of a use point and (i^w, j^w) be the iteration of a definition point. Then we get the definition point for each use of elements of the array a as follows.

- If $j^r = 1$, the value to be used is defined before entering the loop. We use ⊥ to represent this situation.

```
for i = 1, n do
  for j = 1, n do
    ... = a(i, j-1)
    a(i, j) = ...
  endfor
endfor
```

Fig. 2. A loop with a pair of definition and use points

- If $2 \leq j^r \leq n$, the value to be used is defined in the iteration $(i^r, j^r - 1)$. This data dependence is carried by the inner loop.

The LWT for this loop nest is shown in Fig. 3. For a pair of a use point and a definition point that defines the value to be used, the iteration of the definition point is expressed using the iteration of the use point (i^r, j^r). Dependence levels in the loop nest are also shown in LWTs.

Dependence level is the nest level of the most outer loop which carries data dependence. In the example shown in Fig. 3, the dependence level is 2, which is the nest level of the inner loop.

If the dependence is carried by the outer loop, the dependence level is 1. If the definition point is located outside the loop(that is the case $\perp$), the dependence level is 0. If there is no loop carried dependence, the dependence level is defined as ∞.

In general, for a n-deep loop nest, the LWT is constructed as follows.

Let $i^w = (i_1^w, i_2^w \ldots, i_n^w)$ be an iteration of a definition point in a n-deep loop nest and $i^r = (i_1^r, i_2^r \ldots, i_n^r)$ be an iteration of a use point.

Let F_r and F_w be subscripts of use and definition of array elements in the loop. If there is a data dependence, there exist a use iteration i^r and a definition iteration i^w such that $F_r i^r = F_w i^w$.

The LWT for this loop is constructed as follows:

1. Express i^w using i^r. If F_w is invertible, $i^w = F_w^{-1} F_r i^r$.
2. Restrict the range of the use iteration variable i^r according to the loop bounds constraints of the iteration variable i^w. For a use point i^r that does not satisfy this restriction, the corresponding definition point is $\perp$.
3. For each loop nest level which satisfies $1 \leq l \leq n$, select one from the following steps according to the relationship between i_l^w and i_l^r.
   - If $i_l^w > i_l^r$, the definition point is lexicographically after the use point. The definition point is $\perp$.
   - If $i_l^w < i_l^r$, there is a loop carried dependence. The definition point is $i^w = F_w^{-1} i^r$.
   - If $i_l^w = i_l^r$, go down to the nest $l+1$ and repeat the selection from these steps.
4. If $i_l^w = i_l^r$ is satisfied for all l nest levels, the data dependence is determined by the execution order of the definition and the use in the loop body of

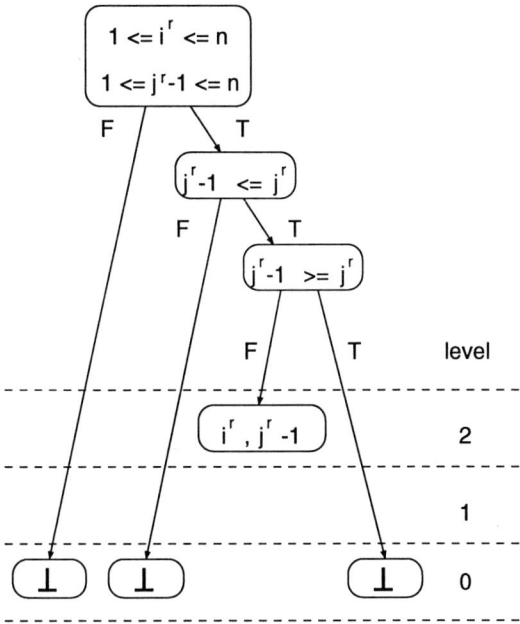

Fig. 3. LWT

the inner most nest. If the definition point precedes the use point, there is a loop independent dependence and the definition point is expressed using the iteration i^w. If the use points precedes the definition point, there is no dependence and the definition point is $\bot$.

In the example shown in Fig. 2, there is a relationship between a use point and a corresponding definition point as $(i^w, j^w) = (i^r, j^r - 1)$. This relationship is calculated as follows.

Firstly, restrict the range of the use iteration according to the loop bounds of the definition point. Substituting the replationship $i^w = i^r$ and $j^w = j^r - 1$ shown above to the loop bounds conditions $1 \leq i^w \leq n$ and $1 \leq j^w \leq n$, we get two inequalities $1 \leq i^r \leq n$ and $1 \leq j^r - 1 \leq n$. Rearranging the latter inequality gives $2 \leq j^r \leq n + 1$. For $j^r = 1$, the corresponding definition point is located out of the loop($\bot$). This means that the value defined before entering the loop is read at the use point $j^r = 1$.

Then the outer loop is examined and the two nodes corresponding to $i^w \leq i^r$ and $i^w \geq i^r$ is created in the LWT. Because these conditions are always true, these nodes are not added in Fig. 3.

For the inner loop nest, the condition $j^w \leq j^r - 1 \leq j^r$ is always true while the condition $j^w = j^r - 1 \geq j^r$ is always false. Thus the iteration $(i^r, j^r - 1)$ is the definition point for the use point (i^r, j^r). Because the dependence between the definition point and the use point is carried by the inner loop (level 2), the node

($i^r, j^r - 1$) which represents the definition point is added in the level 2 row in the LWT.

3.2 Merging Multiple LWTs

In our method, a use point u which requires communication is selected and then its corresponding definition point d to the use point can be obtained by tracing the edges of the LWT. The sequence of the def-use chain is constructed by finding all use points corresponding to the definition point d and put these define and use points in the lexicographically order of the iterations. Arranging the LWTs corresponding to these use points in the def-use chain in the lexicographical order, the sequence of the LWTs can be constructed. Then the def-use chain of each nest level is constructed by tracing the same nest level of the sequence of the LWTs.

4 Implementation Issue

In this section, we show an optimization technique at compile-time and how it works at run-time to realize message coalescing.

Depending on whether the nest level of a definition point is equal to the nest level of a use point or not, we use different techniques to realize message coalescing. If the both nest levels are equal to each other, a dynamic message coalescing technique using run-time libraries is applied. We call this technique as *message caching*. If the nest levels are different, that is, the definition point is out of the loop in which the use point located, a static message coalescing is performed by reordering the communication code and making temporary array variables to keep the received data.

4.1 Message Caching

Basic Principle Data elements previously received by PE r are kept in the temporary buffer(called *receive buffer* for short) maintained by the run-time library routines for communication in PE r. For a receive request for a data element in PE r, if the valid value of the data element r is kept in the receive buffer, this valid value is supplied from the receive buffer. Decisions of the duration of the value kept in the receive buffer and the way to ensure the validness of the value of the maintained data element are the implementation dependent issues.

Management of Receive Buffer The information about "which data is used multiple times" and "where the definition point and the last use point of the data are" is obtained by the examination of the LWT. With this information, allocation and release of an area in the receive buffer and the consistency control for the data which is defined multiple times are performed.

In order to ensure the consistency, a *valid flag* is provided to indicate the validness of the value of the data kept in the receive buffer. A *sent flag* is also provided in the sender PE to indicate whether the data is sent after the value of the data is defined.

Because communication is performed with synchronized send/receive messages, the sent flag that corresponds to the valid flag of the receive buffer must be provided in the sender PE.

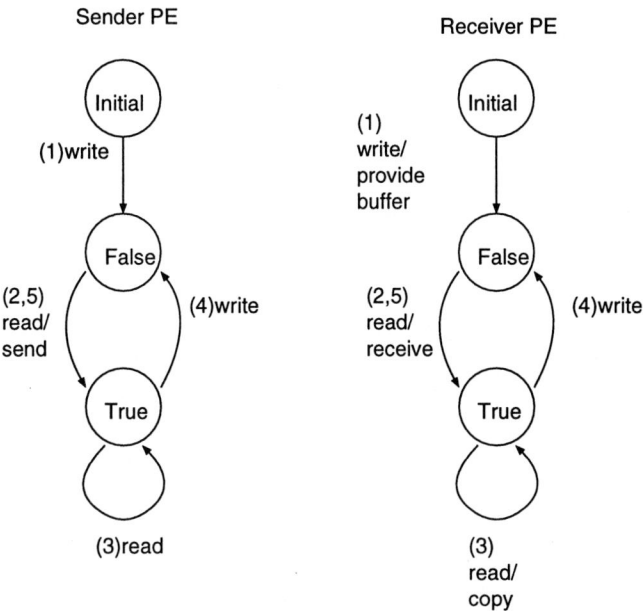

Fig. 4. State diagram of control of receive buffer

The consisitency of the data kept in the receive buffer is maintained as shown in Fig. 4. The details is shown as follows.

1. When the data to be sent is defined in the sender PE for the first time, the sent flag turns to false. In the receiver PE, area is allocated in the receive buffer and the valid flag turns to false.
2. When the data is used in the receiver PE, the data is transferred from the sender PE and the sent flag turns to true. In the receiver PE, the data is received, and the valid flag turns to true. The received data is saved into the receive buffer and the value to be used is supplied from the receive buffer.
3. While the data is not defined in the sender PE, the data is supplied from the receive buffer for the use of the data in the receiver PE.
4. When the data is defined in the sender PE again, the sent flag turns to false. In the receiver PE, the valid flag corresponding to the redefined data turns to false and the area in the receive buffer is released.

5. When the data is used in the receiver PE again, the updated value of the data is transferred from the sender PE and the sent flag is turned to true. In the receiver PE, the data is received and the valid flag is turned to true. The received data is saved into the receive buffer and the value to be used is supplied from the receive buffer.

The receive buffer can be regarded as a cache area controlled by software. When the value of the data is updated, this cache area is invalidated without control signal by making use of the the result of the analysis at compile-time. The control of these flags is equivalent to the coherence control by the self invalidation on a compiler aided cache coherence scheme.

Optimization The area in the receive buffer and the both valid and sent flags corresponding to that area are deleted when they become unnecessary. This deleting point in the execution is the last use point examined by the data-flow analysis. After the use at this point, any use and definition to the data is not performed. It is therefore guaranteed that the area of the receive buffer and the both valid and sent flags can be deleted. This prevents the excessive use of the memory for the receive buffer.

If there exist some control statements such as if statements in a loop and the data-flow is not precisely analyzed, it is enough for optimization at compile-time to find a relationship between a use points and a definition point in the loop. At run-time, the consistency of the data is guaranteed by turning the both sent and valid flags to false at the definition points and turning these flags to true at the use points.

If the valid and the sent flags are prepared for each element of the array, the send/receive status of the element can be presented accurately by these flags. However, this array of flags amounts to huge tables and the cost to define and use of element in the table may not be negligible.

To reduce the size of table, the buffer area and the flags are provided for a part of the array represented by a rectangular region. All elements in this rectangle must be defined and used in succession so that the definition points and the use points can be regarded as one definition point and one use point respectively.

Each dimension of the rectangular region in the array is represented by a tuple $[l : u : s : b]$ where l, u, s, b represent lower bound, upper bound, step size and block size respectively.

If the size of the rectangle maintained together is changed at run-time, the rectangle is divided into smaller rectangles and flags are provided for each rectangles.

4.2 Static Message Coalescing

The static message coalescing is applied if the definition point is located out of the loop. The code to receive data is moved out from the loop and the received values are saved in a temporary array variable. In the loop nest, the use of the data is replaced by the use of the temporary variable. If the data transferred can be vectorized, a temporary array variable is provided.

This code replacement technique can be applied when the data reference accompanied with communication can be separated into communication and reading of the value of the variable. Thus a temporary array variable for the data to be received can be provided at compile time. This leads to reduce the overhead of allocation and release of buffer area managed by the run-time library routines.

```
for i = 1, n do
  for j = 1, n do
    for k = 1, n do
      c(i,j) = c(i,j) + a(i,k) * b(k,j)
    endfor
  endfor
endfor
```

⇓ Insert send/recv codes

```
for l_i = 1, n/P do /* P: Number of PE */
  for j = 1, n do
    for k = 1, n do
      send b(k,j) to owner(a(i, j))
      recv tmp_b = b(k,j)
      c(l_i,j) = c(l_i,j) + a(l_i,k) * tmp_b
    endfor
  endfor
endfor
```

⇓ Move send/recv codes

```
for pe=1, P (pe ≠ p) do
  send b(n(p-1)/P+1..n(p-1)P,1..n) to pe
  recv tmp_b(n(pe-1)/P+1..n(pe-1),1..n) from pe
endfor
for l_i = 1, n/P do
  for j = 1, n do
    for k = 1, n do
      c(l_i,j) = c(l_i,j) + a(l_i,k) * tmp_b(k,j)
    endfor
  endfor
endfor
```

Fig. 5. Moving send/recv codes in the matrix multiply program

In the matrix multiply program shown in Fig. 5, all the arrays a, b, c are divided in the first dimension and distributed to PEs. Then elements of the array b excluding the locally located ones must be used accompanied with communication. Because all values of the array b are not updated in the loop nest, the communication code can be moved out from the loop.

5 Conclusion

In this paper, we have shown the technique for parallelizing compilers to reduce the redundant communication of data. The data-flow analysis to detect the redundant communication and the dynamic/static message coalescing technique to reuse the transferred data are described.

In the current implementation, the static message coalescing can be performed when there is no definition point in the loop nest. We got parallelized program of problem size 128 × 128 shown in Fig. 5 and measured the execution time of it on Fujitsu AP1000 multiprocessor. We achieved the 4 times speedup compared to the program parallelized without message coalescing.

In the future work, we will examine more detailed data-flow analysis across loop nests. In addition, we will evaluate the effectiveness of the techniques we proposed by incorporating them into the parallelizing compiler TINPAR.

Acknowledgments

We would like to thank Fujitsu Laboratory Ltd. for offering their AP1000. We also wish to thank all members in Tomita Laboratory of Kyoto University for their contribution through daily discussion. This work is supported in part by the Grant-in-Aid #00093007 from the Ministy of Education, Science, Sports and Culture.

References

1. High Performance Fortran Forum. High Performance Fortran language specification (ver. 2.0). Technical report, January 1997.
2. Seema Hiranandani, Ken Kennedy, and Chau-Wen Tseng. Evaluating compiler optimizations for fortran D. *J. Parallel and Distributed Computing*, 21:27 45, 1994.
3. Atsushi Kubota, Ikuo Miyoshi, Koji Maeyama, Shin ya Goto, Shin ichiro Mori, Hiroshi Nakashima, and Shinji Tomita. TINPAR: A parallelizing compiler for message-passing multiprocessors. In *Proc. of Int'l Symposium on Parallel and Distributed Supercomputing*, pages 214–223, September 1995.
4. Dror E. Maydan, Saman P. Amarasinghe, and Monica S. Lam. Array data-flow analysis and its use in array privatization. In *20'th Annual ACM Symposium on Principles of Programming Languages*, pages 2–15, January 1993.

An Automatic Vectorizing/Parallelizing Pascal Compiler V-Pascal Ver. 3

Tetsutaro UEHARA[1], Yoshitoshi KUNIEDA[2] and Takao TSUDA[3]

[1] Center for Information Science, Wakayama University, Japan
[2] Faculty of Systems Engineering, Wakayama University, Japan
[3] Faculty of Information Sciences, Hiroshima City University, Japan

Abstract. This paper descrives the design and implementation of the automatic vectorizing and paralellizing compiler named V-Pascal Version 3. The compiler is designed as a workbench on which various vectorizing and parallelizing techniques are evaluated. Now this compiler has the ability of vectorizing/parallelizing multiply-nested loops as reduced single loops, vectorizing **while**-loops and recursive calls, analyzing aliases caused by pointers, detecting dynamic data-structures such as linked-lists and so on. These special techniques can be applied not only for Pascal.

1 Motivation

With the recent progress in technology of semi-conductor devices and architectures, the performance of computers has been remarkably advanced. Now large-scale numerical computing can be processed at quite high speed by using supercomputers such as vector computers and parallel computers. Today numerical simulation using supercomputers is a fundamental tool for various fields of science and engineering. However, it is still not popular to develope the application programs in parallel programing languages for most programmers, therefore supercomputer manufacturers are forced to provide automatic vectorizing and parallelizing compilers (or *supercompilers*) which can generate suitable object codes from programs written in sequential languages.

Most of these compilers are for FORTRAN77 which has been commonly used for numerical applications. However, recently, block-structured languages, in particlar C, have become quite polular and the demand for supercompilers for such languages is increasing. Moreover, with the arrival of Fortran90, we can now utilize more versatile control and data structures like C — for example, while-loops, recursive calls and data structures constructed with pointers. Since these structures have been considered to be fairly helpful for writability and readability of programs, most of the application programs written in Fortran90 will contain these control/data structures in the future. Therefore automatic vectorization and parallelization techniques to deal with these structures have become a very important issue for supercompilers.

We have already proposed various vectorizing and parallelizing techniques for these versatile structures and some workbench is needed to verify the efficiency of the compilation methods. For this goal, we started to develop an automatic vectorizing and parallelizing compiler named "**V-Pascal**". This paper describes the design and implementation of the V-Pascal Version 3 compiler.

2 Overview

2.1 The target language

The target language chosen for our compiler is Pascal. Pascal is a typical block-structured language which is widely spread for general purpose. It is equipped with more versatile control/data structures than FORTRAN77, which has been the only language for which supercomputer manufacturers have provided useful vectorizing/parallelizing compilers. To extend the horizon of vector supercomputer usage, we were interested in vectorizing/parallelizing programs that manipulate data structures other than arrays. Even in the presence of control structures such as while-loops and recursive calls, the programs should be vectorized/parallelized; certainly the language Pascal is eligible for our purpose. Pascal has been used not only for educational purposes in classrooms but also for real-world problems, for example, the early TeXprocessor by D.E. Knuth, the vectorizing Fortran compiler construction by an American supercomputer manufacturer, and the implementation of an operating system for some Japanese minicomputers. Although the target language is Pascal, the various basic techniques that have been and will be developed for our compiler will also be useful for advanced vectorizing/parallelizing FORTRAN77 and Fortran90 compilers.

Since we have added no language extensions to the syntax of Pascal to express any parallelism, the target programs for V-Pascal are exactly written in traditional sequential manner. To realize "fully automatic" vectorization/parallelization for sequential programs, we didn't add any syntactic extension to our compiler to express parallelism. We only allowed to use compiler directive facilities to indicate data-dependence information which can never be known by the compiler.

There have been four versions of V-Pascal : Versions 1.0, 1.5, 2.0, and 3. The first three will be called "early" versions. The early versions mainly aimed at the study of advanced vectorization techniques, while Version 3, although dovetailing with the early versions in many aspects, is specifically meant for parallelization of Pascal programs for Japanese vector multiprocessors such as Hitachi's S-3800 and highly parallel machines like Fujitsu AP1000.

This paper describes the most recent version, V-Pascal Version 3. For the detail of the early versions, see [1] and [2].

3 Organization of the Compiler

Figure 1 gives the organization of V-Pascal Version 3 which is now being developed. It consists of three phases and the second phase can be also broken into

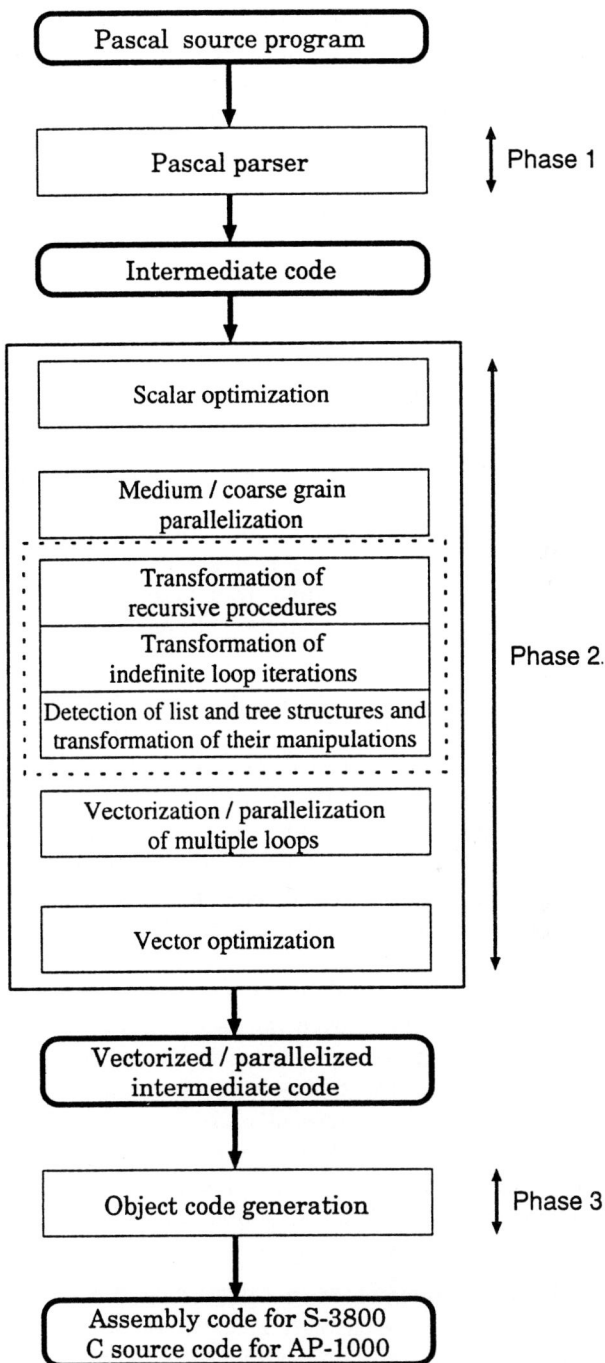

Fig. 1.: The organization of V-Pascal Version 3

small analyzing and optimizing modules.

The first phase, Phase 1, parses the source program and performs syntactic and semantic analysis. After the process, the program will be converted into intermediate code which is independent of the target language. It means that we can support other languages by constructing Phase 1 for them.

The second phase, Phase 2, is the optimization phase which receives sequential intermediate code from Phase 1 and outputs optimized, vectorized and parallelized intermediate code. This phase consists of a number of small optimization modules which are the implementations of vectorization/parallelization methods we have proposed. All of the modules has an uniform input/output interface; they work as filter programs to receive and output intermediate code which has the unified format. This allows us to reconfigure the compiler and to add new optimizing facilities easily.

The last phase, Phase 3, converts machine-independent intermediate code to machine-dependent execution code. Machine-dependent optimization is also performed in this phase. Now two versions of Phase 3 are operational, one of which generates vectorized/parallelized assembly code for S-3800 and the other yields parallelized C source code for AP1000.

As described in above, intermediate code is used as a unified input/output data format over all phases and modules in Phase 2. The design goal of intermediate code is as follows.

- Scalar instructions are simple instructions having load-store RISC-style architecture with unlimited number of registers. Vector instructions are also simple instructions of a typical Cray-style vector processor which has unlimited number of vector registers with infinite length. The vector-macro operations are also defined to cover all of vector-macro instruction sets of supercomputers which are provided by domestic manufacturers.
- Parallel execution primitives are defined in general terms for tightly coupled vector multiprocessors. Differences between target machines are absorbed by run-time libraries.
- Control structures are carefully designed. In addition to **for**-loops, **forever**-loops with exit instructions from the middle of the loops are defined to express **while**-loops, **repeat-until**-loops and loops implicitly constructed by **if-goto-statements**. Although these structures can also be expressed implicitly by **goto**-statement of intermediate code, we added these control structures to direct which part should be vectorized/parallelized in the intermediate code. Special branch nodes are defined so that multiple alternatively optimized codes for the same instruction sequence can be retained until the choice can be made later in Phase 3.

V-Pascal Ver.3 is implemented in C++ while the early versions are written in Pascal. The main reason of this changeover is that we found that the object-oriented approach is fairly helpful for us to handle quite complicated data structures used in compilers. In particular, many graph-shaped data structures are used in our compiler and we had to write a sequence of pointer manipulation statements to handle them in Pascal, but using classes of C++, we can neatly express the sequence in a simple manner. The other reasons to use C++ are that we wanted to use various utilities to help the development, such as compiler-compilers (lex and yacc) and useful source-code-level debugger (gdb) which cannot be found for Pascal.

4 Facilities of Analysis

Each of the modules in Phase 2 calls various common analyzing modules to perform optimization, vectorization and parallelization. This section describes the analyzing facilities with which V-Pascal Version 3 is equipped.

4.1 Alias Analysis

When two different variables refer to the same location of the memory, each of them is called an alias of the other and they are called an alias pair. In cases alias pairs may exist, an assignment of a variable may change value of the other variable and it causes misjudgments in data-flow analysis. Therefore, for precise data-flow analysis, we have to determine all possible alias pairs. In other words, precise alias analysis is indispensable before data-flow analysis.

```
procedure F(var a,b : integer);
   begin
      ...
   end;
begin
   ...
   { (a,b) is an alias pair }
   F(c,c);
   ...
end;
```

```
var p,q : ↑integer;
begin
   ...
   new(p);
   { (p↑,q↑) is an alias pair }
   q := p;
   ...
end;
```

Fig. 2.: An example of alias pairs caused by call-by-reference parameters

Fig. 3.: An example of alias pairs caused by pointer references

In Pascal, there are two cases which cause aliases. One of them is the case caused by call-by-reference parameters. When a function or procedure has two or more **var** arguments and actual parameters for them are the same, these parameters are an alias pair, as shown in Figure 2 for example. There are many methods to analyze this kind of aliases, and Cooper[3] proposed an algorithm for precise analysis of this problem. We implemented the algorithm on our compiler.

The other case, in which an alias occurs in Pascal, is by pointer references. When two of pointer variables have the same content as shown in Figure 3, they point to the same location of the memory and that causes aliases. Since the values of these pointer variables cannot be estimated in most cases, there have been few methods to overcome this problem and almost all existing compilers give up analysis and assume that all possible pairs of such variables may be aliases. We have proposed a method to detect data structures such as linear linked lists and trees at compile time and through this process we can eliminate some possibilities of existence of pointer-aliases precisely. The details of the algorithm are described in [4]. After the analysis, we can also detect pairs of pointer variables which always refer to different data structure (for example, as R and S in Figure 4), and which cannot be an alias pair.

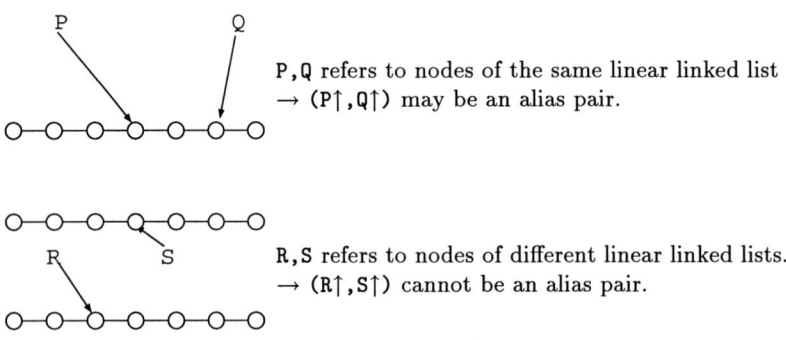

Fig. 4.: An example of the detection of aliases of pointer variables

4.2 Control/Data-flow Analysis

Control and data-flow analysis in V-Pascal Version 3 is performed by a traditional method, but the result is more precise than others because of the existence of advanced alias analysis facilities.

The results are represented in the **D-matrix**[1]. It takes a hierarchical structure so that dependences between procedures/functions, those between loops plus basic blocks, and those between intermediate code statements are represented by the D-matrix. In other words, the parallelism of various granularities are designated by the same data structure. An example is shown in Figure 5.

4.3 Value-Range Estimation of Integer Variables

It is a well-known problem that dependence analysis of arrays is obstructed with the existence of an array subscript expression which contains some variables whose values cannot be determined statically. Therefore it is important to estimate the value range of an array subscript expression at each step of execution to raise the compiler's power for dependence analysis for vectorization and parallelization. This information acquisition is done with data-flow analysis, and the relevant information is appended to integer variables and temporaries of the intermediate-code statements. The detail will be described by a paper in preparation.

```
procedure P;
    procedure Q;
    begin
        A;
        { block B}
        for ... do
    begin
            S₁;
            S₂;
            S₃;
        end;
        { block B end}
        C;
    end;
    begin
        Q;
end;
```

(a) A sample program.

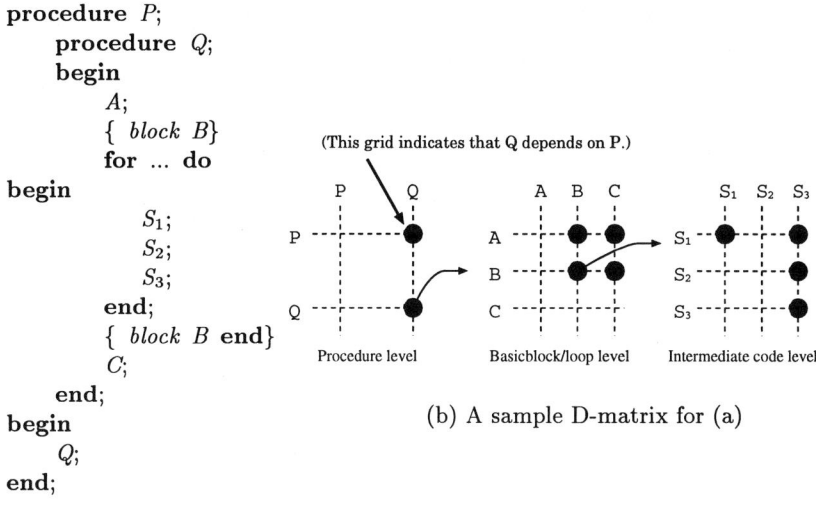

(b) A sample D-matrix for (a)

Fig. 5.: An example of D-Matrix

4.4 Dependence Analysis of Array Subscripts

Dependence analysis of array subscripts is essential for automatic vectorization and parallelization. There have been many proposals of methods to perform this analysis precisely and efficiently; one of the most famous method is the Omega test by W.Pugh[5]. We also have an original approach for this analysis which is implemented on our compiler[6]. Our method can analyze nested loops accurately, and in most cases, in polynomial time of the number of variables. When loop bounds and array subscripts are linear expressions of the surrounding loop control variables, our method is exact, i.e., the power of analysis is as same as that of Omega-test. When loops have symbolics, our method tries to find the range of the symbolics that may cause dependences preventing vectorization or parallelization, as the result compilers can produce conditions to judge the possibility of vector or parallel execution at the execution time. We also proposed a method to analyze loops with loop bounds and array subscripts which contain non-linear expressions.

The outline of our method is as follows. The dependence analysis of arrays is equivalent to the problem of obtaining integer solutions of simultaneous Diophantine equations and inequalities. We can easily check the existence of real solutions by linear programming which takes only the polynomial time of the number of the variables, and after that we obtain a convex hull which expresses a space of the real solution and may contain integer solutions. Then we perform exhaustive search of an integer solution around each apex of the convex. In the worst cases, this process takes exponential time in the number of variables. However, we think that it is not a disadvantage of our method. The reason is shown using Figure 6. This is a typical solution space where both integer solutions and real solutions exist. As shown in the figure, in most cases when a real solution exists, integer solutions also exist, and at least one of them is expected to be near an apex of the convex (P in the figure), or the apex is an integer solution itself (Q). These solutions will be detected in short time of which the order is about the polynomial time of the number of the apexes. Moreover, even when no integer solutions exist while the real solution exists, which is a rare case, we can also expect that the convex is small enough to check all possible integer solutions quickly.

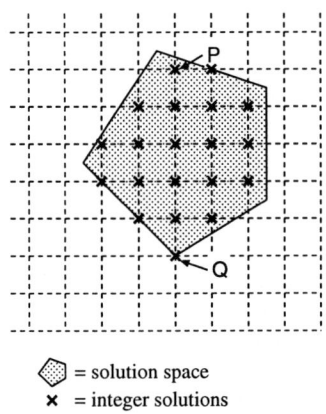

= solution space
× = integer solutions

Fig. 6.: An example of solution space

5 Vectorization/Parallelization Facilities

As mentioned before, Phase 2 of V-Pascal Version 3 consists of many small optimizing modules which realize our proposed parallelizing/vectorizing methods. This section describes some of the methods we proposed and implemented.

5.1 Vectorization/Parallelization of Multiply-nested Loops

Most of the compilers offered by supercomputer manufacturers can only vectorize inner-most loops of multiply-nested loops, and parallelize the outer loops. However we have already proposed a method to reduce multiply-nested **for**-loops into equivalent single loops which are then vectorized by extensive use of vector indirect addressing [1]. An example is shown in Figure 7. As we have shown in [7], recent high-speed vector supercomputers tend to have large N_{half}[8] values, therefore the reduction is efficient to make vectors longer and obtain higher performance.

```
            for i:=1 to 5
              for j:=1 to i
                ..a[i,j]..
```

⇓reduce

```
vi[] := {1,2,2,3,3,3,4,4,4,4,5,5,5,5,5};
vj[] := {1,1,2,1,2,3,1,2,3,4,1,2,3,4,5};
for k:=1 to 15
  ..a[ vi[k], vj[k] ]..
```

Fig. 7.: Reduction of a multiply-nested loop into a single loop

5.2 Vectorization of While-Loops

While traditional FORTRAN 77 has only DO-loops to express loop structures, Pascal has not only **for**-loops but also **while**-loops and **repeat-until**-loops. Since we cannot determine the number of iteration for **while**-loops and **repeat-until**-loops (therefore call them *indefinite loops*) even at the entrance of the loop in execution — in other words, we cannot determine the vector length even in execution time — we have to make some tricks to vectorize these loops. Wolfe proposed two approaches to cope with this problem : one is called loop distribution and the other is strip-mining [9]. We have proposed a hybrid method of them to treat these loops more neatly[10] and obtained high speedups for more generalized cases. Wolfe also discussed how to vectorize nested loops of which the inner loop is an indefinite loop and the outer loops is a **for**-loop, by interchanging these loops and vectorizing only the **for** loop, but this method is not efficient when the vector lengths are too short. We also proposed a method to cope with this problem interchanging these loops dynamically to keep the vector length long enough [11].

5.3 Vectorization and Parallelization of Recursive Procedures

Generally, recursive procedures (and functions) can be reduced into non-recursive procedures by simulating stack operations explicitly in the source code. In most cases the obtained procedures consist of several loops where the first-order recurrences along each branch of call trees of the original procedure dominate. When the loops can be simplified not to contain control statements except the exit, some of them can be vectorized by vector-macro operations, and when the vector length is estimated to be long enough, a fast vector algorithm [12] can be used. We call this approach the *depth-first method*. However, in many cases this method cannot be applied because of the complexity of the loops.

The second approach is called the *breadth-first method*. The basic idea is as follows; when a recursive procedure contains two or more recursive calls and it

function $F(arg: arg\text{-}typ): retval\text{-}typ;$
begin
    S_{head}; { Obtain $arg_1 \cdots arg_n$ }
    if P_1 **then** $retval_1 := F(arg_1);$
    if P_2 **then** $retval_2 := F(arg_2);$
    if P_3 **then** $retval_3 := F(arg_3);$
                     $\vdots$
    if P_n **then** $retval_n := F(arg_n);$
    S_{tail} { Obtain the return value }
end;

Fig. 8.: A normalized recursive procedure for breadth-first method.

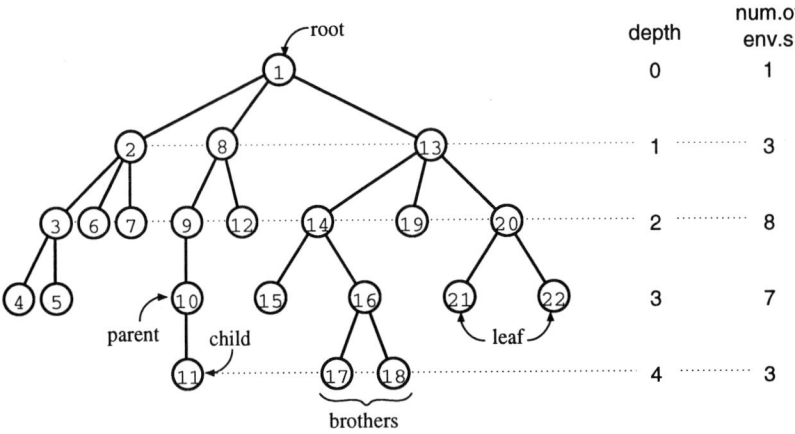

Fig. 9.: A sample of an environment tree.

can be normalized in the form as shown in Figure 8, we try to execute the recursive calls $F(arg_1) \cdots F(arg_n)$ in parallel. This is called *breadth-first execution* of a recursive procedure. We have introduced a concept of "environment" analysis to find recursive procedures to which this method can be applied. Environments are dynamically generated instances of the procedure for each recursive calls, and the relation between environments can be expressed in a form of a tree like a call-tree (Figure 9). When environments at the same depth of the tree are always independent to each other, these environments can be executed in parallel and the breadth-first method can be applied. To determine the independence of environments is quite easy when the procedure contains no writing statements for global variables; when none of $arg_1 \cdots arg_n$ and $P_1 \cdots P_n$ depends on $retval_1 \cdots retval_n$ in Figure 8, environments at the same depth are independent. However when the procedure performs assignments to global variables, this analysis is difficult and we leave it for the future work.

6 Conclusion

An overview was given of the current status of the V-Pascal Version 3 compiler. In the V-Pascal Version 3 compiler, special techniques have been developed even for Algol-like features, so that they may also be useful for the full-fledged vectorizing/parallelizing compilers for the languages Fortran 90 and C. Although the initial target machine has been a tightly coupled vector multiprocessor like Hitachi's S-3800, a slight reconfiguration of compiler components allows parallelization for distributed-memory parallel computers. To demonstrate this, we have provided a compiler called **V-Pascal/DM** [13] for Fujitsu AP1000, to which the compiler output is fed in the form of parallelized C source code. V-Pascal/DM performs automatic parallelization based on an SPMD model.

This compiler is still under developement and we are now working to complete it and show the effectiveness of the vectorization and parallelization methods we have proposed.

Acknowledgements

Developing a native compiler from a scratch is exciting but hard job. This work would never be accomplished without the great works and effors by the following people, which are gratefully acknowledged here : Kenji Umeda, Junji Sakai, Toji Okayama, Kenji Fujita, Shinya Nakagawa, Ikuo Miyoshi, Hideki Kagemoto, Hitoshi Murai, Masayuki Yamamoto, Masakazu Nomoto and Hiroyuki Wada. Thanks are also due to Yasuo Okabe (Data processing center, Kyoto Univ.) for helpful discussions and comments.

References

1. T. Tsuda and Y. Kunieda, "V-Pascal: An Automatic Vectorizing Compiler for Pascal with No Language Extensions," *Journal of Supercomputing*, 4: 251-275, 1990.
2. T. Tsuda, "Design and Implementation of a Vectorizing Compiler for the Block-Structured Language Pascal," *Supercomputer*, 46 (VIII-6): 12-21, 1991.
3. Cooper, K.D. *Analyzing Aliases of Reference Formal Parameters*, Conf. Rec. Twelfth ACM Symposium on Principles of Programming Languages(Jan.), pp. 281-290. 1985.
4. A. Matsumoto, Han D.S., and T. Tsuda, "Alias Analysis of Pointers in Pascal and Fortran 90, Part I. Dependence Analysis between Pointer References," in press for *Acta Informatica*, 1995.
5. Pugh, W., "The Omega Test : a fast and practical integer programming algorithm for dependence analysis", *Proceedings of Supercomputing '91*, pp.4-13.
6. I. Mizunuma, T. Uehara, Y. Okabe, Y. Kunieda and T. Tsuda, "Data-Dependence Analysis of Nested Loops Containing Symbolics and Nonlinear Expressions," *Proc. of the 9th National Convention of Japan Society for Software Science and Technology* (Fujisawa, Sept. 1992), pp. 485-488, 1992 (in Japanese).

7. Uehara, T. and Tsuda, T., *Benchmarking Vector Indirect Load/Store Instructions*, Workshop on Benchmarking and Performance Evalutation in High Performance Computing (Tokyo, Japan), pp.16–25, 1993.
8. Hockney, R. W. and Jesshope, C. R., "Parallel Computers 2", *Adam Hilger*, 1988.
9. M.Wolfe, "Optimizing Supercompilers for Supercomputers", *MIT Press*, 1989.
10. K. Suehiro and T. Tsuda, "Automatic Vectorization/Parallelization of WHILE Loops," *Proc. 45th Annual Convention IPS Japan* (Tokushima, Nov. 1992), pp. 5-51 & 52, 1992 (in Japanese).
11. H.Murai, K.Suehiro, Y.Okabe, K.Kunieda, T.Tsuda, "Vectorizing while loops by loop interchange", *Proc. of the 11th National Convention of Japan Society for Software Science and Technology* (Osaka, Mar. 1994), pp.65–68, 1994 (in Japanese).
12. M. Nakamura and T. Tsuda, "A Fast Algorithm for First Order Recurrences on Vector Supercomputers", *Transactios of Information Processing Society of Japan*, Vol.36, No.3, pp. 669–680, 1995 (in Japanese).
13. K. Umeda, T. Uehara and T. Tsuda, "An Automatic Parallelizing Compiler for Distributed Memory Parallel Computer V-Pascal/DM," *Proc. 48th Annual Convention IPS Japan* (Kashiwa, March 1994), paper 5G-1, 1994 (in Japanese).

An Algorithm for Automatic Detection of Loop Indices for Communication Overlapping

Kazuaki Ishizaki, Hideaki Komatsu, and Toshio Nakatani

IBM Tokyo Research Laboratory
1623-14 Shimotsuruma, Yamato, Kanagawa 242, Japan
TEL ++81-462-73-4664, FAX ++81-462-74-4282
Email: {ishizaki, komatsu, nakatani}@trl.ibm.co.jp

Abstract

This paper presents a compiler algorithm that automatically detects the appropriate loop indices of a given nested loop and applies loop interchange and tiling in order to overlap communication with computation. It also describes method of generating communication for the tiled loop on distributed memory machines. The algorithm presented here has been implemented in our High Performance Fortran (HPF) compiler, and experimental results have shown its effectiveness on the RISC System/6000 Scalable POWERparallel System.

1 Introduction

On distributed memory machines, parallelizing compilers [1,2,3,4], such as the one for High Performance Fortran (HPF) [5], make it possible to write programs in the global memory address without dealing with the details of interprocessor communication. To obtain satisfactory performance, it is important to apply message vectorization. This combines a series of short messages in a loop into a single long message, to reduce the startup overhead of communication. It is the key optimization feature in parallelizing compilers for distributed memory machines [6].

When message vectorization is applied, the execution time for the loop comprises the pre-loop communication time, the computation time for the loop, and the post-loop communication time. The execution time decreases by overlapping the communication with the computation. Many researches to reduce communication time by overlapping communication with computation are proposed. In previous methods [13,15], the programmer had to write non-blocking send/receive [7] or PUT/GET directly by hand in order to use the optimization of overlapping communication with computation, as we will explain in a later section. Various methods of executing a loop in a tiled wavefront method have been proposed [8,9] for situations in which the loop contains true dependence. These methods apply tiling to the loop in order to overlap communication with computation and reduce the synchronization cost. The cited papers discuss methods for choosing the best tile size according to the characteristic parameters of the machine; however, they do not discuss methods for detecting the appropriate loop indices to apply tiling to automatically.

This paper presents an algorithm that automatically detects the appropriate loop indices of a given nested loop and applies loop interchange [10] and tiling [10,11] in order to overlap communication with computation. It also describes a method of generating communication for the tiled loop on a distributed memory machine. The algorithm presented here has been implemented in our HPF compiler, and experimental results have shown its effectiveness on the RISC System/6000 SP [12].

Traditionally, for a uniprocessor, tiling is used to improve the data locality by changing the order of data accesses in a loop so that reuse with the same elements occurs in the innermost loop [19]. For the tiling of distributed memory machines, our algorithm applies loop interchange and tiling to a nested loop so that communication by non-local data accesses without reuse occurs in the outermost loop. When only reuse with non-local accesses does not occur in a loop index, our algorithm applies tiling to that loop index in order to prevent redundant communication.

The structure of this paper is as follows. In Section 2, we summarize related work. In Section 3, we introduce definitions of some terms. In Section 4, we present the concept behind our algorithm. In Section 5, we describe our loop transformation algorithm for tiling loops in order to overlap communication with computation. In Section 6, we describe a method of generating communication. In Section 7, we give the performance results that we obtained in our experiments, and discuss the bottlenecks of the current system. Finally, in Section 8, we outline our conclusions.

2 Related Work

We summarize related work in the two categories. One is the overlapping communication with computation. The other category is the loop transformations for distributed memory machines.

2.1 Overlapping communication with computation

There are five areas of previous work related to improving the performance by overlapping communication with computation:

1. Overlapping communication with computation by using pipeline communication in loops where there is true dependence [8,9]; these loops are tiled by a compiler. The cited papers discuss methods for choosing the best tile size.
2. Overlapping the communication for the next computation with the current computation in loops that require prefetch communication [13]. The cited papers discuss the performance of programs optimized by the programmer.
3. Overlapping communication with computation by transforming the original loops with stencil communication into loops with two separate computation processes, one with non-local data and the other with only local data [2,14]. These papers discuss methods for generating code in the limited cases of FORALL statements or one-dimensional distributed arrays.
4. Overlapping the communication for selection of the next pivot with computation of matrix elimination in Gaussian Elimination [15]. The paper discusses a manual optimization for shortening the critical path of the computation.
5. Hiding the communication time and eliminating redundant communication by scheduling sends as early as possible and receives as late as possible [16]. The cited paper discusses a compiler framework for eliminating redundant communications, but it does not give any experimental results.

In areas 1, 2, 3 and 4, the cites papers do not discuss loop transformation algorithms that can be implemented on an automatic parallelization compiler. In areas 1 and 2, our algorithm can apply loop transformations to loops in order to allow a compiler to overlap communication with computation automatically. The nested loop that is transformed by our algorithm is the same as that in previous methods [8,9, 13]. The cited papers do not discuss any automatic loop transformation algorithms, such as the one implemented in our compiler. Our algorithm is robust enough to cover a broad area of loop parallelization. It can apply to the loops that include prefetch and pipeline communication. Thus our algorithm makes it possible for communication to be overlapped with computation in languages that do not permit communication to be specified explicitly, such as HPF.

2.2 Loop transformations

Much research [17,18] has been done on the use of automatic loop transformations to improve the performance of the nested loops on distributed memory machines. These cited papers focus on loop transformations and automatic data alignments to minimize the amount of communication, whereas our algorithm focuses on reducing the communication time by overlapping communication with computation, and can be applied to nested loops transformed by other algorithms.

3 Background

In this section, we introduce the loop representation, data dependence vectors, and tiling in order to explain our algorithm.

3.1 Loop representation

To simplify the discussion, we discuss only perfectly a nested loop in this paper.

Definition 1: We represent a perfectly nested loop as follows. Here, l and r are m-dimensional arrays in an n-nested loop. Each iteration in the loop is identified by a column vector $\vec{i} = (i_1,..., i_n)$. We call this vector a loop index vector, and each element a loop index. Here, i_i is the value of the ith loop index, counting from the outermost to the innermost loop. The lower bound of the loop is denoted by $\vec{1} = (l_1,..., l_n)$, and the upper bound of the loop is denoted by $\vec{u} = (u_1,..., u_n)$. The subscript function $f(\vec{i}) = H\vec{i} + \vec{a} = \vec{j}$ maps the loop index vector $\vec{i}$ to the array index vector $\vec{j} = (j_1,..., j_m)$. All elemental values of a matrix H that is a linear transformation and a vector $\vec{a}$ are integers.

Definition 2: The distribution function $g(\vec{j}) = \vec{p}$ maps the array index vector $\vec{j}$ to the processor index vector $\vec{p} = (p_1,..., p_k)$, where k is the rank of the processor configuration. Let P be a set of processor index vectors for executing the program.

In Fig. 1, n is three, $\vec{l}$ is $(1,1,1)$, $\vec{u}$ is $(8,8,8)$, k is one, and P is $\{(0),(1)\}$. The rank m of the array A is two. The mapping function of a reference A(I, J) is $f(\vec{i}) = \begin{pmatrix} 0 & 1 & 0 \\ 1 & 0 & 0 \end{pmatrix} \vec{i}$. The distribution function of the array A is , $g(\vec{j}) = \left(\left\lfloor \frac{j_2 - 1}{4} \right\rfloor \right)$.

```
        REAL A(8,8), B(8,8), C(8,8)
*HPF$ PROCESSORS P(2)
*HPF$ DISTRIBUTE (*,BLOCK) onto P :: A, B, C
        DO 10 J = 1, 8
          DO 10 I = 1, 8
            DO 10 K = 1, 8
10            A(I,J)=A(I,J)+B(I,K)*C(K,J)
```

Fig. 1: Example of a program

3.2 Dependence vector

A dependence vector shows possible execution orders constrained by their data dependence. Newly, we distinguish between true and anti data dependences to generate prefetch and pipeline communication. In this paper, an anti dependence vector $\vec{d}_a$ shows that an execution order is constrained by anti dependence. A true dependence vector $\vec{d}_t$ shows that an execution order is constrained by true dependence [20].

Definition 3: D, which is a set consisting of $\vec{d}_a$ and $\vec{d}_t$, is defined as follows:

$$D = D_a \cup D_t, D_a = \{\vec{d}_a \mid \vec{d}_a = \vec{i}' - \vec{i}, \text{anti dep. from } \vec{i} \text{ to } \vec{i}'\}, D_t = \{\vec{d}_t \mid \vec{d}_t = \vec{i}' - \vec{i}, \text{true dep. from } \vec{i} \text{ to } \vec{i}'\}$$

$$\vec{d} = (d_1, ..., d_n), d_i = \left[d_i^{\min}, d_i^{\max}\right], d_i^{\min} \in Z \cup -\infty, d_i^{\max} \in Z \cup \infty$$

3.3 Tiling

In general, tiling maps an n-nested loop to a $n+t$-nested loop by adding t inner loops for a fixed number of iterations. Fig. 2 shows the code after tiling of both the loop index variables I and K in the example shown in Fig. 1, using a tile size of 4 x 4. On a uniprocessor, tiling is used to improve the data locality; this reduces the number of iterations between accesses of the same data, which allows the same data to be kept in the data cache, and hence reduces the number of memory accesses.

```
        REAL A(8,8), B(8,8), C(8,8)
        DO 10 I = 1, 8, 4
          DO 10 K = 1, 8, 4
            DO 10 J = 1, 8
              DO 10 KK = K, MIN(K+3, 8)
                DO 10 II = I, MIN(I+3, 8)
10                A(II,J)=A(II,J)+B(II,KK)*C(KK,J)
```

Fig. 2: Example of tiling for a uniprocessor

4 Concept of the algorithm

In this section, we introduce the definition of reuse generally. We then discuss the usage of reuse information for distributed memory machines.

4.1 Types of Reuse

Reuse [19] occurs when the same data is read or written more than once in the loop. A temporal reuse occurs when two references access the same memory location. A spatial reuse occurs when two references access nearby memory locations, such as within the same cache line. A self reuse occurs when a reference in different iterations accesses the same memory location. A group reuse occurs when different references access the same memory locations.

The following terms [19] are used for the above types of reuse. Self-temporal reuse is a reference that accesses the same memory location in different iterations. Self-spatial reuse is a reference that accesses the same cache line in different iterations. Group-temporal reuse is more than one references that access the same memory location. Group-spatial reuse is more than one references that access the same cache line.

4.2 Temporal Reuse

On distributed memory machines, A(I) and A(I+1), which are consecutive in the view of the programming language, may not be assigned consecutive physical memory locations; this depends on the array distribution. Since it is hard to discuss spatial reuse generally in considering the array distribution, we therefore discuss only temporal reuse.

4.2.1 Self-Temporal Reuse

First, we consider self-temporal reuse. Self-temporal reuse occurs if two iterations $\vec{i}_1$ and $\vec{i}_2$ reference the same data element whenever $H\vec{i}_1 + \vec{a} = H\vec{i}_2 + \vec{a}$, that is, when $H(\vec{i}_1 - \vec{i}_2) = 0$. The solution of this equation is $\ker H$, which is called the self-temporal reuse vector R_{ST}.

On a uniprocessor, choosing some index from the zero-value of $\ker H$ as the innermost loop index allows a reference to exploit self-temporal reuse, because reuse occurs when the value of the loop index variable (loop index variables K for the array B in Fig. 2) that does not appear in the subscripts is changed. It then applies tiling to these indices, and moves the tiled loops to the innermost position so that the processor can read the same elements from the data cache many times in the innermost loop. The key to increasing the performance is to improve the data locality in an innermost loop.

On the other hand, we here discuss the case that this method is used on distributed memory machines. If reuse with non-local data access occurs in an innermost loop, the performance will be greatly decreased. This is because many communications are normally required for the same array element to occur in an innermost loop. In Fig. 2, when communication is performed just before the tiled loop (DO 10 KK=...), communication for accessing the same region of array B is required more than once; this is inefficient. The key to increasing the performance is to vectorize communication for non-local data accesses in outermost loops as much as possible. When tiling is applied to the nested loop, we must ensure that a compiler does not generate redundant communications for accessing the same array elements by reuse.

4.2.2 Group-Temporal Reuse

We discuss references that have the same affine subscript expressions $f_1(\vec{i}) = H\vec{i} + \vec{a}_1$ and $f_2(\vec{i}) = H\vec{i} + \vec{a}_2$. There is group-temporal reuse between two such references with distance $\vec{r}$ if there are iterations $\vec{i}_1$ and $\vec{i}_2$ such that $H\vec{i}_1 + \vec{a}_1 = H\vec{i}_2 + \vec{a}_2$, that is, $H(\vec{i}_1 - \vec{i}_2) = H\vec{r} = \vec{a}_2 - \vec{a}_1$. To determine whether such an $\vec{r}$ exists, we solve the system of equations to obtain a particular solution $\vec{r}_p$. The general solution, $\ker H + \vec{r}_p$, is called the group-temporal reuse vector R_{GT}. On a uniprocessor, this reuse reduces the number of memory references.

In order to exploit this reuse with non-local data access, we say that the compiler has to support message coalescing on distributed memory machines. Therefore, the group-temporal reuse information reduces the number of communications.

4.3 Usage of Reuse Information

Our algorithm applies loop interchange to move the loop indices to positions further out than those in which reuse occurs, because this ensures that the compiler does not generate redundant communication with the same array elements by reuse. It then applies tiling to the loop indices in which reuse does not occur in order to overlap communication with computation, tiling only the loop indices that require communication. We therefore specify **Condition 1**.

Condition 1: Tiling is applied to loop indices that require communication and in which neither self-temporal nor group-temporal reuse occurs. These are the indices in which either the self-temporal or the group-temporal reuse vector has a value of zero.

5 Algorithm for loop transformations

In this section, we present an algorithm for loop transformations such as tiling and loop interchange in order to overlap communication with computation.

The algorithm is structured in the following five steps:
1. The In Set [2] is calculated as a set of array index vectors that require interprocessor communication by non-local data accesses.
2. The communication vector [20] is calculated as a vector that indicates on the basis of the loop index whether interprocessor communication is required.
3. A check is carried out to determine whether interprocessor communication can be vectorized.
4. The loop indices for tiling are found by using the reuse vector.
5. Loops are interchanged in order to move these loops indices to outer nested loops, and tiling is done to generate loops for overlapping communication with computation.

Fig. 3 shows the algorithm for finding the tiled loop indices (steps 1-4). In the result of the algorithm, let the tiling vector $\vec{t}$ be a vector showing the tiled loop indices. The indices corresponding to non-zero elements of $\vec{t}$ can be applied tiling to. The variable *top* shows that the outermost loop index can be a target for loop interchange.

We will now explain the algorithm that applies loop transformations to overlap communication with computation, using the program in Fig. 1 as an example.

Example:
The input parameters to the algorithm in Fig. 3 are as follows:

$n = 3, \vec{l} = (1,1,1), \vec{u} = (8,8,8), P = \{(0), (1)\}, l = \{A(I, J)\}, R = \{A(I, J), B(I, K), C(K, J)\}, D = \emptyset$

5.1 Calculating the In Set (step 1)

First, for a loop to be executed in parallel, its iteration space is partitioned and each iteration sub-space is allocated to a single processor. Our algorithm can use any iteration partitioning algorithm to calculate the iteration sub-space. In this paper, we use the owner-computes rule [8]. The elements of the left-hand side array l are determined by the original iteration space $(\vec{l}:\vec{u})$ and the subscript expression $f_l(\vec{i})$ of the array l.

Definition 4: Let the Local Iteration Set $Q(\vec{p})$ be a set of loop index vectors that access array elements on the processor $\vec{p}$.

The compiler then calculates a set of array references that causes non-local data access when Local Iteration Set is executed. The accessed elements of the right-hand side array r are determined by the Local Iteration Set $Q(\vec{p})$ and the subscript expression $f_r(\vec{i})$.

Definition 5: Let the In Set $Y(\vec{p})$ be a set of array index vectors that cause non-local data access when each loop index of the Local Iteration Set $Q(\vec{p})$ is executed on the processor $\vec{p}$.

Example:

In Fig. 1, we assume $l = \{A(I, J)\}$ as the left-hand side operand. The mapping functions f_l and the distribution function g_l for the operand A(I, J) and the Local Iteration Set Q are as follows:

$$f_l(\vec{i}) = \begin{pmatrix} 0 & 1 & 0 \\ 1 & 0 & 0 \end{pmatrix} \vec{i}, \quad g_l(\vec{j}) = \left(\left\lfloor \frac{j_2 - 1}{4} \right\rfloor \right), \quad Q((0)) = (1{:}4, 1{:}8, 1{:}8), Q((1)) = (5{:}8, 1{:}8, 1{:}8)$$

In Sets are calculated for all right-hand side operands in the loop. In this paper, the results of the operands without communication are omitted, because they do not affect the results of loop transformations. We show the result of array B that requires communication. In Fig. 3, we assume $r = R = \{B(I, K)\}$ as the right-hand side operand. Mapping functions f_r and distribution function g_r for the operand B(I, K) and the In Set Y_r are as follows:

$$f_r(\vec{i}) = \begin{pmatrix} 0 & 1 & 0 \\ 0 & 0 & 1 \end{pmatrix} \vec{i}, \quad g_r(\vec{j}) = \left(\left\lfloor \frac{j_2 - 1}{4} \right\rfloor \right), \quad Y_r((0)) = (1{:}8, 5{:}8), Y_r((1)) = (1{:}8, 1{:}4)$$

5.2 Generating a communication vector (step 2)

We now present a method of generating a communication vector to represent communication information based on the loop index. Because, we want to calculate communication information and other information based on the loop index, such as the data dependence vector and the reuse vector. In the kth index of the In Set, a reference to different non-local elements on each processor indicates that interprocessor communication occurs on the kth array index.

According to **Definition 1**, H_r in $f_r(\vec{i})$ maps the loop index vector to the array index vector. For mapping communication information in the array index vector to it in the loop index vector, the algorithm uses the transpose matrix of H_r as a mapping function.

If the depth n of the nested loop is smaller than the rank m of the array r, the communication information is reduced when it is mapped to the loop index. This is because the matrix does not completely project an array index vector into a loop index vector. Practically speaking, if the ith value of $\ker H_r$ is non-zero, it shows that the ith loop index variable does not appear in the subscripts. If the index variable for the ith nested loop does not appear in the subscripts of the right-hand side array, the compiler has to add the information to the ith index conservatively in order to show that communication occurs. The result vector is defined as follows:

Definition 6: The communication vector $\vec{b} = (b_1,...,b_n)$, b_i = non-zero or zero, is defined as a vector in which each elemental value represents whether processors require communication, on the basis of each loop index.

Example:

In Fig. 3, each processor reads a different range of only the second dimension of the In Set. As a result, the In Set shows that communication is required in the second dimension of array B. The compiler then calculates the loop index to access the second dimension of array B. It is calculated from the following expression. We use the vectors $\vec{e}_1$, $\vec{e}_2$, and $\vec{e}_3$ to represent (1, 0, 0), (0, 1, 0) and (0, 0, 1), respectively.

$$\vec{b}_r = \begin{pmatrix} 0 & 1 & 0 \\ 0 & 0 & 1 \end{pmatrix}^T \begin{pmatrix} 0 \\ 1 \end{pmatrix} = span\{\vec{e}_3\}.$$

The result shows that the loop index variable K accesses the second dimension of array B. The compiler adds the vector $span\{\vec{e}_1\}$ that communication requires to the loop index whose variable J, does not appear in the subscript B(I, K). Here, the communication vector is $\vec{b}_r = span\{\vec{e}_1, \vec{e}_3\}$.

5.3 Checking whether communication can be vectorized (step 3)

We define a vector to determine an appropriate mode of communication:

Definition 7: Let the communication dependence vector $\vec{c}$ be a vector in which the value of each element represents whether communication occurs, as a result of anti dependence $\vec{d}_a$ or true dependence $\vec{d}_t$ for the operand. The communication dependence vector consists of a true communication dependence vector $\vec{c}_t$ and an anti communication dependene vector $\vec{c}_a$. We derive them as follows:

$$\vec{c}_t = (c_{t1}, c_{t2}, ..., c_{tn}), c_{ti} = \begin{cases} d_{ti} & (\text{if } b_i \neq 0) \\ 0 & (\text{if } b_i = 0) \end{cases}, \vec{c}_a = (c_{a1}, c_{a2}, ..., c_{an}), c_{ai} = \begin{cases} d_{ai} & (\text{if } b_i \neq 0) \\ 0 & (\text{if } b_i = 0) \end{cases}$$

Then, the compiler can apply message vectorization if **Condition 2** is satisfied [20].

Condition 2: Message vectorization is applied if an operand satisfies exactly one of the following conditions:

1. It has only an anti communication dependence vector $\vec{c}_a$.
2. It has only a true communication dependence vector $\vec{c}_t$.
3. It has no data dependence.

If the operand has only a true communication dependence vector $\vec{c}_t$, it requires vector pipeline communication. Otherwise, the operand requires vector prefetch communication.

Example:

$D = \emptyset$ shows that any operand in the loop has no dependence, and that it thus satisfies **Condition 2**. The communication for array B can be vectorized by using prefetch communication.

5.4 Finding appropriate loop indices for tiling (step 4)

To find loop indices that satisfies **Condition 1**, the compiler calculates the bitwise **and** of the communication vector $\vec{b}$ and the logical negative of the reuse vectors, such as the self-temporal the group-temporal reuse vector. The ith zero value of the reuse vector shows that reuse does not occur in the ith loop index. The ith non-zero value of the communication vector shows that communication occurs in the ith loop index. If the ith value of the reuse vector is zero and the ith value of the communication vector is non-zero, the compiler sets ith non-zero value of the vector as a result. If the ith value of the vector is non-zero, the compiler determines that the ith loop index can be applied tiling to. It then determines the scope of the tiled loop by examining whether the nested loops are fully permutable [11].

Example:

In the example, we focus only on self-temporal reuse, because not more than one reference is made to the same array with communication. Thus, we assume that there is no group-temporal reuse. In Fig. 1, the self-temporal reuse vector R_{ST} for the operand B(I,K) is $\text{span}\{\vec{e}_1\}$. The bitwise **and** of $\vec{b}_r = \text{span}\{\vec{e}_1, \vec{e}_3\}$ and $\text{span}\{\vec{e}_2, \vec{e}_3\}$ is $\vec{t} = \text{span}\{\vec{e}_3\}$.

The compiler then finds the scope of the nested loop that is fully permutable, to ensure the legality of the loop transformations. Since $D = \emptyset$ in the loop, the whole nested loop is fully permutable. The range of the loop index is between (*top* =) 1 and (*inner* =) 3. As a result, the algorithm returns $\vec{t} = \text{span}\{\vec{e}_3\}$ and *top*=1 in Fig. 3.

5.5 Apply loop interchange and tiling (step 5)

First, the compiler chooses the loop index to which tiling is to be applied on the basis of the communication method in the loop.

If the operands in the loop require only vector prefetch communication, tilling is applied to all loop indices, because the loop can be executed without synchronization after communication is completed, and thus the parallelism is not varied if tiling is applied to any loop index. If the operands in the loop require vector pipeline communication, tiling is applied to the loop index that accesses the array dimension which is distributed among the fewest processors, because amount ofr each communication should be as small as possible to increase the effectiveness of the overlapping.

Secondly, to perform communication for a tiled loop at the outermost loop possible, the compiler interchanges the found loop indices with the outermost loop indices that is fully permutable. The compiler then applies tiling to the loop index that was moved into the outermost loop.

Example:

The operand in the loop has only prefetch communication. Since only the third index has a non-zero value in the vector $\vec{t}$, the compiler interchanges the loop index variable K with the outermost loop index. Finally, it applies tiling to the new outermost loop.

Algorithm

IN : ($\vec{l}$: loop index vectors, $\vec{u}$: loop index vectors, n: **integer**, l: left hand side operand,
　　R: **set of** right-hand side operands, F: **set of** access functions for operands,
　　G: **set of** distribution functions for operands,
　　P: **set of** processor index vectors, D: **set of** dependence vectors)

OUT : ($\vec{t}$: index vector /* tiling vector */, top: **integer**)

i, m, top, inner: **integer**;
is_comm: **boolean**;
$\vec{i}$, $\vec{d}$: loop index vector;
$\vec{y}$, $\vec{z}$: array index vector;
$Q(P)$: **set of** loop index vectors;
r : right-hand side operand;
$Y_r(P)$: **set of** array index vectors for operand;
$\vec{b}_r$: loop index vector for operand r ;
$f_r(\ )$: access function for operand r ;
H_r : mapping matrix for operand r ;
$g_r(\ )$: distribution function for operand r ;

/* step 1: calculate In Set */
foreach $\vec{p} \in P$ **do**
　$Q(\vec{p}) := \emptyset$;
　foreach $\vec{i} \in [\vec{l}:\vec{u}]$ **do**
　　if $g_l(f_l(\vec{i})) = \vec{p}$ **then** $Q(\vec{p}) := Q(\vec{p}) \cup \vec{i}$;
　end foreach
end foreach

is_comm := **false**;
foreach $r \in R$ **do**
　foreach $\vec{p} \in P$ **do**
　　$Y_r(\vec{p}) := \emptyset$;
　　foreach $\vec{i} \in Q(\vec{p})$ **do**
　　　if $g_r(f_r(\vec{i})) \neq \vec{p}$ **then**
　　　　$Y_r(\vec{p}) := Y_r(\vec{p}) \cup f_r(\vec{i})$;
　　　endif
　　end foreach
　　if $Y_r(\vec{p}) \neq \emptyset$ **then** is_comm := **true**;
　end foreach
end foreach
if not is_comm **then return** (o , 0);

/* step 2: generate Communication Vector */
foreach $r \in R$ **do**
　$\vec{z} := \mathbf{o}$;
　m = rank of *operand r*;
　for i := 1 **to** m **do**
　　if $\left(\bigcap_{\vec{p} \in P} \{y_i | \vec{y} = (y_1,...,y_m), \vec{y} \in Y_r(\vec{p})\} \right) = \emptyset$ **then**
　　　$\vec{z} = \vec{z} \cup \text{span}\{\vec{e}_i\}$
　　endif
　end for
　$\vec{b}_r := H_r^T \vec{z}$;
　if $\vec{b}_r \neq \mathbf{o}$ **then**
　　for i := 1 **to** n **do**
　　　if $\exists \text{span}\{\vec{e}_i\} \in \ker f_r$ **then**
　　　　$\vec{b}_r = \vec{b}_r \cup \text{span}\{\vec{e}_i\}$;
　　　endif
　　end foreach
　endif
/* step 3: check condition 2 */
if $_{\exists \vec{d} \in D:} (\vec{d} \text{ is anti dep. by opr. } r \text{ and } \vec{d} \cap \vec{b}_r \neq \emptyset)$ **and**

$\exists \vec{d} \in D: (\vec{d} \text{ is true dep. by op } r. r \text{ and } \vec{d} \cap \vec{b}_r \neq \varnothing)$ **then**
    return (o , 0);
  end if
end foreach

/* step 4: calculate $\vec{t}$ */
$\vec{t} = \mathbf{o}$;
foreach $r \in R$ **do**
  for i := 1 **to** n **do**
    if $\left(\left(\ker f_r \cap span\{\vec{e}_i\} = \varnothing \right) \text{ or} \right.$
    $\left. \left(\left(\ker f_r + \vec{r}_p \right) \cap span\{\vec{e}_i\} = \varnothing \right) \right)$
    and $\vec{b}_r \cap span\{\vec{e}_i\} \neq \varnothing$ **then**
      $\vec{t} := \vec{t} \cup span\{\vec{e}_i\}$;
    end if
  end foreach
end foreach

if $\vec{t} = \mathbf{o}$ **then return** (o , 0);

/* find fully permutable nest from top to n */
top := 1;
if $D = \varnothing$ **then return** ($\vec{t}$, top);
for inner := n **to** 1 **by** -1 **do**
  if $t_{inner} \neq 0$ **then break;**
end for

while (top < inner) **do**
  if $\forall \vec{d} \in D: \left(\left(d_1, ..., d_{top-1} \right) \succ 0 \text{ or } \forall top \leq i \leq n : d_i \geq 0 \right)$ **then**
    break;
  top := top + 1;
end while
if (top = inner) **return** (o , 0);
for i := 1 **to** top-1 **do** $t_i := 0$;
return ($\vec{t}$, **top**);

Fig. 3: Algorithm for detecting the tiled loop indices (steps 1-4)

6 Code Generation

In this section, we describe methods for generating communication in such a way that it is overlapped communication with computation. The methods depends on the communication method of each operand, which may be prefetch communication or pipeline communication. We assume that a non-blocking communication interface is implemented in distributed memory machines. In this paper, SEND/RECEIVE means a function that issues a non-blocking send/non-blocking receive. WAIT means a function that waits for all data to be received.

6.1 Overlap for Prefetch Communication

An array that requires prefetch communication must reference elements before they are written. To overlap communication with computation of the tiled loop, the processor performs prefetch communication alternately with exchanging the two receive buffers. First, the processor sends array elements that will be read in the next tiled loop to another buffer before the execution of the tiled loop. Then, it starts to calculate the tiled loop with the current buffer. Here, the processor can overlap communication with computation. Before starting to calculate the next tiled loop, the processor waits for the arrival of the data sent to the buffer, in order to synchronize itself with the other processor. After exchanging the old buffer for the received buffer, it starts to calculate the tiled loop.

6.2 Overlap for Pipeline Communication

The array that requires the pipeline communication to be written at iteration $\vec{i}$ is read at iteration $\vec{i} + \vec{d}$. The processor sends the array elements to another processor that references them after the calculation of the tiled loop is completed, and starts to calculate the next tiled loop during the communication. Thus, the processor can overlap communication with computation.

7 Experiments

In this section, we give the experimental results obtained by using our algorithm, which we have implemented in our HPF compiler [20]. To measure the effectiveness of the algorithm, we compiled two applications written in HPF. One is a matrix multiplication code that requires prefetch communication. The other is a Successive Over Relaxation (SOR) code that requires prefetch and pipeline communication.

For our experiments, we used an IBM SP with 32 thin nodes. The high-performance switch provides a DMA mechanism between the network and the system buffer.

7.1 Matrix Multiplication

We ran the matrix multiplication code with 1600 x 1600 arrays distributed in (*, BLOCK). The performance results are shown in Fig. 4 (a). The vertical axis shows the speedup ratio normalized by the execution time without communication overlap, while the horizontal axis shows the size of tile used by the compiler. In this experiment, a tile size between 200 and 500 would be the best choice.

7.2 Successive Over Relaxation Method

For the second experiment, we ran the SOR code with 2000 x 2000 arrays distributed in (BLOCK, *). The performance result is shown in Fig. 4 (b). The vertical axis shows the speedup ratio normalized by the execution time of the sequential code on a single processor, while the horizontal axis shows the size tile used by the compiler. The results without our optimization are also shown for the processor configurations of 3 x 3 and 4 x 4 in the array distribution (BLOCK, BLOCK). In this experiment, a tile size between 100 and 200 would be the best choice.

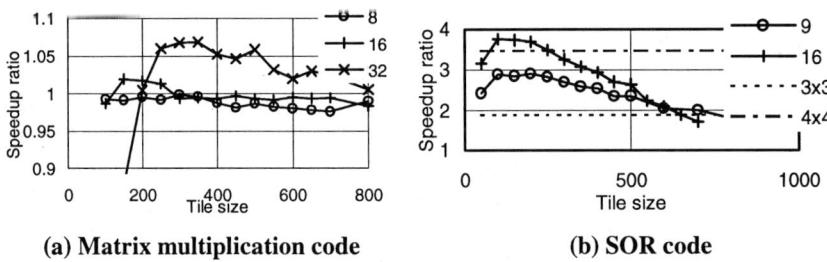

(a) Matrix multiplication code   (b) SOR code

Fig. 4: Performance results of the experiments

7.3 Discussion

In the matrix multiplication code, prefetch communication is generated. As the number of processors increases, the effectiveness of the algorithm becomes apparent for a wider range of the tile size. The performance gains also increase when more processors are used. In the SOR code, both prefetch and pipeline communication are generated. The program is executed in a wavefront method. The performance with our algorithm is better than when the array distribution is done by the user.

The best tile size is determined by the relationship between the communication time among processors and the execution time of the tiled loop. In the case that there is prefetch communication, the communication time takes up less of the execution time if the communication time is hidden by the calculation time of the tiled loop. The best tile size gives a ratio of 1:1 for the communication time and the calculation time of the tiled loop. When there is pipeline communication, the parallelism increases and the amount of each communication decreases if the tile size is smaller, but the communication overhead increases with the total number of communications. The performance decreases remarkably when the tile size is bigger than the best tile size in the case of matrix multiplication by 32 processors and SOR, because the communication among processors waits for the computation of the tiled loop to be completed.

In the current SP system, we can overlap computation with the transfer through the network, but we cannot overlap computation with the transfer between the system buffer and the user buffer. This is because a DMA transfer in an SP system only supports copying of data between the network and the system buffer. In our SP system, the following two overheads exist when communication and computation are overlapped by using a non-blocking communication interface :

1. An interrupt to the processor, generated by the DMA mechanism at the completion of the data transfer from the network to the system buffer
2. A copy between the system buffer and the user buffer by the processor

Roughly speaking, the first overhead takes $60\,\mu s$, and the second overhead takes $28\,\mu s$/KB. These overheads also bring unnecessary data for computation into the data cache, and thus decrease the data locality.

Despite these overheads of the current SP mechanism, our algorithm improves the performance by choosing an appropriate tile size. If a future SP system supports direct data transfer between the network and the user buffer on the processor, we would be able to obtain even a better speedup.

8 Conclusions

We have presented a compiler algorithm that hides the communication time by automatically overlapping communication with computation. Previously, this type of optimization was possible only by hand-coding programs. We have also described how to generate communication for tiled loops. We have implemented this algorithm in our HPF compiler, and experimental results have shown its effectiveness.

We are currently working on a heuristic algorithm to find an appropriate tile size for overlapping communication with computation.

9 Acknowledgments

We are grateful to Osamu Gohda, Gyo Ohsawa, Toshio Suganuma, and Takeshi Ogasawara of Tokyo Research Laboratory for implementing our HPF compiler and for participating in helpful discussions.

References

1. Stanford SUIF Compiler Group: "SUIF: A Parallelizing and Optimizing Research Compiler," Technical Report, Stanford University, CSL-TR-94-620, 1994
2. C. W. Tseng: "An Optimizing Fortran D Compiler for MIMD Distributed-Memory Machines," PhD thesis, Rice University, CRPC-TR93291, 1993
3. Z. Bozkus, A. Choudhary, G. Fox, T. Haupt, and S. Ranka: "Fortran90D/HPF Compiler for Distributed Memory MIMD Computers: Design, Implementation and Performance Results," in Proceedings of Supercomputing '93, pp. 351-360, 1993
4. P. Banerjee, J. A. Chandy, M. Gupta, J. G. Holm, A. Lain, D. J. Palermo, and S. Ramaswamy: "The PARADIGM Compiler for Distributed-Memory Message Passing Multicomputers," in Proceedings of the First International Workshop on Parallel Processing, pp. 322-330, 1994
5. High Performance Fortran Forum: "High Performance Fortran Language Specification, Version 1.0," Technical Report, Rice University, CRPC-TR92225, 1992
6. S. Hiranandani, K. Kennedy, and C. W. Tseng: "Compiling Fortran D for MIMD Distributed-Memory Machines," Communications of the ACM, Vol. 35, pp. 66-80, 1992
7. T. Horie, K. Hayashi, T. Shimizu, and H. Ishihata: "Improving AP1000 Parallel Computer Performance with Message Communication," in the 20th Annual International Symposium on Computer Architecture, pp. 314-325, 1993
8. A. Rogar and K. Pingali: "Process Decomposition Through Locality of Reference," in Proceedings of the SIGPLAN '89 Conference on Program Language Design and Implementation, 1989
9. D. J. Palermo, E. Su, J. A. Chandy, and P. Banerjee: "Communication Optimizations Used in the PARADIGM Compiler for Distributed-Memory Multicomputers," In Proceedings of the 23rd International Conference on Parallel Processing, pp. II:1-10, 1994
10. M. Wolfe: "High Performance Compiler for Parallel Computing," Addison-Wesley Publishing Company, 1995
11. M. E. Wolfe and M. S. Lam: "A Loop Transformation and Theory and an Algorithm to Maximize Parallelism," IEEE Transaction on Parallel and Distributed Systems, Vol. 2, No. 4, pp. 452-471, 1991
12. T. Agewara, J. L. Martin, J. H. Mirza, D. C. Sadler, D. M. Dias, and M. Snir: "SP2 System Architecture," IBM Systems Journal 344, No.2. pp. 152-184, 1995
13. A. Lain and P. Banerjee: "Techniques to Overlap Computation and Communication in Irregular Iterative Applications," in Proceedings of the International Conference on Supercomputing, pp. 236-245, 1994
14. C. Koelbel, P. Mehrotra, and J. V. Rosendale: "Supporting Shared Data Structures on Distributed Memory Architectures," in Proceedings of the ACM SIGPLAN '90 Symposium on Principles and Practice of Parallel Programming, pp. 177-186, 1990
15. S. Hiranandani, K. Kennedy, and C. W. Tseng: "Preliminary Experiences with the Fortran D Compiler," in Proceedings of Supercomputing '93, pp. 338-350, 1993
16. R. Hanxlenden and K. Kennedy: "GIVE-N-TAKE: A Balanced Code Placement Framework," in Proceedings of the ACM SIGPLAN '94 Conference on Program Language Design and Implementation, pp. 107-120, 1994
17. A. W. Lim and M. S. Lam: "Maximizing Parallelism and Minimizing Synchronization with Affine Transforms," Conference Record of the 24th Annual ACM SIGPLAN-SIGACT Symposium on Principles of Programming Languages, 1997
18. J. M. Anderson, S. P. Amarasinghe and M. S. Lam: "Data and Computation Transformations for Multiprocessors," in Proceedings of the Fifth ACM SIGPLAN Symposium on Principles and Practice of Parallel Processing, 1995
19. M. E. Wolfe and M. S. Lam: "A Data Locality Optimizing Algorithm," in Proceedings of the ACM SIGPLAN '91 Conference on Program Language Design and Implementation, pp. 30-44, 1991
20. K. Ishizaki and H. Komatsu: "A Loop Parallelization Algorithm for HPF Compilers," 8th Workshop on Language and Compilers for Parallel Computing, pp. 12.1-15, 1995

NaraView: An Interactive 3D Visualization System for Parallelization of Programs

Mariko SASAKURA[1], Kazuki JOE[2], and Keijiro ARAKI[3]

[1] Department of Information Technology, Okayama University, 3-1-1, Naka, Tsushima, Okayama 700, JAPAN
[2] Department of Computer and Communication Sciences, Wakayama University, Sakaedani, Wakayama 640, JAPAN
[3] Graduate School of Information Science and Electrical Engineering, Kyushu University, Kasuga Koen 6-1, Kasuga, Fukuoka 816, JAPAN

Abstract. For effective use of parallelizing compilers, an interactive environment which allows users to instruct the way of parallelization is needed. As the first step to build such an environment, we have developped a program visualization system named NaraView. The system provides two powerful methods for 3D visualization of program structure and data dependence. 3D visualization of program structure illustrates a hierarchical loop structure of given programs and suggests which parts of the program have been parallelized. 3D visualization of data dependence explains each data dependence on any variable or array element which is accessed at a specific loop. By using these methods, users can easily understand which part of the program should be more parallelized. We also show several examples to demonstrate the efficiency of these methods.

1 Introduction

Parallelizing compilers are a tool which analyzes sequential programs and restructures them to parallel programs. Making maximum use of parallelizing compilers, old but sophisticated and bug-free programs may be restructured into parallel forms with keeping their original semantics. Regarding development of new programs for parallel computers, since typical users tend to be an old hand at writing conventional sequential programs, the users want compilers to parallelize them instead of contriving new parallel algorithms with great efforts.

Although parallelizing compilers try to transform any sequential program to a parallel program automatically, they often fail to extract some kind of parallelism of given sequential programs. One of the most significant problems is that the strategy of choosing the best transformation methods is not known. Therefore, existent parallelizing compilers require the instructions from users to optimize the program codes: which transformation methods should be applied to which loops.

However, it is hard for users to decide transformation methods just from taking a glance at the source program. To determine them, users need detail information about control and data dependence of the program. Control and data dependence whose analyses are typically obtained via parallelizing compilers are difficult for users to interpret.

Therefore, we are developing a visualization system named NaraView, which provides an interactive compilation or programming environment for parallel computer users. The main idea for NaraView is to present the detail information procured by parallelizing compilers to users visually and, hence, intuitively. Such intuitive comprehension helps users to suggest the way of parallelization to the compilers interactively.

In this paper, we focus the visual aspects of NaraView by proposing two methods for 3D visualization: program structure and data dependence. The visualization of program structure is a map for given program with 3D axes of program flows, loop and function hierarchies, and palallelizm. We will give a definition to the *program structure* in Section 3. The visualization of data dependence is a relational map of given loops and variables or array elements, which may have data dependences.

There have been many studies concerning program visualization for parallel programming. Most of them visualizing trace data obtained through execution of parallel programs to improve their performance or debug the programs[3],[5],[9].

However, our motivation is different from these works. We are interested in exploring and detecting the parts of a given program, which have potential parallelism, by a glance of their visualized exteriors. This motivation is close to [6]. However, our target is a source code level while [6] targets an instruction level.

In Section 2, we give an overview of NaraView. A method for 3D visualization of program structure is explained in Section 3. In Section 4, a method for 3D visualization of data dependence is described. We will show some example figures which are generated by proposed methods in Section 3 and 4. Finally we give a conclusion in Section 5. All examples in this paper are written Fortran, but we believe the main idea of visualization of program structure and data dependence can be applied to any imperative program languages.

2   Overview of NaraView

NaraView is a visualization system for parallelizing compilers, which is aimed at assisting users to detect parallelism from source codes by a glance of structure, control and data dependence of given programs.

Especially NaraView can be regarded as a visual user interface tool for a parallelizing compiler. It visualizes the information which is extracted from given source programs by a parallelizing compiler. NaraView collaborates with Parafrase-2[8], which is one of the most popular parallelizing compilers in academic interests, in order to furnish an interactive system for more conspicuous parallelization.

On the other hand, NaraView supplies users with the visualized and intuitive information of source programs with parallel forms generated by parallelizing compilers or written by the users without their execution. This information is desirable from the viewpoints of scalability and portability for effective parallel processing.

NaraView consists of several components, *views*. A *view* gives a layout for the extracted information; which parts are chosen to display and how they are placed. Users can understand the characteristics of the information through appropriate *views*. The views assist them to investigate the best strategy for parallelization of the original programs.

Currently, NaraView provides the below four views:

Program structure view which shows a kind of map of given programs.
Source code view which shows source codes of an indicated part of the original programs.
CFG view which shows a control flow graph of an indicated part of the programs.
Data dependence view which shows data dependences of an indicated part of the programs.

Parallelization with NaraView is like this: first, NaraView invokes program structure view to give users an abstract and overall impression of the given sequential or partially parallelized programs as well as clues to further investigation. Next, users specify a loop on which they want to focus. According to their indications, NaraView invokes corresponding views of CFG and/or data dependence in the loop as well as the original source program of the loop. Using the information given by the views, they may find unnecessary data dependent parts intuitively, and can improve the source program with other parallelization methods.

3 Visualizing program structure

3.1 Overview

Program structure view shows *program structure* of given programs to select a part of the program which should be focused. Here, we use *program structure* as following meanings:

- how many and where loops are located in the given program,
- how many and where function calls are,
- and which loops have been already parallelized.

We also use *hierarchical structure* with the same meanings for nested loop structure. We describe this in Section 3.2.

All information about program structure is included HTGs (Hierarchical Task Graphs) proposed by Girkar[4], which are an intermediate representation in Parafrase-2. Program structure view gets information from HTGs, and visualizes it. In program structure view, three kinds of information are selected to visualize, *program flow*, *a measure of parallelism* and *hierarchical structure*.

3.2 Explanatory notes

Program structure view generates 3D *visible objects* which consists of a set of nodes. A node is indicated by a colored cube in the view. Each node corresponds to a line of source codes or a chunk of source codes. We define six kinds of node.

A root node represents the root of a hierarchical tree of HTGs and is displayed in red. There is only one root node in a *visible object*.

A basic node corresponds to a statement in source codes which is executed sequentially, and is displayed in sky blue.

A parallel node represents a loop body which is executed in parallel, and is displayed in green.

A loop node represents the existence of a loop, and is also displayed in red.

A call node represents a function call, and is displayed in dark blue.

An if node represents an *if* statement, and is displayed in yellow.

A node is expressed by a triplet (x,y,z). The exterior of *visible objects* changes as the correspondences of program information to x, y and z are changed. The correspondences are given as below so that users can understand program structure intuitively to extract parallelism from the exterior of the *visible object*.

x: program flows.
y: measures of parallelism.
z: levels of loop hierarchy structure.

Program flows The x-axis represents a program flow, which indicates the order of execution of each node labeled with some number. A node with a big number is executed after the termination of the node with a smaller number.

Usually, program flows contain conditional branches. Since we are interested in having users grasp program structure of given parallel programs intuitively, just the critical part of such flows ought to be displayed.

Levels of loop hierarchical structure The z-axis represents a level of hierarchical structure based on loops. Since we are interested in parallelizing methods related to loops, we focus hierarchical structure obtained from HTGs. By this hierarchical structure, each hierarchy level represents a loop, so that we can easily know if the loop has been parallelized.

When a loop node α is included in the body of a loop node β, we say α is a deeper node or in deeper hierarchy than β. Similarly we say β is a shallower node or in shallower hierarchy than α.

Values of the z-axis for shallower hierarchy are smaller than deeper one. The top level of hierarchy, where the value of the z-axis is 0, consists of just a root node. Any nodes in the second level, where the values of the z-axis are 1, do not belong any loop.

Measures of parallelism Towards the y-axis, nodes are placed as many as the number corresponding to a measure of parallelism so that users can find intuitively which parts of the program have been parallelized.

Usually such a measure of parallelism is given as the number of instructions which can be executed in parallel. Since our target is source level information, we define our measure of parallelism here.

We assume the loops which can be executed in parallel are expressed by DOALL in source codes.

1. the measure of parallelism at the root node is 1.
2. the measure of parallelism at the outside of any loops is 1.
3. when the measure of parallelism at a loop which belongs the hierarchical level $n(1 \leq n)$, is $w(1 \leq w)$, the measure of parallelism of nodes which belongs to the body of the loop is,
   - $w \times p$, when the loop is DOALL, and the total number of the loop iterations is p.
   - otherwise, w.

Program structure view displays nodes in y-direction by the number corresponds to the measure of parallelism. According to this measure, a sequence of nodes which spreads toward the y-axis has highly parallelism while another sequence of nodes which does not spread toward y is not executed in parallel. Thus, users may pay attention to the part which has no spread toward the y-axis. In practice, it is hard to compare the measure of parallelism when loop bounds are given as variables. In this case, program structure view asks users for concrete values to the variables interactively.

3.3 An example for program structure view

Figure 3 represents a program structure view of a parallel program for Gaussian elimination, which is restructured by Parafrase-2. Characters, arrows, and a curved line are added by hand for explanation.

In the figure, the flow of the program (the x-axis) starts from the upper left and ends at the lower right, the measure of parallelism (the y-axis) is expressed as the horizontal width and the level of hierarchical structure (the z-axis) is indicated from almost up to down. We show dotted lines which indicate the axes.

We can easily find two loops can be executed in parallel. Since those parallelized loops have no data dependence, Parafrase-2 could parallelize them without users' indications. However, other loops were not parallelized because they have some data dependence.

By using program structure view, we can select the loops which may have the possibility of parallelization. For example, if-statement is one of the obstacles to be parallelized. Therefore, we may investigate the possibility of parallelization of the loops which have no if-statement. We indicate such loops in the figure by arrows.

For further investigatoin, users select a loop which should be focused. Then NaraView can provide other views to visualize the details of the loop nodes.

4 Visualizing data dependence

4.1 Overview

We want to visualize data dependence to assist users to select transformation methods. We think the following issues are important to visualize data dependence.

- Make it clear that there exist data dependences in a given loop.
- If there are data dependences, clarify the possibility of parallelization by transforming the loop.
- Develop a tool which visualizes data dependence from the given source program automatically.

Data dependence is one of the most important information to estimate a loop can be parallelized or not. So far, data dependence graphs(DDG)[7] or iteration graphs [2] are used to show data dependence. But, both are not powerful enought for our purpose because they aim to explain data dependence to compiler resercheres.

Therefore, we propose data dependence view which automatically visualizes data dependence of any loop without conditional branches. The basic idea is to visualize data dependences from the viewpoint of data. Data means array and scalar variables which are accessed in the loop. A set of data is mapped on the x-y plane. We want to display data accesses with making the existence of data dependence clear. In addition, we define *access time* to clarify the distance of dependence, which corresponds to the z-axis. To recognize the possibility of parallelization well, we display *loop grids* which point to places where each loop is started.

4.2 How to get information

Data dependence view shows users the following information.

- data which are accessed in a loop.
- iterations at which the accesses occur.

These information can be acquired from trace data on which execution time of data accesses are recorded. We get the information from the HTG because we do not want to insert extra codes to the source program for recording trace data. A process to get information is:

1. Select a loop which should be focus by using program structure view.
2. Indicate the number of iterations. (start and end of the iterations, for example.)

3. Generate data accesses information. Data dependence view calculates when and where data accesses occurrs via HTG. If there are variables of which value is unknown, the view asks users for concrete values.

Data accesses information consists of:
- data which is accessed.
- the iteration when the access occurrs.
- the source code statement when the access occurrs.
- the kind of access (Read and/or Write).

4.3 Explanatory notes

Data dependence view consists of cubes, poles, *loop grids*, and *AVD map*.

Cubes A data access is displayed as a colored cube. The coordinate of each cube is expressed by a triplet (x,y,z). Each axis and color has the following meanings.
    x,y   location of the data
    z   access time
    color the kind of the access

The layout of data on the x-y plain is defined by users. *AVD map* which is described in Section4.3 shows the current data layout.

We define *access time* to record the time when accesses occur. It is mapped on the z-axis. In this view, we provide two kinds of access time and users can choose whichever they like.

Statement access time: Assign access time by each statement.
Iteration access time: Assign access time by each iteration.

By using statement access time, we can show all data dependences in detail. By iteration access time, we can focus loop carried data dependences. We can set the above access times to individual loop. For example, statement access time for outer loops and iteration access time for inner loops is possible.

The kind of access is represented as color of a cube.

A blue cube represents a read access.
A red cube represents a write access.
A purple cube represents read and write accesses.

Poles A pole connected between two cubes indicates data accesses to the same data, and represents the existence of data dependences. There are three kinds of poles which correspond to flow dependence, anti dependence and output dependence.

Life time pole: If there are more than one consecutive accesses after a write access, a life time pole connects from the write access to the last read access. The life time poles are represented in green and it also shows the existence of flow dependence spontaneously.

```
      do 10 i = 2, n-1
        do 20 j = 2, m-1
          a(i, j) = (a(i-1, j) + a(i, j-1)
                   + a(i+1, j) + a(i, j+1))/4
20      continue
10    continue
      end
```

Fig. 1. Example program 1.

Anti-dependence pole: If there is a read access and the following access to the same data is a write, an anti-dependence pole connects these two accesses. Anti-dependence poles are represented in yellow.

Output-dependence pole: If there are more than one consecutive write accesses to the same data, an output-dependence pole connects these accesses. Output-dependence poles are represented in pink.

Users can make each pole visible or invisible.

Loop grids Loop grids are semi-transparent planes, which are placed perpendicularly to the z-axis. Users can display loop grids to indicate the beginning of each iteration of loops or specified iterations.

AVD map There is another plane placed perpendicularly to the z-axis, namely, AVD(Array-Variable disposition) map. It shows the layout for arrays and variables with easy comprehension of the view for data dependence. Characters on AVD-map stands for the names of variables or arrays which are mapped there.

4.4 Examples for data dependence view

In this section, we show two examples for data dependence view. One is an example of loop skewing and the other is scalar expansion[7]. These examples show data dependence view helps users to select transformation methods.

An example of loop skewing We show an example which can be parallelized by loop skewing[1]. Loop skewing is a transformation method which handle wavefront computations, so called because the updates to the array propagate like a wave across the iteration space.

Figure 1 is a program which have a typical wavefront computation. Figure 4(a) shows data dependences of the program, whereas $n = 10$ and $m = 10$. Loop grids which correspond to the outer loop are displayed.

The feature of wavefront computations is that there are two read accesses after a write access; write one occurs in the execution of the inner loop and the other occurs in the execution of the outer loop. After a write access to an

```
      do 1600 k = i + 1,n
      m = aary(k,i) / aary(i,i)
      do 1550 j = i + 1,n + 1
      aary(k,j) = aary(k,j) - m * aary(i,j)
1550  continue
1600  continue
      end
```

Fig. 2. Example program 2.

element of array a, there are two read accesses before and after the loop grid. Life time poles go across the loop grid. Therefore, we can know the loop is a wavefront computation easily.

Figure 4(b) shows data dependences of a program which is obtained by applying loop skewing. In the figure, all read accesses and a write access to the same data occurs in different region partitioned by loop grids. Thus we can estimate the loop can be executed in parallel.

An example of scalar expansion This example shows how users use combination of program structure view and data dependence view, and how users know scalar expansion, which is one of transformation methods, can be applied.

In Figure 3 shown in previous section, we could find five loops which are pointed by arrows. Now, we focus one of them which is labeled "Focused loop" in the figure. This loop is a double loop. We know the inner loop have already been parallelized because it consists of only green nodes which are lined up parallelly to the y axis. When users *cut* the focused loop node, Figure 5 appears.

Figure 5 shows data dependence view of the focused loop. The source code of this loop is represented in Figure 2. This figure displays data accesses when $n = 5$ and $i = 1$. Loop grids which correspond to the outer loop are displayed too. The AVD map is placed at the bottom of the figure. The z-axis goes from almost down to up, which is shown by a dotted line.

The figure tells us the obstruction which prevents this loop from being parallelized is data dependence on variable m. m has been rewritten and read several times in the outer loop. We can know it easily by life time poles on m.

In this case, we can apply scalar expansion to m, and obtain a more parallelized program. Its data dependence is shown in Figure 6. In the figure, each element of array (which is produced by scalar expansion) m has only one write access, and read accesses to the element occurs in the same region partitioned by the loop grids. Thus we know the outer loop can be executed in parallel.

5 Conclusion

NaraView is a visualization system for parallelizing compilers. In this paper, we proposed two views of NaraView: program structure view and data dependence view.

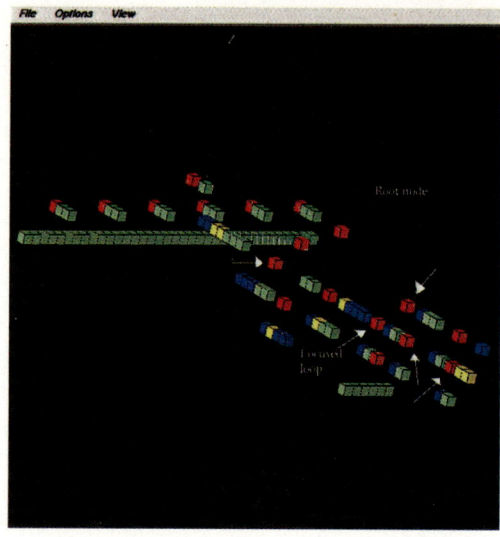

Fig. 3. A program structure view of a progam of Gaussian elimination.

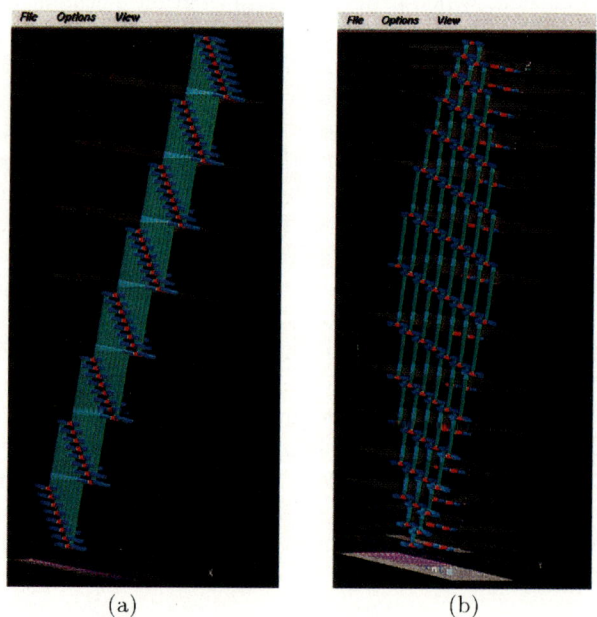

Fig. 4. (a) A data dependence view of example program 1. (b) A data dependence view of the example program 1 with loop skewing.

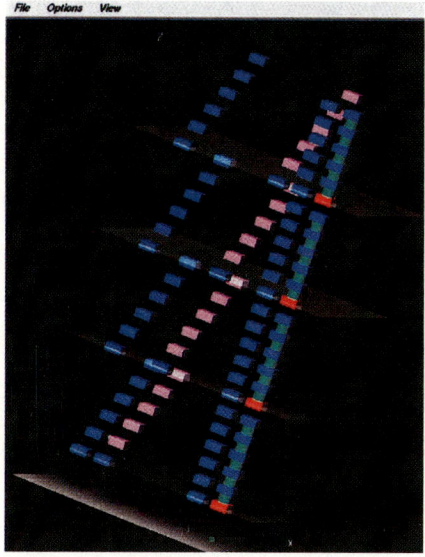

Fig. 5. A data dependence view of example program 2.

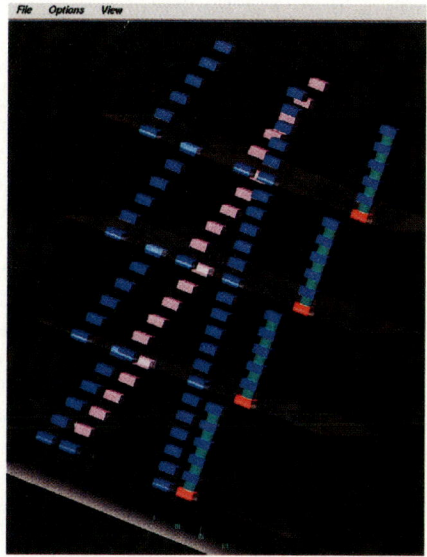

Fig. 6. A data dependence view of the example program 2 with scalar expansion.

Program structure view shows users 3D *visible objects* which explain the outline of given programs for intuitive understanding, and this leads them to go down to investigate each loop which does not seem to have parallelism at a glance.

Data dependence view shows data dependence in the loop focused by using the program structure view. We explained by examples that this view is useful to select transformation methods to parallelize the given programs.

Although we do not mention in this paper, we think data dependence view has more various uses. For example, we can investigate how data should be partitioned and distributed by mapping data to the appropriate layout on the x-y plain. Also the view is helpful to education and debugging.

We have a plan to extend NaraView to have the function that users can modify given source programs by using visual interface. For example, when some redundant dependences are detected in a view, users can remove the dependences by cutting the dependence poles without modifying the source program. If this operation succeeds to find parallelism well, the source program is updated by NaraView automatically.

References

1. Bacon, D. F., Graham, S. L. and Sharp, O. J. "Compiler Transformations for High-Performance Computing", ACM Computing Survey, vol. 26, no.4, pp. 345-420, 1994.
2. Banerjee, U. "Loop Transformations for Restructuring Compilers: the foundations ", Kluwer Academic Publishers, 1993.
3. Dykes, S. G., Zhang, X., Shen, Y., Jeffery, C. L. and Dean, D. W. "*GRAPH: A Tool for Visualizing Communication and Optimizing Layout in Data-Parallel Programs", 1995 International Conference on Parallel Processing, II pp.121-129, 1995.
4. Girkar, M.B. and Polychronopoulus, C. D. "The Hierarchical Task Graph as a Universal Intermediate Representation", International Journal of Parallel Programming, Vol.22, No.5, pp.519-551, 1994.
5. Heath, M. T. and Etheridge, J. A. "Visualizing the Performance of Parallel Programs", IEEE SOFTWARE Vol.8, No.5, pp.29-39, September, 1991.
6. Novack, S. and Nicolau, A. "VISTA: the visual interface for scheduling transformations and analysis", Languages and Compilers for Parallel Computing. 6th International Workshop Proceedings, pp.449-60, xi+655, 1993.
7. Polychronopoulos, C.D. "Parallel Programming and Compilers", Kluewer Academic Press, 1988.
8. Polychronopoulos, C.D. et al. "Parafrase-2: An environment for parallelizing, partitioning, synchronizing, and scheduling programs on multiprocessors", Proceedings of the 1989 International Conference on Parallel Processing, 1989.
9. Reed, D. A., Shields, K. A., Scullin, W. H., Tavera, L. F. and Elford,C. L. "Virtual Reality and Parallel Systems Performance Analysis", IEEE Computer vol.28, no.11, pp.57-67, Nov. 1995.

Hybrid Approach for Non-strict Dataflow Program on Commodity Machine

Kentaro Inenaga, Shigeru Kusakabe, Tetsuro Morimoto, and Makoto Amamiya

Dept. of Intelligent Systems, Kyushu University
Kasuga Fukuoka 816, Japan

Abstract. Dataflow-based non-strict functional programming languages have attractive features for writing concise programs. In order to avoid performance penalties on non-dataflow stock machines, we speculatively use a stack frame instead of a heap frame for a fine grain function instance, which may require dynamic scheduling. As a static approach, we introduce a merging policy to a thread partitioning algorithm in order to find functions with a potentially strict call interface. To complement this static analysis, we provide a hybrid runtime mechanism which can dynamically change a suspended stack frame into a heap frame. The results of the performance evaluation indicate that we can reduce superfluous heap frame management and achieve practical performance even on stock machines.

1 Introduction

Dataflow-based functional programming languages have attractive features for writing concise programs. Functional paradigm offers a high level programming, and dataflow paradigm can abstract away the timing problems such as synchronization control. Especially, languages that support non-strict and eager evaluation provide high expressive power, as well as high degree of implicit parallelism.

Although special architectures have been proposed to efficiently implement dataflow-based functional languages[6, 10], low performance of such languages on stock machines prevents them from being a popular language. In order to show the feasibility of such a language, we are implementing a dataflow-based non-strict language on stock machines with commodity processors, ranging from engineering workstations to a distributed memory parallel machine.

Our early implementations based on a naive dataflow scheme showed poor efficiency. In order to support non-strict execution and extract maximum parallelism, our abstract machine supported fine-grain execution and we generated a very fine-grain thread code. Non-strict and fine-grain eager execution requires frequent dynamic managements such as thread scheduling and synchronization at a very fine-grain level. On stock machines, which have no hardware support for fine-grain execution, fine-grain dynamic managements are implemented by software and will incur heavy overhead.

In order to achieve high efficiency, we exploit strictness in non-strict dataflow language programs. Many functions in non-strict language programs may have a strict call interface. We use an abstract machine which supports both fine-grain non-strict evaluation and stack-base strict evaluation. As a static approach, we introduce a merging policy to a thread partitioning algorithm, in order to find function calls which may be able to use a strict call interface.

To complement this static approach, we develop a hybrid runtime mechanism. While non-strictness of function call is rarely used in lenient programs, non-strictness of data-structure is used extensively[7]. Even if a function can be invoked in a strict way, the function may be suspended by a non-strict data-structure access, and may require dynamic scheduling. For such a function call, we speculatively use a stack frame while providing a hybrid runtime mechanism which can dynamically change the suspended stack frame into a heap frame, so that it can be dynamically scheduled. [1]

We also show the preliminary performance evaluation of our implementation. The results indicate that we can reduce superfluous heap frame management overhead and achieve practical performance even on stock machines by the static thread merging and the hybrid runtime system.

In this paper, section 2 discusses the problem of the naive implementation. Then, section 3 examines a thread merging algorithm to find strict function call interfaces. In section 4, we discuss the runtime systems to speculatively use stack frames for suspensive function calls. We also show the performance evaluation of our implementation in section 5.

2 Fine-grain lenient execution

In order to support non-strict execution and extract maximum parallelism, we generated very fine-grain thread codes of the abstract machine, whose target machines include a special fine-grain architecture[1, 5].

2.1 Fine-grain function call interface

A function consists of threads, each of which is an exclusively executable instruction sequence. A function call is executed as a split-phase transaction. Instance creation and parameter passing are performed separately. To realize fine-grain lenient execution, each function argument is passed respectively to the corresponding thread at the callee site. A function call can start if any of the arguments is prepared, thus subcomputation in the caller can be overlapped with the callee side.

Fig.1 exemplifies an overview of a function call, f(Exp1, Exp2). The argument expressions Exp1 and Exp2 can be evaluated in parallel, and the callee computation can proceed if either of Exp1 or Exp2 is prepared. This function call mechanism is possible to exploit fine-grain parallelism and realize non-strict computation.

In the figure, the "**call** f r" creates an instance of a function f, and store the pointer to the instance in r. The "**rins**" releases the current instance. The "**link** r reg $slot$" sends the value of reg as the $slot$-th parameter data to the callee instance specified by r. The "**rlink** r reg $slot$" sends the continuation, register

---

[1] Other than the non-strict data-structure access, a remote invocation and a remote access will make a function suspended. However, in this paper we focus on how to manage suspension caused by non-strict operation.

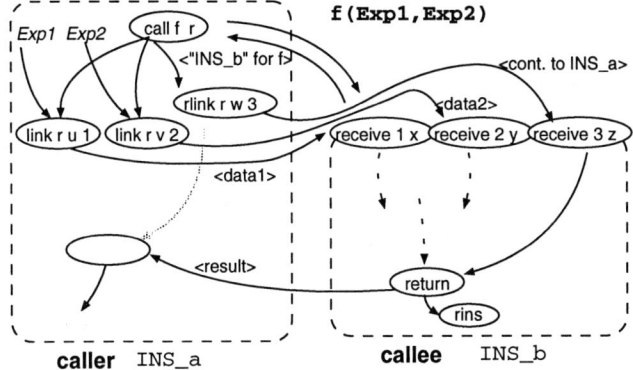

Fig. 1.: Overview of a fine-grain function call for f(Exp1, Exp2).

reg for the returned value and the continuation threads in the callee instance, as the *slot*-th parameter data to the callee instance specified by *r*. The "**receive** *slot reg*" receives the data transferred from the caller instance through the *slot*-th slot, and store the data into *reg*. The "**return** *rp reg*" triggers the continuation threads specified by the rlink operation after setting the value of *reg* to the content of *rp*.

2.2 Fine-grain runtime model

Our runtime model adopts a multi-thread execution model. Each function instance is implemented by using an instance frame. Threads within a function share the context on an instance frame for the function. Thread level scheduling and context switch mechanism realize non-strict and eager execution (lenient evaluation). Operations with unpredictably long latency are realized as a split-phase transaction. The runtime model is latency tolerable and expected to achieve high throughput. Since the implementation target here is stock machines without special support for fine-grain processing, we develop the software runtime system as shown in Fig. 2 on stock machines.

Instance frames hold the local variables and synchronization variables. Each thread has a synchronization counter variable, whose initial value is determined at compile time. When a thread is triggered, the counter is decremented. If it reaches zero, the thread becomes ready and pushed into the thread queue.

The thread queue is provided in order to dynamically schedule intra-instance threads. If the execution of the current thread terminates, the next ready thread is popped from the thread queue for the next execution. If the thread queue is exhausted before the instance terminates, the instance becomes a "suspended instance" and is pushed into the idle pool before the execution switches to another frame. If the instance terminates, the runtime releases the instance frame. Runtime system manages runnable instance frames by using instance queue. Runtime system picks up one of the runnable frames and passes the execution control to the corresponding code to activate the instance frame.

When a frame in the idle pool (suspended instance frame) receives data

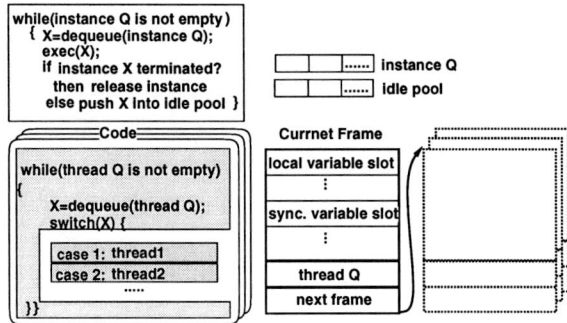

Fig. 2.: Overview of Runtime system.

from another active frame, the corresponding thread in the suspended frame is triggered. If the thread becomes ready and is pushed into the thread queue, the instance frame also becomes ready and is moved to the instance queue.

2.3 Experiment

On special machines dedicated to fine-grain execution, fine-grain abstract machine codes can execute efficiently. However, on stock machines without fine-grain execution support, fine- grain execution control will incur heavy overhead. As an example, we consider the following program which calculates the summation form `low` to `high`(See Fig. 3). Fig. 4 shows the fine-grain abstract machine code of the program. In the figure, an arrow shows a direct continuation arc, a wavy arrow indirect arc. A solid box indicates a thread, and a dashed box an function.

```
function summ(low,high:integer) return(integer)
= if low = high then low
  else{let mid:integer = (low+high)/2
       in summ(low,mid) + summ(mid+1,high)};
```

Fig. 3.: Summation program

We implemented the runtime system shown in Fig. 2 on an engineering workstation, and measured the elapsed time of the code shown in Fig. 4. We also measured the elapsed time of C program code. [2]

The elapsed time of the fine-grain code is about 0.8 seconds, and that of the C code is about 0.007 seconds. The difference of the elapsed time is very large. Although `summ` does not require any non-strictness, function instances of the fine-grain code requires frequent heap frame management, and arguments `low` and `high` are passed individually. Peak memory volume used for function instances amounts to 801KB, while on-chip data-cache size is 15KB. Such execution controls incur heavy overhead. This execution style can be regarded as an over-labored version from the the program semantics point of view.

---

[2] Compiled by gcc without no option, and executed on SS10 (clock:50MHz, Memory:64MB) with `low=1, high=1000`.

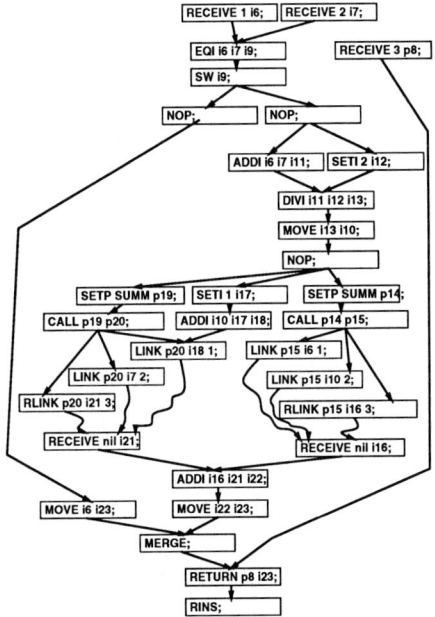

Fig. 4.: Fine-grain abstract machine code of summ.

3 Strict call interface
3.1 Thread merging

If the target machine is a special machine dedicated to fine-grain parallel processing, fine-grained code is useful to achieve high throughput, as well as to realize non-strictness, by means of dynamic scheduling at a fine-grain thread level. However, on a general purpose stock machine, if a function activation does not require non-strictness, full lenient style execution using an expensive heap frame may be ineffectual as examined in the previous section, and should be avoided in order to achieve high efficiency.

On a stock machine with commodity processor, a function call is expected to be performed with a stack frame. As the first step to realize a stack-based execution of function call in a non-strict program, we examine the thread merging method, which generates coarser grain threads from fine-grain codes while keeping necessary non-strictness. In the thread merging process, we can find function calls which can use a strict call interface.

As a thread generation algorithm, separation constraint partitioning[9] is proposed. The rule is that two nodes must reside in different threads if there exists any indirect dependency between them. Indirect dependences such as a certain indirect dependence by non-strict data access and a potential indirect dependence by function call may require dynamic scheduling. In order to find as many function calls with strict interfaces as possible, we add a merging priority to the thread partitioning algorithm:

Merge threads containing instance creation instructions and inter-instance data transfer instructions, then merge other threads.

Fig. 5 shows the code after thread merging for the summ program. As shown in Fig. 5, threads for parameter sending (or receiving) at the call boundary are merged up into a single thread. In this summ case, arguments can be sent in one thread, thus the function can use a strict call interface. Furthermore, since the only fan-out point in the graph code is a conditional branch SW, the actual number of execution threads is one. The callee has no suspension point, thus the function can be invoked with stack frame.

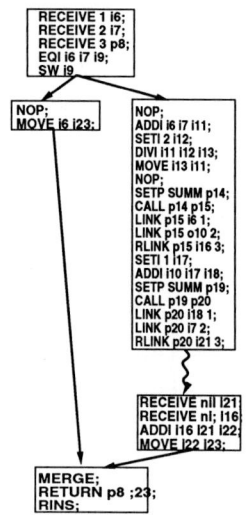

Fig. 5.: Code after thread merging.

3.2 Introducing strict call interface

As stated above, we can generate a strict code from a program of non-strict language. Furthermore, we can generate a single-threaded function code, which can be efficiently executed with a stack frame on stock machines.

We translated the code shown in Fig. 5 to use a strict call interface, and executed it with stack frames on SS10. The elapsed time was about 0.015 seconds. This is a big performance improvement, and only two times slower compared to C version.

We also measured the elapsed time on a distributed parallel machine, Fujitsu AP1000. It consists of 8 × 8 Sparc processor (25MHz). We ran two versions of summ 100,000: heap frame version and stack frame version. The elapsed time of former was 0.77 seconds and the latter 0.022 seconds. We can also see a big performance improvement on the distributed parallel machine.

In order to efficiently execute a non-strict functional program, calling a function with a stack frame is worth introducing as long as it does not violate the semantics.

4 Hybrid runtime for stack/heap activation

As stated in the previous section, we can improve the performance of non-strict functional programs on stock machines, if we can generate function codes with a strict call interface and execute the codes with stack frames. While functional non-strictness is rarely used, non-strictness regarding data-structures is used in lenient programs for scientific computing, sorting and search problems, symbolic computing, NAS parallel benchmarks, and small kernels[7]. Non-strict data-structures such as I-structures are effective to abstract the timing problem. Read requests and a write request to a data-structure element (I_FETCH and I_STORE) can be issued asynchronously. The read for an element is deferred until the location is filled. See Fig. 6 for state transitions of I-structures[2].

This means that functions which handle non-strict data-structures may require dynamic scheduling. Even if a function can be invoked with a strict call interface, the function may be suspended by a non-strict data-structure access. In this section, we discuss how to execute a function which can use a strict call interface, but may be suspended by a non-strict data-structure access.

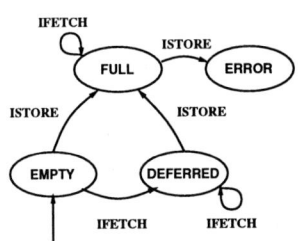

Fig. 6.: State transitions of I-structures.

4.1 Frame allocation

In order to complement static analysis, we use two types of stack-based function invocation with a strict call interface, while we use a heap frame for a function invocation with a non-strict call interface.

1. Regular stack invocation for a function which can be statically determined not to be suspended.
2. Suspensive stack invocation for a function which cannot be statically determined not to be suspended.

Suspensive stack invocation speculatively uses a stack frame for a function call, and the behavior is almost the same as regular stack invocation if the function terminates without suspension. Otherwise, if it is suspended, the stack frame will change into a heap frame. The generated heap frame is managed as the same as the heap frame created for function with a non-strict call interface(see section 2.2).

As an example, consider a function which includes an I_FETCH. Fig. 7 exemplifies a function call, which does not suspend, using the suspensive stack invocation mechanism.

1. A caller invokes a callee function which includes an I_FETCH.
2. The callee issues an I_FETCH request.
3. The I_FETCH successfully reads the value and the callee continues its computation using the result of the I_FETCH.
4. The callee terminates its computation, and returns the result and a "terminate" signal to the caller.
5. The caller checks the signal, and confirms the termination of the callee.
6. The caller continues its computation.

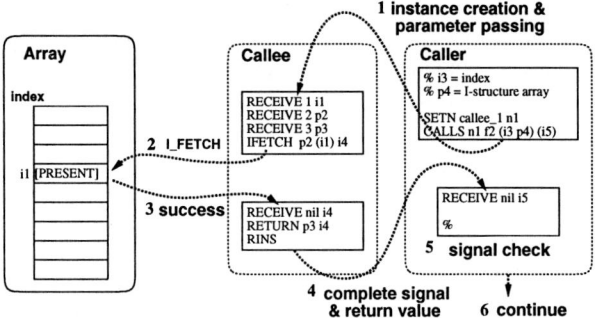

Fig. 7.: Suspensive stack invocation without suspension.

On the other hand, Fig. 8 exemplifies a suspended function call under the suspensive stack invocation mechanism.
1. A caller invokes a callee function which includes an I_FETCH.
2. The callee issues an I_FETCH request.
3. The I_FETCH fails and is deferred.
4. The callee is suspended, waiting for the result of the I_FETCH, and the context is swapped to heap.
5. The callee returns the pointer to the heap context and a "suspended" signal to the caller.
6. The caller checks the signal, and recognizes the suspension of the callee. By the propagation of suspension from the callee, the caller is also suspended.
7. The context of the caller is also swapped to heap.

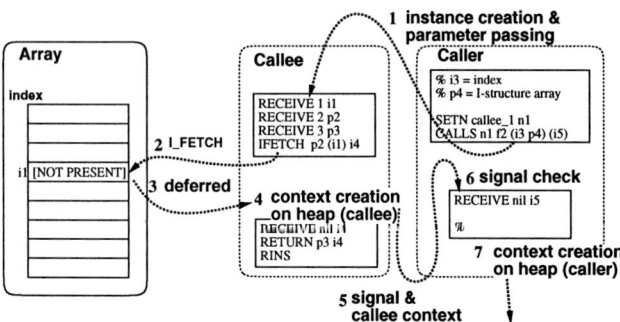

Fig. 8.: Suspension of suspensive stack invocation.

4.2 Compilation issues

In order to maximize the opportunity to use a stack-based invocation mechanism, we use a rather relaxed rule for thread merging. The rule of the separation constraint partitioning is that two nodes cannot be merged (i.e., they must reside in different threads) if there exists either a certain indirect dependence (CID) or a potential indirect dependence (PID) between them[9]. On the other hand, we relax the rule concerning certain indirect dependences caused by data-structure access as long as merging threads does not violate non-strict semantics of the program. This relaxation brings the potential suspension points into code, while reducing the merging complexity and boosting the thread granularity. However, the runtime mechanism described above manages occasional suspensions.

5 Evaluation
5.1 Simple program

As an example, we consider a simple matrix multiplication program:

```
function matmul(a,b:array of real;n:integer) return (array of real)
= mkarray (i,j) in ([1..n],[1..n])
      body foreach k in [1..n]
           sum a[i,k]*b[k,j];
```

Fig. 9.: Matrix Multiplication

In this program, within the body of mkarray and foreach, instances of filling functions and reduction functions can use strict call interface, but may request non-strict data-structure access. Table 1 shows the experiment result.

- In "heap" version, all functions use a non-strict call interface, and data-structure non-strictness is assumed. All instances are created and managed as heap frames.
- In "suspensive" version, all functions use a strict call interfaces, while data-structure non-strictness is assumed: functions with array access are invoked with suspensive frame.
- In "stack" version, all functions and all data-structures are assumed as strict. The code is manually serialized. All functions are single-threaded and invoked with regular stack frame even if the function includes data-structure access.

	heap	suspensive	stack
elapsed time(sec)	9.4	4.1×10^{-2}	3.7×10^{-2}

Table 1.: Elapsed time of matmul(size20).

From the elapsed time of the heap version and the suspensive version, we can see that suspensive mechanism can efficiently invoke functions. Although the best case is that the functions can be invoked with regular stack frames without suspension, suspensive call stack frame is effective even if the function has non-strict data-structure access. From the elapsed time of the suspensive and the stack version, the penalty to ensure dynamic scheduling is about 10% in this case. By providing this suspensive stack frame mechanism, static analysis can decide to speculatively use a strict-based function call even if it cannot be determined whether the function call will be suspended or not.

5.2 Examination of suspension

We examine the cost of heap frame management for the following cases. As a sample program, we use an artificial program to create an array A[]:
- A function is invoked on heap from the beginning.
- A function instance is allocated on heap after its suspensive stack invocation is suspended.

```
A:array of integer =
   mkarray i in ([0..n-1])
   body if i = s then 1
        elsif i = 0 then A[n-1]
        else A[i-1]
```

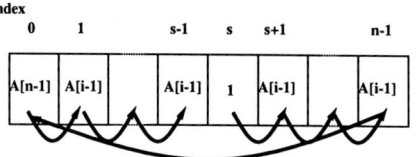

Fig. 10.: Array with self-reference

In this example, the filling function has self-reference to A. Since the dependence may have the backward direction depending the parameter s, the execution order of the filling functions cannot be determined statically. In this program, if s=0 then no instance will be suspended, while if s=9999 all instances will be suspended.

Table 2 shows the results of program execution where instances of the filling function are invoked with suspensive invocation. Since an array access is implemented by an I_FETCH, it depends on the behavior of the I_FETCH whether an instance may be suspended or not.

	parameter s			
	0	1000	4000	9999
suspensive invocation (suspended or not)	0/10000	1000/9000	4000/6000	9999/1
IFETCH (deferred or not)	0/10000	1001/8999	4001/5999	10000/0
elapsed time(sec)	1.0×10^{-1}	3.6×10^{-1}	4.0	2.4×10^{1}

Table 2.: Non-strict array generation (suspensive invocation)(size10000)

Table 3 shows the results of program execution where all instances of the filling function are invoked with heap frame invocation from the beginning.

From these tables, we can see that the number of suspended instances, and the elapsed time are almost the same for both suspensive and heap invocation versions when s=9999. Since the elapsed time is the same, the cost of function invocation on heap from the beginning and that of function invocation whose instance is swapped to heap after its suspensive stack invocation is suspended can be said as almost the same. We can conclude that suspensive invocation will not degrade the performance of functions which may be suspended, compared to heap invocation.

	parameter s			
	0	1000	4000	9999
heap invocation (suspended or not)	0/10000	1000/9000	4000/6000	10000/0
IFETCH (suspended or not)	1/9999	1001/8999	4001/5999	10000/0
elapsed time (sec)	2.4×10^1	2.4×10^1	2.4×10^1	2.4×10^1

Table 3.: Non-strict array generation (heap invocation)(size10000)

6 Discussion and related work

In the context of fine-grain object oriented programming language, hybrid stack-heap runtime system is also proposed[8, 11]. These seem to be provided for programs with irregular and potentially long latency communications, not for programs with non-strict semantics.

TAM[4] is also developed to implement fine-grain dataflow programming languages. Compiler controlled threading manages synchronization and scheduling reflecting a cost hierarchy. Although frame allocation of TAM is performed on heap by default similar to our previous abstract machine, TAM does not support hybrid invocation mechanism.

While some of the strict dataflow languages, such as Sisal, have achieved high performance on stock machines and demonstrated the feasibility of dataflow functional programming language[3], non-strict dataflow languages require more sophisticated techniques to achieve high efficiency. Table 4 shows the elapsed time of C program, Sisal program, and our program of `matmul` on an engineering workstation.

	C	Sisal	$V_{suspensive}$
Elapsed time(sec)	1.6×10^{-2}	3.7×10^{-2}	4.1×10^{-2}

Table 4.: Elapsed time of matmul(size20)

As seen from the table, by speculatively using stack frame, `matmul` program of non-strict language can execute almost as fast as a strict Sisal program. Since C is not so high level language as dataflow functional languages, C program runs about two time faster than ours. [3]

7 Conclusion

A lot of fine-grain function instances which may require dynamic scheduling are created while executing a non-strict dataflow program. The performance of the naive implementation using a heap instance frame is degraded on stock machines. The study described in this paper tried to achieve high performance of non-strict dataflow programs on stock machines. We discussed compilation issues

---

[3] Even if V is a non-strict language, it is possible to serialize all filling functions and translate them into a loop form for this `matmul`. Although this transformation technique is out of scope of this paper, further translation can make V code run as fast as C program.

to generate a code to be suitable for the execution style of commodity machines. We also developed a runtime system which speculatively uses stack frame invocation while assuring dynamic scheduling. As a static approach, we introduced a relaxed merging policy to a thread partitioning algorithm in order to find more functions which can use a strict call interface in non-strict dataflow language programs. Although a lot of function calls can exploit strict call interface, some of them may suspend and require dynamic scheduling. As a runtime support, we provided a hybrid runtime mechanism which can dynamically change a suspended stack frame into a heap frame. The results of the performance evaluation indicated that we can reduce superfluous heap frame management and achieve high efficiency for non-strict dataflow language even on stock machines.

References

1. M.Amamiya and R.Taniguchi : "Datarol: A Massively Parallel Architecture for Functional Language", Proc. IEEE 2nd SPDP, pp. 726-735, 1990.
2. Arvind, R. S. Nikhil, and K. K. Pingali "I-strutures:Data Structures for Parallel Computing" Technical Report TR-87-810,Cornel University, Ithaca, New York 14853-7501, Feb, 1987.
3. D. C. Cann "Retire Fortran? A Debate Rekindled", CACM, Vol.35, No.8, p.p.81-89, 1992.
4. D. E. Culler et al. "Fine-grain Parallelism with Minimal Hardware Support: A Compiler-Controlled Threaded Abstract Machine" In Proc. of 4th ASPLOS, 1991
5. T. Kawano, S. Kusakabe, R. Taniguchi, and M. Amamiya. "Fine-grain multi-thread processor architecture for massively parallel processing"In *Proc. of HPCA '95* pp.308–317, Jan. 1995.
6. B. Lee and A. R. Hurson, "Dataflow Architectures and Multithreading," IEEE Computer, Aug. 1994, pp.27-39.
7. K. E. Schauser and S. C. Goldstein. "How Much Non-strictness do Lenient Programs Require?" In Proc. FPCA, 1995
8. J. Plevyak, V. Karamchet, X. Zhang, A. A. Chien "A Hybrid Execution Model for Fine-Grained Languages on Distributed Memory Multicomputers," In Proc. of Supercomputing'95. 1995.
9. K. E. Schauser, D. E. Culler, and S. C. Goldstein "Separation Constraint Partitioning - A New Algorithm for Partitioning Non-strict Programs into Sequential Threads" In Proc. of POPL, Jan. 1995.
10. A. Shaw, Arvind, R. P. Johnson, "Performance Tuning Scientific Codes for Dataflow Execution" Proc. of PACT96, pp.198–207, 1996.
11. K. Taura, S. Matsuoka, and A. Yonezawa, "*StackThreads*: An Abstract Machine for Scheduling Fine-Grain Threads on Stock CPUs" Proc. of JSPP94, pp.25–32, 1994.

Resource Management Methods for General Purpose Massively Parallel OS SSS–CORE

Yojiro Nobukuni, Takashi Matsumoto, Kei Hiraki

Department of Information Science, Faculty of Science, University of Tokyo
7-3-1 Hongo, Bunkyo-Ku, Tokyo 113, Japan

Abstract. We propose two resource management methods; a scheduling policy that reflects resource consumption states and a memory-replacement strategy based on page classification under distributed shared memory architecture. The performances of the two mechanisms are evaluated by a probabilistic simulation. An instruction-level simulator simulates variety of process sets with finite resources on proposed resource-management methods. The results show the superiority of the proposed resource management mechanisms.

1 Introduction

NUMA architecture [10, 1] is widely accepted as basic architecture for very high-performance computers because huge systems can be built by simply connecting many pairs of processing elements and local memories.

Current parallelizing methods for on NUMA systems [11, 2, 8] and optimization techniques for parallel applications[12, 7] optimize execution of a single parallel application on a fixed system configuration. Therefore, running multiple parallel applications with competing resource allocation in general-purpose environment is much less efficient than that is expected. Dynamic partitioning of parallel system is a necessary feature of a general-purpose parallel systems but it may further reduce the performance. Operating system level optimization is required to coordinate resources to run each process efficiently.

We propose two resource management mechanisms for efficient running of multiple parallel processes. One is a process scheduling mechanism that utilizes information on physical page usage. The other is a memory replacement mechanism based on page attribute used for distributed-shared-memory management. The performances of these mechanisms are evaluated by detailed probabilistic simulation.

Previous operating systems[5, 9] that allowed gang scheduling with dynamic repartitioning did not use resource informations and has limited scheduling flexibility. DHC[4] designed for UMA uses management structure close to ours but only uses load informations[3].

Section2 describes the resource management mechanisms. Section3 shows outline of an operating system, $SSS - CORE$ and proposes scheduling polisies and page replacement policies. The methodology and the results of the simulation are given in section4 and 5 respectively. We conclude in section6.

2 Resource Management Mechanism

2.1 Scheduling Policy

Resource Management Tree To take resource consumption state into scheduling policy, resource related information must be managed. Our approach is to construct a data structure called *resource management tree* (RMT) to maintain system-wide resource usage and each process's resource consumption state.

RMT is hierarchically structured to be scalable. In addition, variants of parallel systems can be supported by adopting real structure of RMT to each specific systems, from workstation clusters to parallel super computers with flat networks. Scheduling decisions based on the structure naturally reflects the distances and hence the access costs between distributed resources. RMT has further advantages. It reduces the quantity of required physical resource for storing resource information. Bottlenecks of accessing the information is avoided.

Each node of the resource management tree logically holds information for resources seen below the node. They are number of processors and physical pages, number of total free processors and physical pages, and number of processors, physical pages and ID of each process. The root node additionally has priority, scheduling constraints and home node of each process. Figure 1 shows an example image for a four processor system with RMT.

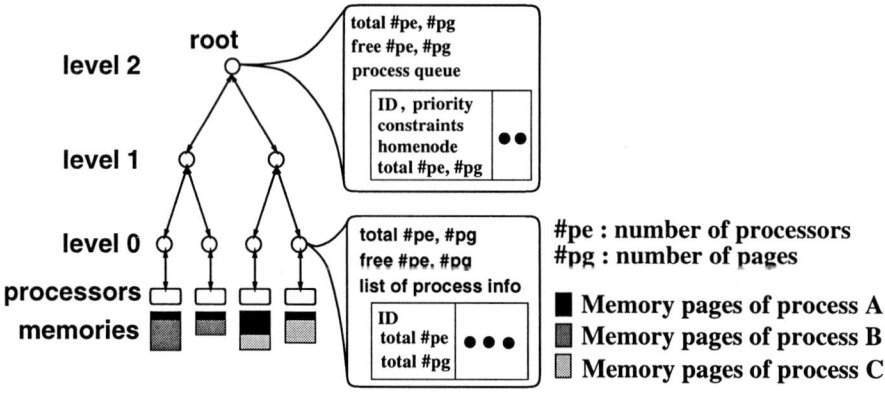

Fig. 1. Resource Management Tree

A process can gain performance by freely using allocated resources and by making the use of application level optimization. This can be done by using 2-level scheduling. The kernel allocates resources to each process by looking into the *resource management tree*. This way, resources that most fit for the use of a process can be allocated. The resource allocation within a process is left to user level scheduler, which freely re-allocate resources into internal threads.

Scheduling Constraints A parallel optimizing compiler[12] generally assumes that system resources are used by a single application. Our goal is to run object codes efficiently in a multiple user/multiple process environment. In such an environment, a process can achieve higher speedup when allocated resources satisfy its requirements and preferences. Scheduling constraints are used by processes to specify these informations to the kernel. The kernel follows the given constraints so that the requirements of a process can be satisfied as much as possible.

A process can use scheduling constraints to specify its requirements of and preferences for the number of processors to use, communication cost between processors, memory access costs, and process migration. The *fixed processors constraint* expresses a constant number of processors a process requires. The *variable processors constraint* enables a process to be allocated variable number of processors.

Priority Computation Since resources are allocated to satisfy each process requirements, a mechanism must be arranged to coordinate fair sharing of resources. Fairness can be achieved by managing priorities according to the amount of used resources and strength of given scheduling constraints and scheduling in priority order. Aging priorities according to these terms realizes fairness.

Priority is based on following values; (1)amount of used resources: U_r, (2)strength of scheduling constraints: R_c, (3)degree of constraints satisfaction: $S_c = 0\ or\ 1$, (4)amount of wasted resources: W_r, (5)presence of waiting process: $f_w = 0\ or\ 1$
The aging value of a process is computed from next expression.

$$aging = (U_r R_c S_c + W_r) f_w * t - C_r(1-t)$$

Smaller the value, higher the priority. C_r is the aging coefficient. To prevent priority values to divert, the sum of the aging values of processes that were running before the time slice is equally divided and distributed to waiting processes.

2.2 Memory Replacement Strategy

Under general environment where multiple processes execute simultaneously, system must be designed to bear situations when physical memories are exhausted. Selecting victim pages from those that are less frequently accessed and have less re-reference cost will enhance system performance.

Pages that belong to the currently running process are usually more referenced. Generally, local pages are more referenced than those shared by threads. Suppose a shared copy page has been replaced, it can be obtained with small cost through network from remote cluster. However, replacing pages that invoke disk accesses on re-reference can be a source of slow down.

3 SSS–CORE

The mechanisms proposed in the paper are planned to be built into an operating system called *SSS–CORE*. The performance of the resource management mechanisms introduced are evaluated by simulating system with *SSS–CORE*.

SSS–CORE[6] is a general purpose massively parallel operating system for NUMA parallel distributed systems. It provides multiple user/multiple job environment with timesharing and space partitioning. The main objective of *SSS–CORE* is achieving maximum performance from each parallel application.

SSS–CORE provides a mechanism that allows information transfer between kernel and user level as described in the previous section. The information includes those used by the resource management mechanisms, e.g. usage of physical pages and number of processes on each node.

3.1 Scheduling Policy

The performances of five kernel-level scheduling policies are compared through evaluation. Every policy computes process priorities according to resource consumption state at each time slice and schedules processes with highest priorities. Processors are looked for within a particular area and allocated to a process if sufficient number of processors are found in the are. The policies are described below.

Policy0 allocates randomly selected requested number of clusters
Policy1 allocates requested number of continuous clusters in a fixed order
Policy2 first allocate clusters in home-node area where pages of target process exist, then clusters in whole area will be tried on failure.
Policy3 same as **Policy2**, but only home-node area is tried
Policy4 same as **Policy3**, but clusters that actually has target process's pages are allocated

The home-node of a process represents a subtree of the resource management tree that includes its requested number of processors. It somewhat corresponds to the area where the process was previously scheduled. Figure 2 gives a home-node example. Process A in the figure, which has pages at marked clusters in area4, takes a node as its home-node which represents the subtree in area3. Area3 is called *home-node area* of process A.

Policy2, 3, 4 use resource management tree. The difference among these three policies is in how much they persist in allocating processors from clusters where process's currently using physical pages are located. The difference is in the action taken when sufficient number of processors cannot be prepared by those clusters. **Policy2** looks for processors for all clusters. **Policy3** tries to allocate from clusters in home-node area; the subtree below the home-node of target process. **Policy4** gives up scheduling the target process. Figure 2 shows the area where each of policies look for processors to allocate. Area 2, 3, 4 corresponds to the area for **Policy2, 3, 4** respectively. **Policy4** mostly schedules a process

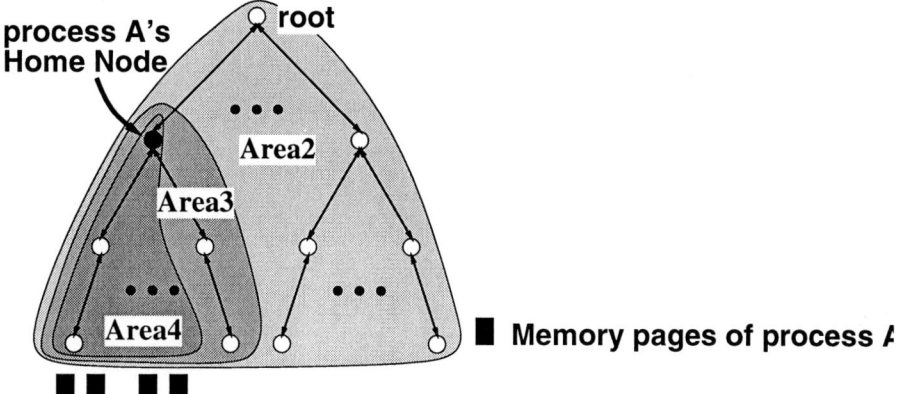

Fig. 2. Scheduling Target Area

to the same processors time to time. Chances that processes will be scheduled to clusters where they hold physical pages are greater in **Policy4, 3, 2** order. More processors may be utilized in reverse order.

Defining as many user-level schedulers as the number of processes to simulate is not possible. A single policy is defined and used by all processes. It schedules the identical threads to the processors that were also allocated to the process at previous allocation by the kernel level scheduler.

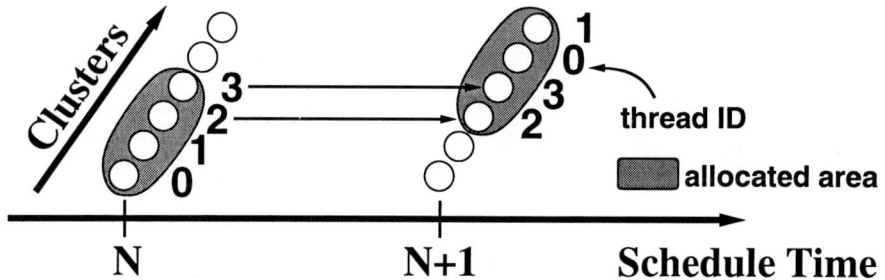

Fig. 3. An Example of User Level Scheduling

Figure 3 shows an example of thread scheduling. In the example, processors that are allocated to the process by $(N + 1)$st scheduling as well as (N)th scheduling will run threads 3 again. Processors that are newly allocated to the process at $(N + 1)$st scheduling will run remaining threads (threads 0, 1, 2) in thread ID order. When thread is scheduled to different clusters, its local pages must be transferred through the network. Distributed shared memory system is responsible for properly transferring the shared pages of the thread. Clearly, the more overlaps in allocation area, the lesser the amount of page transfers.

Note that when time quantum is sufficiently larger than the time required for context switching, required time for computing scheduling itself is relatively small. SSS–CORE will use larger value for time quantum. The time required for scheduling is ignored in the simulation.

3.2 Page Replacement Strategies

Two page replacement strategy is evaluated and compared with each other.

Strategy0 Simple LRU without page classification
Strategy1 Uses page class. Processes are scanned in reverse priority order.

Assuming distributed shared memory system, memory pages can be classified into 6 groups by pointing whether a page; (a) belongs to currently running process or not, (b) is shared page or local page, and (c) has other copy pages or not. The page classes for **Strategy1** are; (1) copy page of not running process, (2) copy page of running process, (3) last one page of not running process, (4) local page of not running process, (5) last one page of running process, and (6) local page of running process. "Last-one" page in the list means a shared page without any copy thus requires a disk access on next access. Ordering between classes 4 and 5 cannot be given trivially. Class 4 is prior to 5 in the list to maximize the efficiency of currently running process.

Since processes are scheduled in priority order, pages of lower priority process are possibly less referenced. **Strategy1** utilizes this characteristic. Both strategies will not select coherency processing shared pages as the victim page for replacement.

Table 1. Process set for simulation

Process set	Number of proc's	Parallelism (Processes with the same parallelism)	Total parallelism	local mem size [pages]	shared mem size [pages]	VR ratio [VR ratio]	Sync Interval [clocks]
A	12	16,36,48,50(2),64,70 96,100,128,192,200	1050	35500	10120	1.00	10000–20000
K	12	48(5),96(5),208,256	1184	49920	6360	1.30	10000
L	11	64(5),128(2),192(2),256(2)	1472	68480	7280	1.68	10000
M	23	16(16),50(6),256(1)	812	35480	30320	1.70	10000

Table 2. Parameters and Costs

Parameters	Values
Number of processors	256
Average memory access interval	10 clk
Disk access cost	100000 clk
Page transfer startup cost	500 clk
Communication startup cost	50 clk
Pages per cluster	400 pages
Page size	4096 Byte
Total memory	409.6 Mbyte
1 quantum	1000000 clk

4 Simulation Model

Operating system level simulation of parallel architecture by highly detailed model is practically impossible. Executing a particular suite of applications on the simulator is not sufficient for evaluating a general purpose operating environment. Therefore, even instruction level simulation does not fit for the objective. We use a detailed probabilistic model for simulation. Probabilistic model is good for simulating variety of processes. However, a stream of process activities cannot easily be given the meaning from program point of view.

A pair of a processor and a memory constructs a cluster. Clusters are connected by tree structured doubly linked interconnecting network. The value a message actually takes for moving one-hop is computed from basic transfer cost of each type of messages and the bandwidth of the network connection where it is passing.

A process has as many number of threads as the number of processors it requests. It uses the *fixed number of processors scheduling constraint*, and thus its parallelism never changes through its life time. Threads here denotes the execution context of a parallel process at a cluster. A thread of a process has own local memory space and a shared memory space shared among threads of the process. Both memory spaces are provided with reference frequency tables that describe how frequently each page of the space is accessed.

Pages of shared space are managed by distributed shared memory system. Sequential consistency memory model with an update protocol is used. Every write access starts update processing by sending update messages to every copy. The processor stops until it collects all acknowledges. To model NUMA systems, memory access cost and basic communication costs are set as to satisfy "local access $\ll$ inter-cluster access$\ll$ disk access". When threads change clusters on which it executes, its local pages are moved on-demand through network. Shared copy pages that do not reside on currently allocated clusters are removed without any cost. Accesses to unloaded virtual pages will cause disk accesses.

Process execution is clock-based probabilistic model. Processes make memory reference actions at each clock if possible. With given interval of effective execution clocks, randomly selected threads of a process synchronize by a simple barrier. Effective execution means the time or clocks spent for other than waiting for synchronization to complete or for memory access processing to end.

5 Simulation Results

The parameters used in simulation are shown in Table 2. System with as many as 256 processors is evaluated. The topology of the network is three and four leveled tree structure. The former expands at root level into 4-way, then 8-way, and 8-way at the bottom level (**w488**). The latter expands 4-way at each level of the network(**w4444**). Table 1 describes the sets of parallel processes simulated.

Each experiment is carried on until one of the processes in a set stops execution or 100 time slices has passed. The results are shown in Figure 6, 4, 6 and

6. Graphs on the left columns are results of **w488** system, and on the right are of **w4444** system.

Each group in a graph is the results of scheduling **Policy0** through **Policy4**. Three left bars of a group are the results of **Strategy0** and the others are of **Strategy1**. Three evaluated values are plotted; (1)net effective execution rate, (2)calibrated effective execution rate. (3)maximum effective execution rate, and (1) is total efficiency of processors when processes are scheduled. Idle processors to which no processes are scheduled are not included. All processor idle times are accumulated to (3) and (2) is computed by following expression, (1) * (1.0 + idle time rate). Processor idle times are accumulated with the ratio of net effective execution rate.

5.1 Kernel Level Scheduling

Policy4 shows the best performance even when compared by net effective execution rate, which is disadvantageous for **Policy4** because processor idle times are not included. When compared by other evaluations, the difference becomes larger. On real systems, processor idle time can be reduced by following two methods.

1. using *variable number of processor scheduling constraint* (will be introduced to *SSS–CORE*), processors can be flexibly utilized for number of processors.
2. un-allocated spaces caused by scheduling oriented processor fragmentation can be utilized by another process not included in a particular set of processes.

In case 1, the calibrated performance can generally be expected because processes will utilize the newly allocated processor space by *variable number of processor scheduling constraint* as much efficiently as they used the same space when allocated by *fixed number of processors scheduling constraint*. When an additional process is assumed for a particular set of processes, it can use the processors in formerly fragmented space as much efficiently as maximum effective execution rate, depending on its characteristics as a parallel process. Thus maximum effective execution rate can be expected for case 2. Evaluating cases for variant process sets other than those experimented is inevitable and important for describing the performance of general purpose operating system and prospecting how the performance of *SSS–CORE* will be. Comparison by calibrated or maximum effective execution rate is validated from this point of view.

Policy4 is the best in efficiency by any of the three estimations. The quantity of page transfer is larger among **Policy2**, **3**, **4** in the order. Table 3 shows that the time spent in synchronization or communication get larger for policies in the same order. Changing processor allocation space time to time cause each memory to be filled with pages from many processes (Disk acssessing time is not included). When **Policy2** or **Policy3** is used, processes scramble for the physical pages and result in lower in efficiency than **Policy4**.

5.2 Memory Replacement Strategies

Strategy1 always outperforms **Strategy0**. Table 4 is the breakdown of replaced counts for each class of pages. Results of the scheduling **Policy4** on **w488** system is shown.

As for **Strategy1**, mostly copy pages are replaced. Process sets A and K, which impose small physical memory requirement, see only copy pages of not running processes victimized. But for **Strategy0**, local pages and last-one shared pages are replaced.

The results show that selecting victim pages according to the classification enhances system performance. Even when the system is somewhat highly loaded, efficiency is preserved by not kicking local pages that are more frequently referenced out of memory.

5.3 Considerations on Realizability of General Environment

Maximum efficiencies of the results are roughly between 65% to 85% for the experiment of scheduling policy **Policy4** and **Strategy1** replacement strategy pair. The lower results come from sequential consistency memory model. The time waiting for preceding accesses to complete is very large. Update processing time is not very large compared to the waiting time. Cooperating more relaxed memory model and lighter consistency managing system solves the problem. In addition, performances of parallel applications with large shared access frequencies can be enhanced with various compilation techniques and by introducing useful communication techniques, such as hierarchical multicasting and acknowledge combining. Thus, lower simulation results does not negate general environment on parallel distributed system.

6 Conclusion

The paper has described kernel level scheduling policy that uses information of resource consumption state and memory replacement strategy that uses page

Table 3. Execution Time Breakdown (%) (w488, Strategy1 replacement)

Set	Type	algo0	algo1	algo2	algo3	algo4
A	Max	57.89	57.36	60.41	64.22	84.02
	Sync	19.98	19.59	20.84	20.03	12.45
	Comm	22.10	23.01	18.72	15.70	3.49
K	Max	42.65	56.57	59.53	59.53	74.91
	Sync	25.87	22.08	23.24	23.24	17.82
	Comm	31.46	21.32	17.20	17.20	7.22
L	Max	38.81	48.93	54.98	54.98	65.85
	Sync	28.52	24.07	23.58	23.58	22.23
	Comm	32.64	26.96	21.39	21.39	11.87
M	Max	24.79	35.73	47.42	61.47	64.37
	Sync	22.83	19.03	18.68	16.57	16.06
	Comm	52.36	45.21	33.86	21.90	19.51

Table 4. Replaced times for each class of pages (**Policy4, w488**)

Page Replace Strategy	Proc Sets	Page Class					
		other's copy	own copy	other's last	other's local	own last	own local
Strategy 0	A	10944	4770	723	4763	3031	130
	K	33105	10487	830	22598	653	803
	L	92253	18063	4911	63885	1369	978
	M	46394	23581	6873	11725	4054	6577
Strategy 1	A	38976	0	0	0	0	0
	K	77665	0	0	0	0	0
	L	414287	0	0	0	0	0
	M	245705	11578	0	0	0	0

classification upon distributed shared memory system. The performances of various methods for these mechanisms are evaluated by simulating on detailed probabilistic model.

As for the kernel level scheduling, the simulation results showes the superiority of those policies that use resource management data structure and allocate processors of clusters with memory affinity to a process. Replacing pages according to the page classification is found much superior than replacement policy by simple LRU order without the classification. When these two mechanisms are in-cooperated together, the effective execution rate of the system is higher than 65% for highly loaded cases.

Acknowledgement

The work is supported by IPA Advanced Information Technology Program (AITP) of Information-technology Promotion Agency (IPA), Japan.

References

1. Intel Supercomputer Systems Division. *Paragon User's Guide*, order number 312489-003 edition, June 1994.
2. F. Douglis and J. K. Ousterhout. Process Migration in Sprite Operating System. *Proc. of the 7th Inter. Conf. on Distributed Computer Systems*, September 1987.
3. D. G. Feitelson. Packing schemes for gang scheduling. In *Proc. IPPS'96 Workshop on Job Scheduling Strategies for Parallel Processing*, pages 54–66, April 1996.
4. Dror G.Feitelson and Larry Rudolph. Distributed Hierarchical Control for Parallel Processing. *IEEE Computer*, 23(5):65–77, May 1990.
5. B. C. Gorda and E. D. Brooks III. Gang Scheduling a Parallel Machine. Technical Report UCRL-JC-107020, Lasrence Livermore NL, December 1991.
6. T. Matsumoto, S. Huruso, and K. Hiraki. General Purpose Massively Parallel Operating System SSS–CORE. *Proceedings of 11th Japan Society for Software Science and Technology*, pages 13–16, October 1994. (in Japanese).
7. Takashi Matsumoto. Synchronization mechanisms and processor scheduling on multiple processors. *IPS Japan SIG report*, pages 1–8, November 1989. (in Japanese).
8. Takashi Matsumoto. Elastic barrier. Generalized barrier synchronization mechanism. *Trans. of IPS Japan*, 32(7):886–896, July 1991. (in Japanese).
9. J. K. Ousterhout, D. A. Scelza, and P. S. Sindhu. Medusa: an experimant in distributed operating system structure. *Comm. ACM*, 23(2):92–105, February 1980.
10. J. Palmar and Jr. G. L. Steele. Connection Machine model CM-5 system overview. *4th Symp. on Frontiers Massively Parallel Comput.*, pages 474–483, October 1992.
11. T.E.Anderson et al. Scheduler activations: Effective kernel support for the user-level management of parallelism. *Proc. of the 13th ACM Sympo. on Operating Systems Principles*, 25(5):95–109, October 1991.
12. R. P. Wilson and et al. SUIF: An Infrastructure for Research on Parallelizing and Optimizing Compilers. *ACM SIGPLAN Notices*, 29(12):31–37, December 1994.

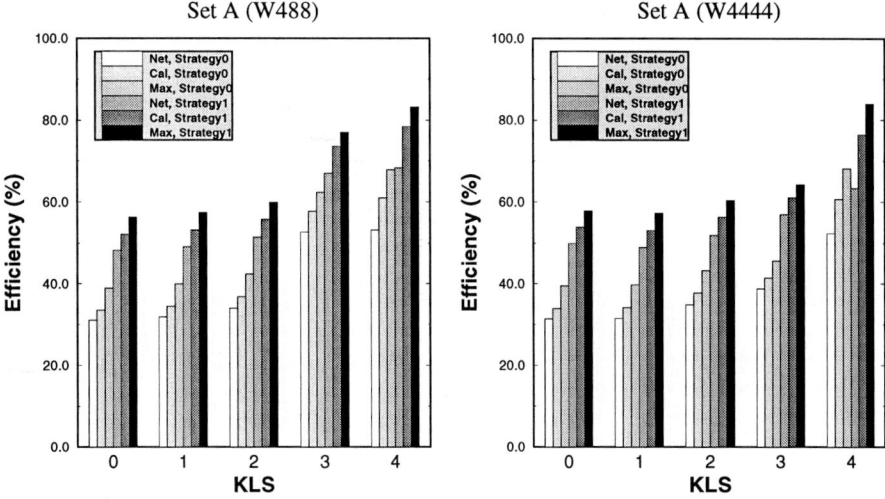

Fig. 4. Results of Process Sets A **w488**(Left), **w4444**(Right)

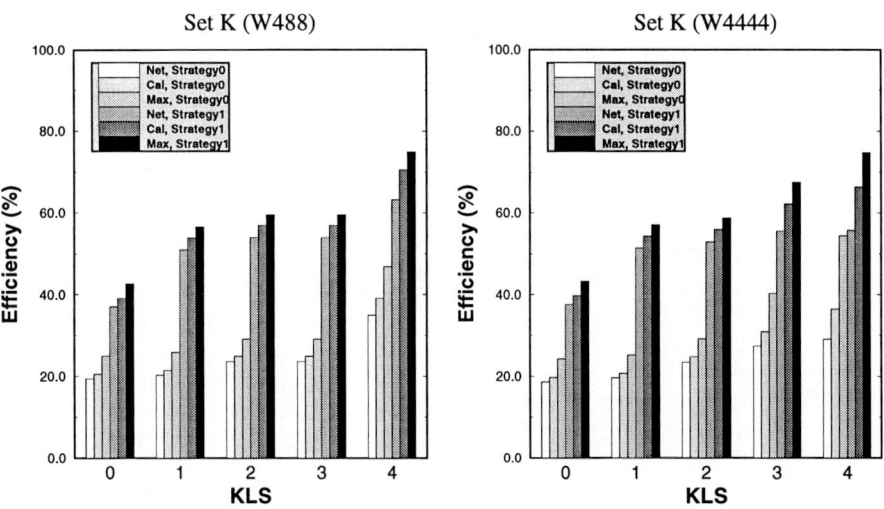

Fig. 5. Results of Process Sets K **w488**(Left), **w4444**(Right)

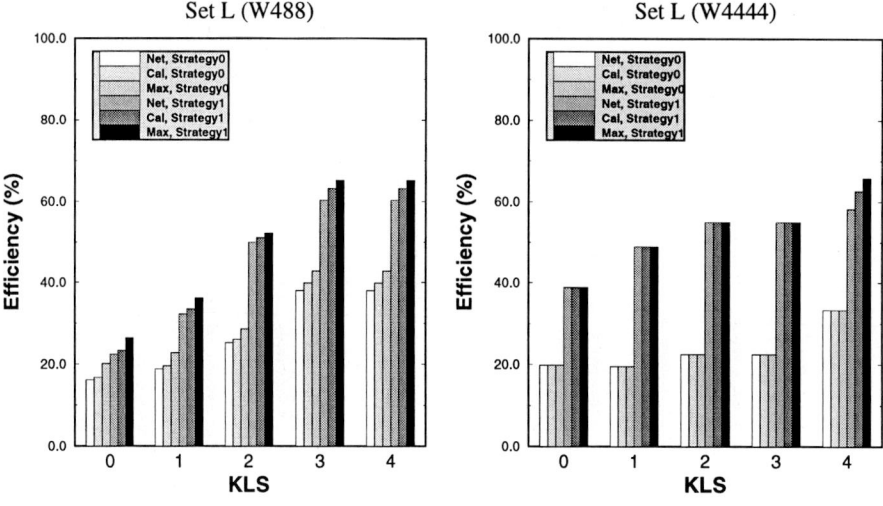

Fig. 6. Results of Process Sets L **w488**(Left), **w4444**(Right)

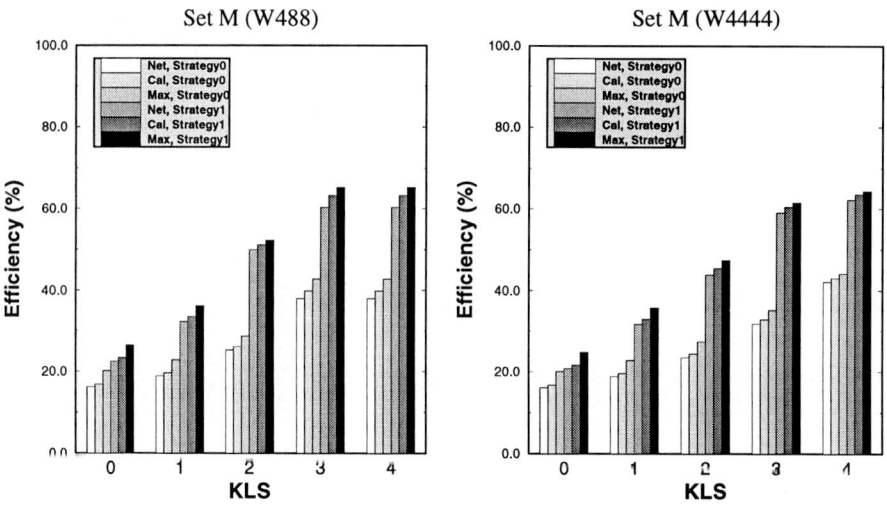

Fig. 7. Results of Process Sets M **w488**(Left), **w4444**(Right)

Scenario-Based Hypersequential Programming: Formulation of Parallelization

Naoshi Uchihira, Hideji Kawata, Fumitaka Tamura

Systems & Software Research Laboratories, Toshiba Corporation,
70, Yanagi-cho, Saiwai-ku, Kawasaki 210, Japan
{uchi,kawata,tamu}@ssel.toshiba.co.jp

Abstract. Hypersequential programming is a new paradigm of concurrent programming in which the original concurrent program is first serialized, then tested and debugged as a sequential program, and finally restored into the target concurrent program by parallelization. Both high productivity and reliability are achieved by hypersequential programming because testing and debugging are done for the serialized version and correctness of the serialized program is preserved during subsequent parallelization. We have proposed a practical embodiment of hypersequential programming based on a sequential execution history, called a *scenario*. This paper formalizes the parallelization step of scenario-based hypersequential programming by introducing a new equivalence relation, *scenario graph equivalence*, and shows a concrete parallelization algorithm.

1 Introduction

The rapid shift toward parallel and distributed computer systems has increased demand for application programmers for concurrent programs. However, concurrent-program development is generally more difficult than is sequential program-development [1]. Especially, testing and debugging is one of the bottlenecks [2]. We have proposed a new paradigm, called *"hypersequential programming"* [3], to remove this bottleneck. Figure 1 shows this paradigm, where the original concurrent program is first serialized, then testing and debugging are done for the serialized version, and finally it is parallelized and the target, reliable, concurrent program is reconstructed. Hypersequential programming makes concurrent programming as easy as sequential programming because testing and debugging are done for the serialized version. Since its correctness is preserved in the subsequent parallelization step, the generated concurrent program is also guaranteed to be reliable.

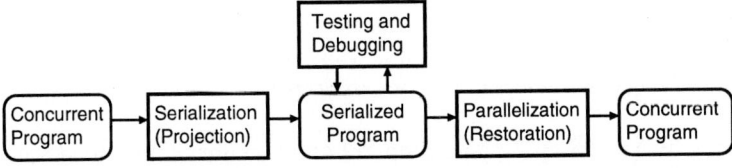

Fig. 1. Block Diagram of Hypersequential Programming

There are a variety of concrete procedures based on this concept. Petri-net-rewriting-based hypersequential programming was presented in [3]. Another practical approach is *scenario-based hypersequential programming*, a basic concept of which was introduced in [4]. In this context, a *scenario* means a sequential execution history intentionally derived by the programmer. In this paper, we formalize the parallelization step of the scenario-based approach.

2 Scenario-Based Hypersequential Programming

In scenario-based hypersequential programming, a programmer develops a concurrent program according to the following steps.

Step 1: Modeling and coding a program
Model the target system as it is, using concurrency and nondeterminacy naturally. Then code it with a concurrent programming language.

Step 2: Testing scenarios and constructing a scenario graph
   Step 2-1: Execute the concurrent program sequentially according to the programmer's test case. The programmer can select interactively which process/statement is executed next with a graphical tool (Fig.2). As a result of a chain of selection and execution, the execution history is constructed. This sequential execution history is called a *scenario*.

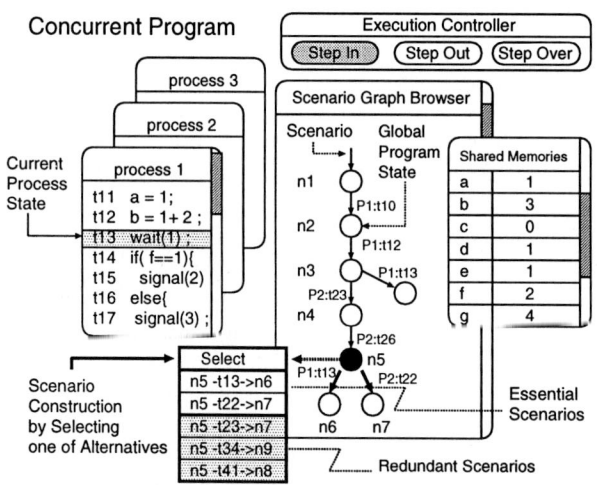

Fig. 2. Scenario Selection and Testing Tool

Step 2-2: Confirm that each scenario satisfies the programmer's requirements in the way of conventional testing and debugging for sequential programs. If bugs are detected, go back to Step 1 and modify the original source code.

Step 2-3: After Step 2-1 and Step 2-2 are done for all intended scenarios, a global state transition system, called a *scenario graph*, is constructed. In the scenario graph, a node represents a global program state, an edge represents an executed program statement in that state, and a path corresponds to a scenario, i.e., a feasible, intended, and debugged behavior of the concurrent program.

Step 3: Parallelization

With dividing the scenario graph onto processes by projection, local scenario graphs are constructed (Fig. 3). This step is a kind of parallelization which consists of the following steps.

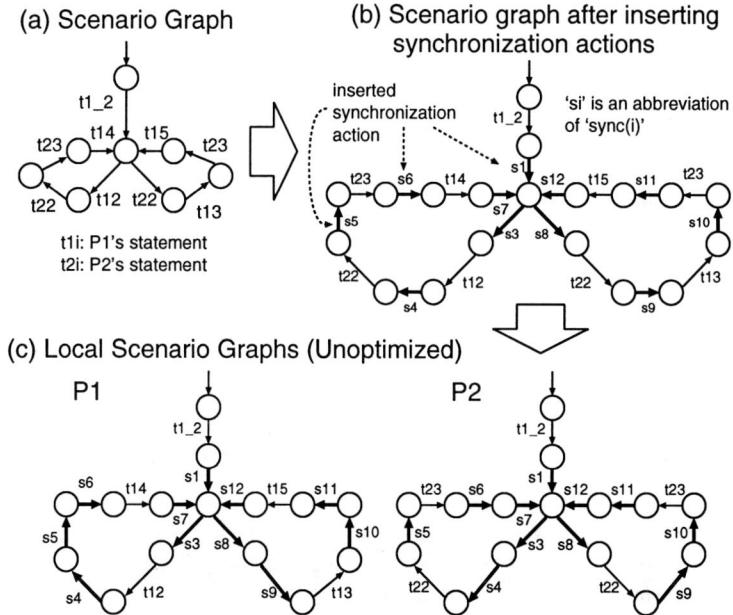

Fig. 3. Parallelization from Scenario Graph SG_C

Step 3-1: Insert dummy actions sync(ID) into the scenario graph which are used for global synchronization after the next projection step. As a basic rule, dummy synchronization actions are inserted immediately after all statements and just before branch statements (Fig. 3(b)).

Step 3-2: Divide the scenario graph onto processes by projection. The divided graph is called a *local scenario graph* in which the statements of each process and all synchronization actions remain. For example, a local scenario graph of P_1 in Fig.3(c) consists of P_1's own statements ($t_{1\_2}, t_{12}, t_{13}, t_{14}, t_{15}$) and synchronization actions ($s_1, ..., s_{12}$) and does not contain P_2's local statements (t_{22}, t_{23}). The local scenario graphs can be regarded as concurrent processes which are synchronized with each other by synchronization actions. Please note that a concurrent program composed of these (unoptimized) local scenario graphs behaves faithfully according to the sequential scenario of the scenario graph because execution of process statements is completely serialized by the inserted dummy synchronization actions.

Step 3-3: Optimize each local scenario graph by removing redundant global synchronization, which means restoration of the original concurrency and nondeterminacy. In this optimization, parallelization techniques of the compiler are utilized so that only precedence constraints among inter-dependent critical statements are preserved and other precedence constraints (unnecessary serialization) are removed as much as possible.

Step 4: Code generation
The final concurrent program is generated from the optimized local scenario graphs and the original program. Each process is reconstructed by inserting additional synchronization statements (semaphore, etc.) into original statements so as to simulate the corresponding optimized local scenario. Furthermore, source code level optimization can be applied.

3 Formalization and Parallelization Algorithm

We formalize the scenario-based hypersequential programming, especially parallelization step (Step 3), by introducing the equivalence relation. First, we define the *communicating transition systems* (CTS) for the target concurrent program.

Definition 1 (Petri Net) $N = (P, T, F, m_0)$ is a Petri net where:

- $P = \{p_1, p_2, ..., p_n\}$ is a finite set of places,
- $T = \{t_1, t_2, ...t_m\}$ is a finite set of transitions,
- $F \subset (P \times T) \cup (T \times P)$ is a set of arcs,
- $m_0 : P \to \{0, 1, 2, ...\}$ is the initial marking, and
- $P \cap T = \emptyset$ and $P \cup T \neq \emptyset$

We assume that a marking $m : P \to \{0, 1, 2, ...\}$ ($m(p)$ means a number of tokens in a place p) of a Petri net is changed according to the conventional transition firing rules. Several notations are introduced. $\bullet t = \{p \mid (p, t) \in F\}$ and $t \bullet = \{p \mid (t, p) \in F\}$ represent input and output places of a transition t, respectively. When $t \in T$ is enabled at a marking m, we denote $m[t >$. After firing t, if m' is a new marking, we denote $m[t > m'$. In the case that $m_1[t_1 > m_2, m_2[t_2 > m_3, ..., m_{k-1}[t_{k-1} > m_k$ for a transition sequence $\theta = t_1...t_{k-1}$, we denote $m_1[\theta > m_k$. A set of reachable markings $R(N)$ of Petri net N is defined as $R(N) = \{m \mid \exists \theta \in T^*.m_0[\theta > m\}$. Here, T^* is a set of finite sequences of T. $\theta[i] = t \in T$ means the i-th element of θ. $c(\theta, i) = (t, k)$ is an i-th element t of θ with a counter k, i.e., $t = \theta[i]$ and t appears k times in $\theta[1], ..., \theta[i]$. Finally, a projection operator $\theta \mid_T$ is introduced by the following definition: $\theta \mid_T = \theta'$ such that $\theta'[i] = \theta[i]$ if $\theta[i] \in T$, otherwise $\theta'[i] = \varepsilon$ (empty sequence).

Definition 2 (Communicating Transition Systems)
$CTS\ C = (N, \Psi)$ is a specific Petri net $N = (P, T, F, m_0)$ with a process structure $\Psi = \{(P_1, T_1), (P_2, T_2), ..., (P_n, T_n)\}$ where

- (P_i, T_i) is a process, and $P_i \subset P$ and $T_i \subset T$ for $1 \leq \forall i \leq n$,

- $P = P_1 \cup P_2 \cup ... \cup P_n$, and $P_i \cap P_j = \emptyset$ for $0 \leq \forall i < \forall j \leq n$,
- $T = T_1 \cup T_2 \cup ... \cup T_n$, and some transitions may be included by multiple processes (i.e., $t \in T_i \cap T_j$), which realize process synchronization, and
- each process is a finite state transition system, i.e.,

$$\sum_{p \in P_i} m(p) = 1 \text{ and } \mid t \bullet \cap P_i \mid = \mid \bullet t \cap P_i \mid = 1 \text{ for } 1 \leq \forall i \leq n, \forall m \in R(N).$$

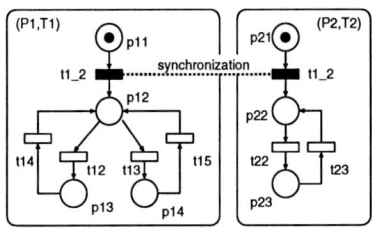

Fig. 4. Communicating Transition Systems (CTS) $\mathcal{C}$

Figure 4 shows an example of CTS $\mathcal{C} = (N, \Psi)$ where $\Psi = \{(P_1, T_1), (P_2, T_2)\}$. (P_1, T_1) and (P_2, T_2) are transition systems with one synchronization transition $t_{1\_2} \in T_1 \cap T_2$.

A state space $SS(\mathcal{C})$ generated from CTS $\mathcal{C}$ is formalized as a global state transition system. $SS(\mathcal{C})$ represents all possible behaviors of $\mathcal{C}$.

Definition 3 (State Space) *A state space $SS(\mathcal{C})$ generated from CTS $\mathcal{C} = (N, \Psi)$ is defined as a labeled transition system $SS(\mathcal{C}) = (S, T, \delta, m_0)$ such that*

- $S = R(N)$,
- $\delta \subset S \times T \times S$ is a transition relation,
- $\forall m, \forall m' \in S, \forall t \in T. (m, t, m') \in \delta$ iff $m[t > m'$.

In scenario-based hypersequential programming, a firing transition sequence θ ($m_0[\theta >$) which the programmer interactively selects is called a *scenario*. A set of scenarios can be also represented by a labeled transition system (a selected subgraph of the original state space). We call it a *scenario graph*.

Definition 4 (Scenario Graph) *A scenario graph $SG = (S, T, \delta', s_0)$ is a subgraph of a state space $SS(\mathcal{C}) = (S, T, \delta, s_0)$ such that $\delta' \subset \delta$.*

In our modeling, transitions correspond to program statements including assignment and comparison statements (e.g., a=b-1, a==b). In the case of a concurrent program with shared memories, it is necessary to consider dependency among transitions which access these memories. Let $D(\mathcal{C})$ be a dependency relation in a CTS $\mathcal{C}$. $(t_i, t_j) \in D(\mathcal{C})$ means that there is a dependency between t_i and t_j. $D(\mathcal{C})$ is a symmetric relation. For a given transition sequence θ, a precedence constraint ($\prec$) among transitions with counters $c(t, k)$ in θ can be defined according to the transition dependency.

Definition 5 (Precedence Constraint) *For a given firing sequence θ of CTS $\mathcal{C}$, $c(\theta, i) \prec c(\theta, j)$ iff $(\theta[i], \theta[j]) \in D(\mathcal{C})$ and $i < j$.*

Then, we introduce the equivalence relation between scenario graphs ($\approx_s$: *scenario graph equivalence*). The scenario graph equivalence is an extended trace equivalence considering precedence constraints. You may wonder why it does not adopt bisimulation equivalence; however, this is unnecessary because a scenario graph is a *deterministic* transition system.

Definition 6 (Scenario Equivalence)
For given scenarios θ_1 and θ_2, $\theta_1 \approx_s \theta_2$ iff

- $\forall i. \exists j. c(\theta_1, i) = c(\theta_2, j)$ and $\forall j. \exists i. c(\theta_2, j) = c(\theta_1, i)$
- $\forall i_1, i_2. \exists j_1, j_2. (c(\theta_1, i_1) = c(\theta_2, j_1)$ and $c(\theta_1, i_2) = c(\theta_2, j_2)$ and $c(\theta_1, i_1) \prec c(\theta_1, i_2) \Rightarrow c(\theta_2, j_1) \prec c(\theta_2, j_2))$
- $\forall j_1, j_2. \exists i_1, i_2. (c(\theta_1, i_1) = c(\theta_2, j_1)$ and $c(\theta_1, i_2) = c(\theta_2, j_2)$ and $c(\theta_2, j_1) \prec c(\theta_2, j_2) \Rightarrow c(\theta_1, i_1) \prec c(\theta_1, i_2))$

Definition 7 (Scenario Graph Equivalence)
For a given scenario graph $SG_1 = (S_1, T_1, \delta_1, s_{01})$ and $SG_2 = (S_2, T_2, \delta_2, s_{02})$, $SG_1 \approx_s SG_2$ iff $\forall \theta_1$ in $SG_1. \exists \theta_2$ in $SG_2. \theta_1 \mid_{T_1 \cap T_2} \approx_s \theta_2 \mid_{T_1 \cap T_2}$ and $\forall \theta_2$ in $SG_2. \exists \theta_1$ in $SG_1. \theta_1 \mid_{T_1 \cap T_2} \approx_s \theta_2 \mid_{T_1 \cap T_2}$.

Using the scenario graph equivalence, we can formalize the scenario-based hypersequential programming as the following problem.

Problem (Scenario Equivalent CTS Synthesis)
For a given scenario graph $SG_{\mathcal{C}}$ of CTS $\mathcal{C} = (N, \Psi)$ where $N = (P, T, F, m_0)$ and $\Psi = \{(P_1, T_1), ..., (P_n, T_n)\}$, synthesize a new CTS $\mathcal{C}' = (N', \Psi')$ where $N' = (P', T', F', m'_0)$ and $\Psi' = \{(P'_1, T'_1), ..., (P'_n, T'_n)\}$ such that

- $T_i \subset T'_i$ for each process (P_i, T_i),
- $SG_{\mathcal{C}} \approx_s SS(\mathcal{C}')$

The synthesized CTS $\mathcal{C}'$ has the same process structure as the original CTS $\mathcal{C}$, but its behavior is restricted by additional dummy synchronization actions sync(ID) for guaranteeing $SG_{\mathcal{C}} \approx_s SS(\mathcal{C}')$. There are a lot of scenario equivalent CTSs. The desirable CTS among them is one such that the fewest additional synchronization transitions are appended, that is, the most original concurrency and nondeterminacy are restored. To evaluate how desirable a CTS is, we introduce a simple measure of concurrency and nondeterminacy.

Definition 8 (Concurrency Rate) *A concurrency rate $\kappa(\mathcal{C})$ of CTS $\mathcal{C}$ is defined as $\kappa(\mathcal{C}) = \frac{|\delta|}{|S|}$ where $SS(\mathcal{C}) = (S, T, \delta, m_0)$.*

$\kappa(\mathcal{C})$ is an average number of branches from one state, which indicates how many processes are active and executable concurrently at the state.

Now, we introduce a concrete parallelization (optimization) algorithm corresponding to Step 3-3 of Section 2, which removes redundant synchronization actions in case of an acyclic CTS (i.e., a scenario graph has no loop structure). This algorithm constructs C_o from C_n (unoptimized CTS) such that $SS(C_o) \approx_s SS(C_n)$ and $\kappa(C_n) \leq \kappa(C_o)$. Before describing the algorithm, several definitions and theorem are introduced.

Definition 9 (Counting Trace) *For a given finite transition sequence θ, a counting trace $ct(\theta)$ is defined as follows.*

$$ct(\theta) = (\langle a\ set\ of\ all\ transition\ counts\rangle, \langle a\ set\ of\ all\ precedence\ constraints\rangle)$$

An transition count (a, k) means that action a occurs k times during θ. For example, $ct(\theta) = (\{(a,2),(b,2),(c,1)\}, \{c(\theta,1) \prec c(\theta,3), c(\theta,3) \prec c(\theta,5)\})$ when $\theta = abcba$ and $(a, c) \in D(C)$.

Definition 10 (Counting Trace Set) *For a given acyclic CTS C, a counting trace set $ctset(C)$ is defined as follows.*

$$ctset(C) = \{ct(\theta) \mid \theta\ is\ a\ firing\ transition\ sequence\ of\ C\}$$

For a acyclic CTS C, $ctset(C)$ can be enumerated because it is finite. And, the following theorem is introduced.

Theorem 1 (Scenario Graph Equivalence of Acyclic CTS)
For given acyclic CTSs C_1 and C_2,

$$C_1 \approx_s C_2\ iff\ ctset(C_1) = ctset(C_2)$$

Proof. When sequences θ_1, θ_2 are finite, $\theta_1 \approx_s \theta_2 \iff ct(\theta_1) = ct(\theta_2)$ from definitions. Then, the theorem can be proved straightforward.

Algorithm 1 (Optimization Algorithm (Acyclic Case)) *For a given acyclic CTS C_n with dummy synchronization actions, redundant synchronization actions can be deleted by the following steps.*

Step 1: *Insert one local non-synchronization dummy action $nsync(P_i, ID)$ for each dummy synchronization action $sync(ID)$ in each process P_i (Fig. 5(b)). Let this extended CTS be C_e. Let a set of all non-synchronization dummy actions which are inserted be ANS.*
Step 2: *Generate a state space $SS(C_e)$ from C_e (Fig. 5(c)).*
Step 3: *Detect a deviation path θ in $SS(C_e)$ such that $ct(\theta) \notin ctset(C_n)$, i.e., some precedence constraint is violated in θ, or deadlock occurs. Then, find the latest branching non-synchronization dummy action $nsync(P_i, ID)$ in θ (Fig. 5(c)). This means that the deviation is caused by introducing $nsync(P_i, ID)$, therefore, a synchronization dummy action $sync(ID)$ is necessary (not redundant).*

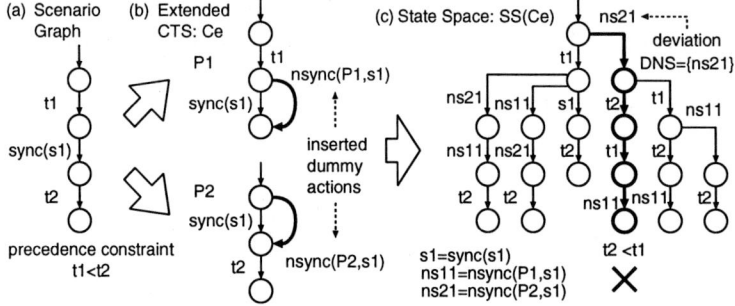

Fig. 5. Simple Example of Parallelization

Step 4: *Let a set of detected non-synchronization dummy actions be DNS. If $nsync(P_i, ID) \in ANS - DNS$, then $sync(ID)$ of P_i is redundant. After all redundant synchronization dummy actions are removed from C_n, the optimized CTS C_o is obtained. Here, $SS(C_o) \approx_s SS(C_n)$ from Theorem 1.*

In case of a CTS having loop structures, first partition the CTS into acyclic subgraphs. Then, each subgraph is optimized according to the algorithm. Finally merge them. This hierarchical approach is similar to task-level parallelization of Girkar and Polychronopoulos [5].

Figure 3(a) is one of the possible scenario graphs SG_C for the CTS C presented in Fig.4. After inserting dummy synchronization actions $(s_1, ..., s_{12})$, Figure 3(b) is generated (Step 3-1). The naive projection of the graph onto two processes P_1 and P_2 (Step 3-2) produces the two local scenario graphs (Fig. 3(c)). We assume $(t_{12}, t_{22}) \in D(C)$ and $(t_{13}, t_{22}) \in D(C)$. Obviously, there are several redundant dummy synchronization actions. Since these local scenario graphs have loop structures, acyclic subgraphs $B_{11}, B_{12}, B_{21}, B_{22}$ are retracted (Fig.6 (a)). Then, the above algorithm is applied to each subgraph (Fig.6(b)). Finally , the optimized local scenario graphs are generated by removing redundant synchronization actions $s_1, s_5, s_6, s_7, s_{10}, s_{11}, s_{12}$ (Fig.7). A CTS C_o reconstructed from these optimized graphs is scenario equivalent to SG_C (i.e., $SG_C \approx_s SS(C_o)$).

4 Example

We now explain scenario-based hypersequential programming by a simple example. The target shared-memory concurrent program P which consists of two processes P_1, P_2, and two shared memories m_1 and m_2.

Step 1: Modeling and coding a target program

The target concurrent program $P = P_1 \parallel P_2$ is described using C language with a multi-tasking library as shown in Fig 8. The multi-tasking library provides semaphore (*signal* and *wait*) as a synchronization mechanism.

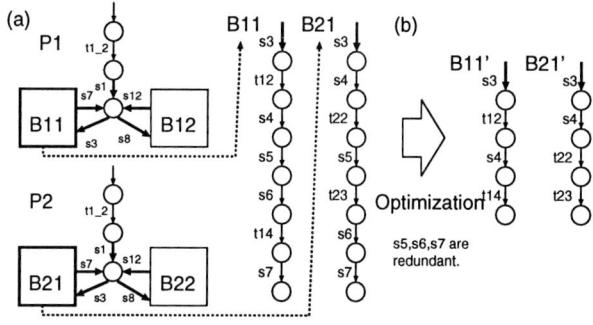

Fig. 6. Partitioning to Acyclic Subgraphs and Optimization

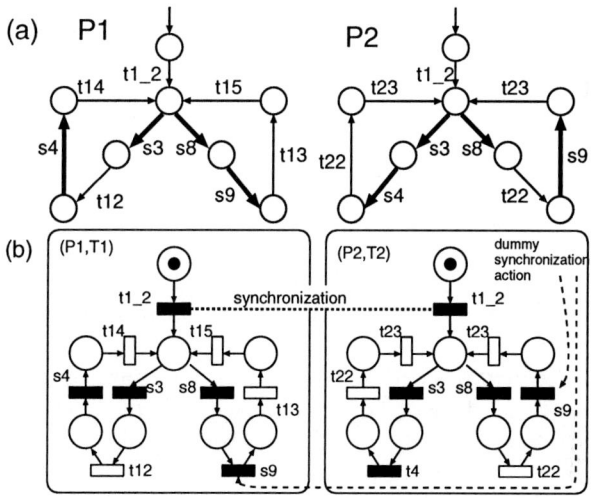

Fig. 7. Optimized Local Scenario Graphs (a) and Their CTS Representations (b)

Step 2: Testing scenarios and constructing a scenario graph

It is supposed that the programmer expects the following test cases (Fig. 9): P_2 must print "hello (good-bye)" after P_1 prints "say hello (good-bye)", respectively. If P_2 prints "hello" just after P_1 prints "say good-bye", it is a bug. We call this bug the "hello-good-bye-bug".

According to the test cases, the programmer executes the program step by step and tests it. If bugs are detected, the source code must be modified. In this example, a bug such that P_2 accesses m_2 (c=m2) before m_2 is initialized in P_1 (m2=0) is detected in the first scenario. Then the programmer debugs it by

```
int m1, m2 ; /* Shared Memories */
task P1()                                task P2()
{                                        {
    int f ;                                  int f, c ;
    m1 = 0 ;                                 f = 0 ;
    m2 = 0 ;                                 while(1){
    while(1){                                    m1 = f ;
        sleep(10) ;                              c = m2 ;
        f = m1 ;                                 if(c==0) {
        if(f==1) {                                   printf("Hello! \n") ;
            printf("Say hello! \n") ;            }
            f=0 ;                                else {
        }                                            printf("Good-bye! \n") ;
        else {                                   } ;
            printf("Say good-bye! \n") ;         f = c ;
            f=1 ;                            }
        }                                }
        m2 = f ;
}}
```

Fig. 8. Original Concurrent Program (Source Code)

Fig. 9. Programmer's Test Cases

inserting signal(1) and wait(1). The modified concurrent program is shown in Fig. 10. Its CTS representation is also shown in Fig. 11. This program still contains the above "hello-good-bye-bug" [1] since the programmer, unfortunately, does not test that case.

The execution histories are represented as a scenario graph SG_C (Fig.12). For example, a path $\theta = P_2 : t_2, \to P_1 : t_1 \to P_1 : t_2 \to P_1 : t_3 \to P_1 : t_4 \to P_2 : t_2 \to P_2 : t_3 \to P_1 : t_5 \to ...$ is a scenario. Since $P_1 : t_1$ and $P_2 : t_3$ access the shared memory m_1 (i.e., $(P_1 : t_1, P_2 : t_3) \in D(C)$), there exists a precedence constraint $c(\theta, 2) = (P_1 : t_1, 1) \prec c(\theta, 7) = (P2 : t_3, 1)$ in θ. This step of testing and constructing a scenario graph can be supported by a visual scenario selection and testing tool (Fig.2). Note that this scenario graph does not include a "hello-good-bye-bug" sequence.

Step 3: Parallelization

After inserting dummy synchronization transitions sync(k), the scenario graph is divided into two processes. Then, redundant synchronization transitions are removed by applying the parallelization algorithm. The optimized local scenario graphs C_o are shown in Fig 12.

---

[1] "hello-good-bye-bug" sequence $= P_2 : t_1 \to P_1 : t_1 \to P_1 : t_2 \to P_1 : t_3 \to P_1 : t_4 \to P_2 : t_2 \to P_2 : t_3 \to P_1 : t_5 \to P_1 : t_9 \to P_1 : t_{10} \to P_2 : t_4 \to P_2 : t_5 \to P_2 : t_6$. $P_i : t_j$ means a statement t_j of a process P_i in Fig. 11.

```
int m1, m2 ;
task P1()                              task P2()
{                                      {
  int f ;                                int f, c ;
  m1 = 0 ;                               f = 0 ;
  m2 = 0 ;                               wait(1) ;     /* Inserted */
  signal(1) ;     /* Inserted */        while(1){
  while(1){                                m1 = f ;
    sleep(10) ;                            c = m2 ;
    f = m1 ;                               if(c==0) {
    if(f==1) {                               printf("Hello! \n") ;
      printf("Say hello! \n") ;            }
      f=0 ;                                else {
    }                                        printf("Good-bye! \n") ;
    else {                                 } ;
      printf("Say good-bye! \n") ;         f = c ;
      f=1 ;                              }
    }                                  }
    m2 = f ;
}}
```

Fig. 10. Debugged Concurrent Program

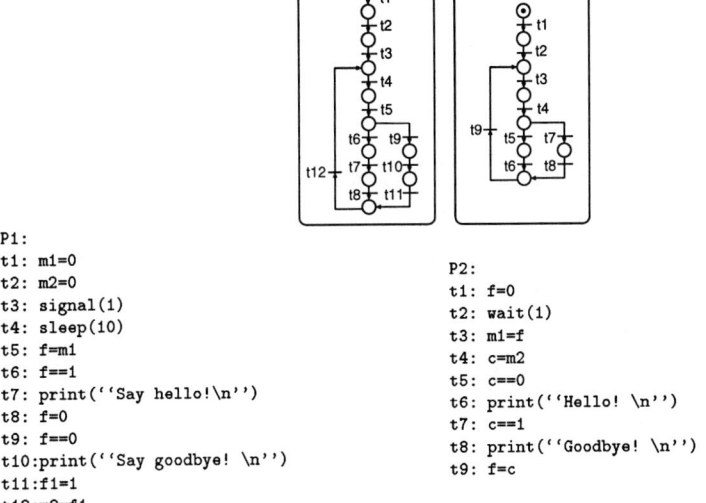

```
P1:
t1: m1=0
t2: m2=0
t3: signal(1)                          P2:
t4: sleep(10)                          t1: f=0
t5: f=m1                               t2: wait(1)
t6: f==1                               t3: m1=f
t7: print(''Say hello!\n'')            t4: c=m2
t8: f=0                                t5: c==0
t9: f==0                               t6: print(''Hello! \n'')
t10:print(''Say goodbye! \n'')         t7: c==1
t11:f1=1                               t8: print(''Goodbye! \n'')
t12:m2=f1                              t9: f=c
```

Fig. 11. CTS Representation $\mathcal{C}$

Step 4: Code Generation

Each process of the concurrent program can be directly reconstructed from these local scenario graphs. Figure 13 shows an example of code generation patterns. Furthermore, since they have several source code duplications, some identical blocks can be folded. In this example, $Block_{11}$ and $Block_{12}$, $Block_{21}$ and $Block_{22}$, $Block_{23}$ and $Block_{24}$ of Fig. 12 can be folded into single blocks, respectively. The final concurrent program is shown in Fig. 14. You can see the "hello-good-bye-bug" is removed in this program. The concurrency rate is rep-

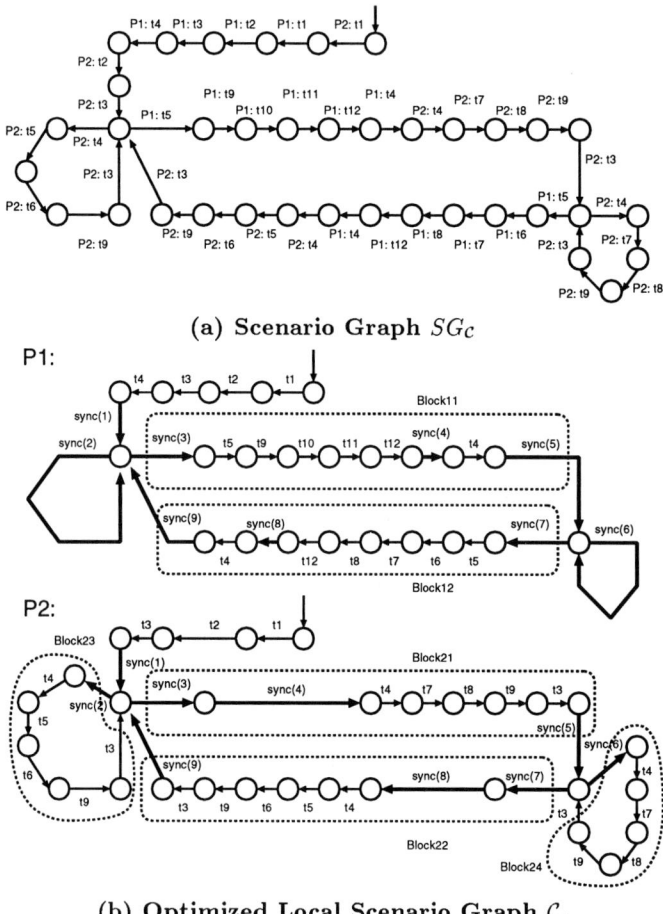

Fig. 12. Scenario Graph and Optimized Local Scenario Graph

resented in Table 1. The concurrency and nondeterminacy are restored to 1.68 in the final program, while that of the original program is 1.99.

5 Concluding Remarks

Hypersequential programming can be applied to a wide range of domains of concurrent programming, including parallel programs for shared-memory parallel computers, multi-task programs for embedded systems, and network protocol programs for distributed systems. We are developing a CASE tool based on this method, where parallelization and optimization algorithms when reconstructing a final concurrent program from a scenario graph are improved by introducing several heuristics. Furthermore, the following techniques should be developed for practical use.

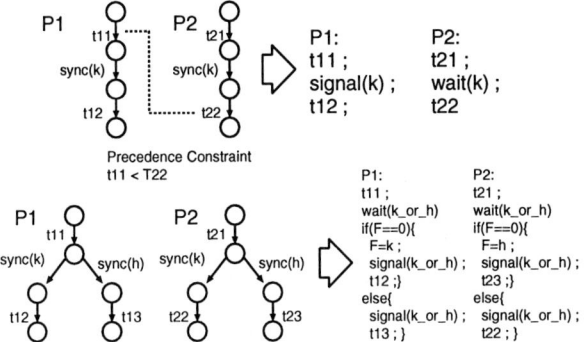

Fig. 13. Code Generation Pattern

```
int m1, m2, sf ;
task P1()
{
  int f ;
  sf = 0 ;
  signal(4) ;
  m1 = 0 ;
  m2 = 0 ;
  signal(1) ;
  sleep(10) ;
  wait(2) ;
  while(1){
    wait(4) ;
    sf=0 ;
    signal(4) ;
    f = m1 ;
    if(f==1) {
      printf("Say hello! \n") ;
      f=0 ;
    }
    else {
      printf("Say good-bye! \n") ;
      f=1 ;
    }
    m2 = f ;
    signal(3) ;
    sleep(10) ;
    wait(2) ;
}}
```

```
task P2()
{
  int f, c ;
  f = 0 ;
  wait(1) ;
  m1 = f ;
  signal(2)
  while(1){
    wait(4) ;
    if(sf == 1) {
      sf = 1 ;
      signal(4) ;
      wait(3) ;
      c = m2 ;
      if(c==0) { printf("Hello! \n") ; }
      else { printf("Good-bye! \n") ; } ;
      f = c ;
      m1 = f ;
      signal(2)
    }
    else {
      signal(4) ;
      c = m2 ;
      if(c==0) { printf("Hello! \n") ; }
      else { printf("Good-bye! \n") ; } ;
      f = c ;
      m1 = f ;
}}}
```

Fig. 14. Final Concurrent Program

Table 1. Concurrency Rate

	Original Program	Scenario Graph	Final Program
Size (edge/node)	473/237	39/37	411/243
Concurrency	1.99	1.05	1.68

- **Global State Abstraction:** In the example of this paper, all memories are considered for identifying global states in the scenario graph. However, memories may be huge in the practical program, then some abstraction is required.
- **Hierarchical Scenario:** Since it is tedious to make all scenarios at the source code (statement) level, hierarchical and top-down construction of scenarios is necessary.

Finally, we briefly mention the related works. Tai, et. al. proposed *deterministic execution debugging* of concurrent Ada programs [6]. This method implements a serialization mechanism based on a given execution history (scenario). Although this serialization is similar to our method, the execution histories are used only for testing and debugging in their method, while we also use them for reconstruction of the final program. Automatic detection of harmful nondeterminacy from the execution histories is also useful, and several researches have been done [7]. However, they are not used for automatic elimination of harmful nondeterminacy. With regard to parallelization of sequential programs, a great deal of research has been done in the domain of compiler optimizations for parallel computers [8]. You might ask why we do not start with a sequential program instead of a concurrent one that is serialized afterwards. The answer is that the topology of concurrent programs is natural for modeling in many target domains. Hypersequential programming preserves the original program's topology during serialization, and restore it during parallelization.

References

1. Pancake, C.M.: Software Support for Parallel Computing: Where Are We Headed?, *Communication of The ACM*, Vol.34, No.11 (1991) 53–64.
2. McDowell,C.E., Helmbold, D.P.: Debugging Concurrent Programs, *ACM Computing Surveys*, Vol.21, No.4 (1989) 593–622.
3. Uchihira, N., Honiden, S., Seki, T.: Hypersequential Programming: A New Way to Develop Concurrent Programs, *IEEE Concurrency*, Vol.5, No.3 (1997).
4. Uchihira, N., Kawata, H.: Scenario-Based Hypersequential Programming: Concept and Example, *2nd International Workshop on Software Engineering for Parallel and Distributed Systems*, Boston, IEEE Computer Society Press (1997) 277–283.
5. Girkar, M., Polychronopoulos, C.D.: Automatic Extraction of Functional Parallelism from Ordinary Programs, *IEEE Trans. Parall. Distrib. Syst.*, Vol.3, No.2 (1992) 166–178.
6. Tai, K.C., Carver, R.H, Obaid, E.E.: Debugging Concurrent Ada Programs by Deterministic Execution, *IEEE Trans. Software Engineering*, Vol.17, No.1 (1991) 45–63.
7. Emrath, P.A., Ghosh, S., Padua, D.A.: Detecting nondeterminacy in parallel programs *IEEE Software*, Vol.9, No.1 (1992), 69–77.
8. Zima, H., Chapman, B.: *Supercompilers for Parallel and Vector Computers*, Addison-Wesley (1990).

Parallelization of Space Plasma Particle Simulation

Yutaka Akiyama[1], Kiyotaka Misoo[2], Yoshiharu Omura[3], Hiroshi Matsumoto[3], Minoru Saito[1], Tamotsu Noguchi[1], Kentaro Onizuka[1], and Makoto Ando[1]

[1] Parallel Application TRC Laboratory,
Tsukuba Research Center, Real World Computing Partnership,
Tsukuba Mitsui Building 16F, 1-6-1 Takezono, Tsukuba 305, Japan
[2] Information and Mathematical Science Laboratory, Inc,
2-43-1 Ikebukuro, Toshima-ku 171, Japan
[3] Radio Atmospheric Science Center, Kyoto University, Gokasho, Uji 611, Japan

Abstract. This paper describes parallelization of the space plasma particle simulation program "KEMPO1" and shows its performance on five different platforms. One of our goals is to solve the Electrostatic Solitary Wave (ESW) problem by intensive computer simulations, which previously took about 1 month for a single experiment (10^7 particles, 10^4 time steps). The parallelized version performs the same calculation in 3 hours and a bigger one (2.7×10^8 particles, 1.6×10^4 time steps) in about 8 hours on our 256-processor Hitachi SR2201 parallel computer. It has made systematic real-world space plasma particle simulations feasible.

1 Introduction

The inter-planet space looks like vacant, however, it is filled with complex electromagnetic phenomena that tell us clues to understand history of the universe and the solar system. The brilliant tail of Comet Hale-Bopp shows the existence of strong solar plasma wind in the space. Our planet Earth is also surrounded by a huge "magnetosphere" which has an extending tail region.

In July 1992, the observatory satellite GEOTAIL was launched and many scientific observations at magnetosphere were obtained, including records of mysterious electrostatic wave emissions. Because cosmic phenomena are enormous in terms of spatial and time scale, such observatory data represent just a very small part of an entire phenomenon. In order to explain the global mechanism of the unexpected electrostatic emissions, the only scientific approach at present is to reproduce the whole phenomenon through computer simulation.

Particle simulations, as well as MHD simulations and hybrid methods, are frequently used in space plasma physics[1]. Plasma particle simulation technique is also useful in engineering purposes, such as designing a free-electron laser and a particle accelerator.

Using PVM[2] and MPI[3], we have parallelized the KEMPO1 simulator[4], which have long been used in space plasma physics. This makes it possible to carry out large-scale computer experiments to clarify the formation mechanism of ESW (Electrostatic Solitary Wave).

2 Space plasma particle simulation

2.1 KEMPO1 simulator

KEMPO (**K**yoto university's **E**lectro-**M**agnetic **P**article c**O**de) was developed in 1985 at Kyoto University, Japan. It calculates the motion of charged particles in electromagnetic fields based on PM (Particle-Mesh) method. KEMPO1[4] is a simplified version of KEMPO for one-dimensional simulations.

PM method consists of two alternating phases: 1) the phase for pushing the particles by the electromagnetic fields, and 2) the other phase for updating the fields influenced by the motion of charged particles. PM method is also called PIC (Particle-In-Cell) simulation[5, 6].

It is ideal to simulate electromagnetic phenomena in three-dimensional calculation. The most well-known three-dimensional code for PM method is TRISTAN by Buneman[7]. TRISTAN has been maintained by NASA researchers and ported onto vector computers and the MasPar parallel computer[8]. However, one-dimensional simulators are more effective with respect to memory usage and calculation speed when the phenomenon will progress along a one-dimensional axis and a large number of particles (*e.g.* 10^5) for each spatial grid are required for a high-resolution experiment.

Parallelization of a plasma particle simulator gives two advantages: 1) **speed up of calculation**, and 2) **extention of simulation size with larger memory**. With tens of Giga bytes of memory on a parallel computer, we can manage billions of particles in simulation, which allows us to build a realistic computational model of non-linear space plasma physics.

2.2 Calculation in KEMPO1

In KEMPO1, electromagnetic fields are defined on one-dimensional grids (typically $M=1,024$ to $8,192$ points), while both electric field $\boldsymbol{E}$ and magnetic field $\boldsymbol{B}$ defined on the grids have three-dimensional elements (planer wave approximation). Each particle, or usually a "superparticle" (a pseudo charge-cloud which preserves charge/mass ratio), can move to any intermediate position x between one-dimensional grids and has velocity vector $\boldsymbol{v}$ with three-dimensional elements (v_x, v_y, v_z).

The PM method ignores direct interaction among particles, and thus each particle is pushed solely by the electromagnetic fields. The motion of a large number of charged particles (*i.e.* electric current) causes field fluctuation. Then the contribution of each particle is divided onto adjacent grids, depending on the particle's position in the mesh. Periodic boundary conditions are used.

The basic equations in KEMPO1 are as follows:

Maxwell equations:

$$\nabla \times \boldsymbol{B} = \mu_0 \boldsymbol{J} + \frac{1}{c^2} \frac{\partial \boldsymbol{E}}{\partial t}$$

$$\nabla \times \boldsymbol{E} = -\frac{\partial \boldsymbol{B}}{\partial t}$$

Initial conditions:
$$\frac{\partial E_x}{\partial x} = \frac{\rho}{\epsilon_0}, \quad \frac{\partial B_x}{\partial x} = 0$$

Motion equations:
$$\frac{d\boldsymbol{v}}{dt} = \frac{q}{m}(\boldsymbol{E} + \boldsymbol{v} \times \boldsymbol{B}), \quad \frac{dx}{dt} = v_x$$

Maxwell equations and the motion equations are differentiated with the central differentiation scheme, defining full-integer grids and half-integer grids on the spatial mesh. Time is also divided into full-integer time steps and half-integer time steps. Based on leap-frog method, the electric field $\boldsymbol{E}$ is updated on full-integer time steps and the magnetic field $\boldsymbol{B}$ is updated on half-integer time steps[4]. Grid length Δx should be shorter than $3\lambda_e$, where λ_e is Debye length (Courant condition). In order to simplify calculation, we define $\epsilon_0 = 1$, and $\mu_0 = 1/c^2$. KEMPO1 supports not only electromagnetic calculations but also fast calculations for electrostatic simulation.

2.3 Large-scale simulation required for the ESW problem

The magnetosphere of the Earth has a long "magnetotail" region blown by the solar wind. The GEOTAIL satellite discovered unexpected emissions of the Electrostatic Solitary Wave (ESW) in the magnetotail which could not be fully explained by the previous theories. We suspect that the ESW is formed in relation to the acceleration of electrons in "magnetic reconnection process" in which partial energy of solar wind flows into the magnetotail. Thus, we expect the ESW problem will give an important clue to understand the dynamics of the magnetosphere.

The particles move along the one-dimensional axis on magnetic flux so that the progress of the electromagnetic fields are described by a one-dimensional model (planar-wave approximation). Inter-particle interactions are negligible in this kind of high-temperature plasma phenomenon.

Omura et al. studied the formation mechanism of the ESW with KEMPO1 [9, 10, 11, 12]. Our hypothesis is that the ESW is composed of sequences of impulsive solitary waves that have grown through nonlinear evolution of electron beam instability, especially bump-on-tail instability, which in turn caused by weak electron beams emitted to thin plasma in the inter-planet space.

In order to simulate the non-linear physics of plasma waves without influences from thermal noise, very high resolution for particle densities is required. The following equation[13] represents the relationship between thermal noise electric field energy $\mathcal{E}_n$ and plasma thermal energy $\mathcal{E}_p$:

$$\frac{\mathcal{E}_n}{\mathcal{E}_p} = \frac{M \Delta x}{2N\lambda_e}$$

where M is the number of spatial grids, N is the number of (super)particles, Δx is the grid length, and λ_e is Debye length. Thermal noise strength is proportional to $\sqrt{M/N}$. To reduce the noise ratio by half, the four-folds number of N is required.

In order to realistically simulate the plasma and the weak electron-beam densities at the magnetotail, more than 10^5 particles are required per Debye length λ_e. When the number of grids is in the range of $M = 1,024$ to $8,192$, the total number of particles required is about $N = 10^8$ to 10^9. At least 10^4 time steps would be needed to see the formation process of the ESW nonlinear whirl patterns.

A single-processor system can simulate at most tens of millions of particles because of both small memory capacity and slow computational speed. In this level of density resolution, the simulated electron beam emission is too strong and thus the excited wave pattern has a very different non-linear nature from the reality. In addition, our previous experiment showed that even a small-scale ESW simulation took one month per experiment on a single processor of Convex C3820GT. Our goal on parallelization is to complete the same simulation in several hours.

3 Implementation of Parallel KEMPO1

3.1 Parallelization scheme

Particle simulation can be parallelized in two ways: 1) spatial decomposition and 2) particle decomposition.

Spatial decomposition (Eulerian method): Commonly used decomposition method, in which spatial data are divided into subregions and are distributed among processors. Suitable when the number of particles per grid is relatively small and/or the mobility of particles is low. Particle data transmission is needed when a particle moves to a different subregion. The fields E and B are updated almost locally. Decyk et al.[5] have implemented one- and two-dimensional PIC simulators (GCPIC) based on spatial decomposition. Ueda et al.[14] have also implemented a PVM-version of KEMPO1 based on spatial decomposition method, exploiting small-scale (1-4 processors) parallelism.

Since naive space-decomposition easily causes load imbalance, the space should better be divided into small fragments and assigned to the processors in block-cyclic manner. Dynamic modification of area division is also considered for cases of unbalanced particle distribution[5, 6].

Particle decomposition (Lagrangian method): In this method, particles are divided into subgroups and distributed among processors. Suitable when the number of particles per grid is very large and no interaction is required among particles.

Processor assignment is never changed even if a particle moves very far from its starting position, so that no particle data transmission is needed between processors. On the other hand, global communication is required to update the fields. Each processor (p) sums up contributions of assigned particles, and calculate partial current densities $J_i^{(p)}$ and partial charge densities $\rho_i^{(p)}$. The actual sums are obtained by reduction operations over all processors, and then updated values (J_i, ρ_i) are broadcasted to processors. Each processor independently perform the entire electromagnetic field calculation using these values delivered, and maintain updated data representing whole regions of the electromagnetic fields.

Advantages of the particle decomposition method includes very simple control and good load balancing. This method is suitable when the number of particles N is large enough with respect to the grids M. Practically, parallelization of field calculation is not necessary if $\frac{M}{N/P} \ll 0.1$, as satisfied in our experiments, where P is the number of parallel processors. If the ratio is not small enough, the computation cost for field update is not negligible and its parallelization should be considered. We chose the particle decomposition method because its superiority is obvious for the high-resolution ESW experiments where a huge number of particles are required.

3.2 Implementation

We have parallelized KEMPO1, using both PVM library[2] (Ver. 3.3.11) and MPI library[3] (MPICH Ver. 1.0.13). Parallel KEMPO1 program creates child processes and perform parallel calculation based on particle decomposition method in the SPMD (Single-Program Multiple-Data) manner (Fortran77, 6×10^3 statements).

Schematic diagram of the program flow is shown in Fig.1. `velcty` (velocity), `positn` (position), and `currnt` (current) subroutines take about 60%, 20%, and 15%, respectively, of the execution time. In the loop, there are two kinds of global communication (reduce and broadcast), however, `charge` (charge) and `ecrrct` (electric field correction) are called every eight loops in the actual code.

Required memory capacity is about $(12M + 7N) \times 4$ bytes, approximately 26GB per one-billion particles.

3.3 Motion video output

KEMPO1's post-processing modules provide graphical drawings of various calculation results, including spatial distribution of fields, wavenumber spectrum, particle distribution in phase-space, and history of fields and energy.

In Parallelized KEMPO1, the AVS format can be generated in addition to the originally supported CALCOMP format. Area-averaging operation is done on each processor in order to reduce data transmission to the host, for drawing a diagram like the $x - v_x$ phase space particle distribution diagram(Fig.2). An automatic VTR recording system has been developed for storing these drawings as a color motion video image.

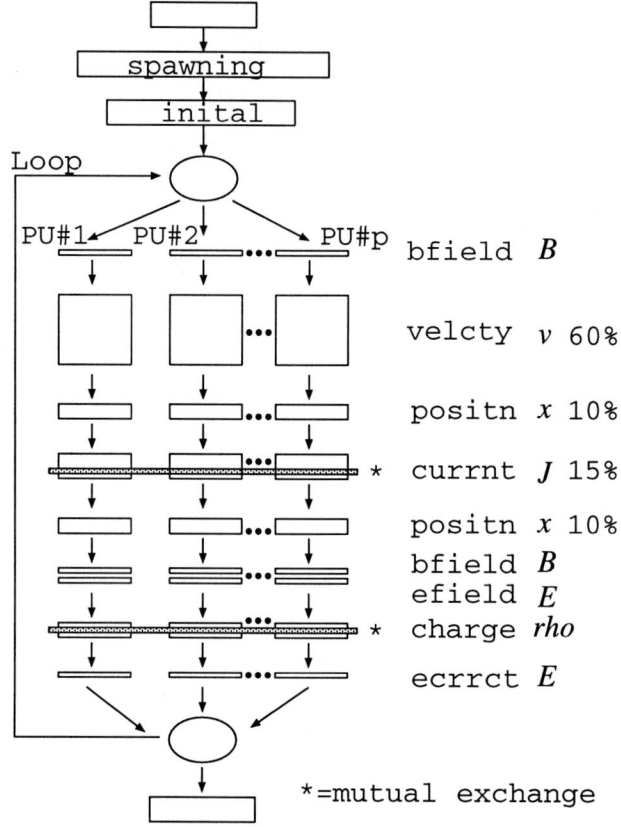

Fig. 1. Schematic calculation flow diagram of Parallel KEMPO1

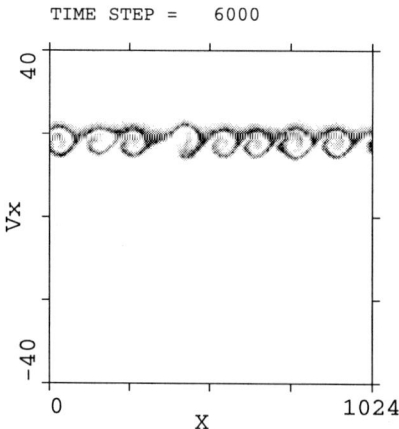

Fig. 2. Example pattern formation in the $x - v_x$ phase space

4 Performance evaluation

Table 1. Systems used in performance evaluation

Machine	Processor chip	#Processors	Memory	Communication
Sun SPARCcenter2000E	SuperSPARC II 85MHz	20	5GB Shared	MPI PVM
SGI Power Challenge	R10000 194MHz	12	2GB Shared	MPI PVM
TMC CM-5	SPARC 32MHz	32	1GB Distrib.	— PVM
Intel Paragon	i860 XR 50MHz	66	1GB Distrib.	— PVM
Hitachi SR2201	PA-RISC1.1+PVP-SW 150MHz	256	64GB Distrib.	MPI PVM

We have measured the performance of parallelized KEMPO1 on the five different platforms listed in Table 1.

Fig.3 through Fig.5 show the execution times for 16 steps (Initialization time excluded. About 16,000 steps needed for a realistic simulation), and Fig.6 through Fig.8 show the parallel efficiency, on SPARCcenter 2000E, Power Challenge, and Hitachi SR2201, respectively. We have used the MPI version for SPARCcenter 2000E and Hitachi SR2201 and the PVM version for others.

The performance on SPARCcenter 2000E for particles $N = 2^{14}, 2^{16}, 2^{20}, 2^{24}$, and 2^{26} is shown in Fig.3 and Fig.6. When the number of particles is greater than $N = 2^{20}(1,048,576)$, efficiency is over 80% up to 20 processors.

The performance on Power Challenge for particles $N = 2^{14}, 2^{16}, 2^{20}, 2^{24}$, and 2^{26} is shown in Fig.4 and Fig.7. Single processor execution time on Power Challenge is three times shorter than that of SPARCcenter 2000E. In the case of $N = 2^{20}(1,048,576)$, parallel efficiency is "super-linear" for more than 4 processors, because data can be stored on the secondary cache.

On CM-5, for particle number $N = 2^{22}$, efficiency does not greatly fall up to the maximum 32 processors. On Paragon, for the same particle number, efficiency does not fall until 32 processors, but then begin to fall because of communication bottleneck. For both machines memory shortage is crucial and we could not use more than $N = 2^{24}$ particles. Because of space constraints graphs for CM-5 and Paragon are not shown.

The performance on Hitachi SR2201 for particles $N = 2^{20}, 2^{24}, 2^{26}$, and 2^{28} is shown in Fig.5 and Fig.8. Since the hyper-crossbar network provides high communication capacity, the parallel efficiency does not fall up to the maximum 256 processors in large-scale problems. With 256 processors, the parallel efficiency is 71% for particles $N = 2^{26}(67,108,864)$, and 82% for $N = 2^{28}(268,435,456)$. A realistic simulation with 16,000 steps can be performed in only three hours for 67 million (2^{26}) particles, and eight hours for 270 million (2^{28}) particles. If we utilize its 64GB memory capacity, even a simulation with 2.4 billion (*i.e.* 64GB/26GB × billion) particles is possible.

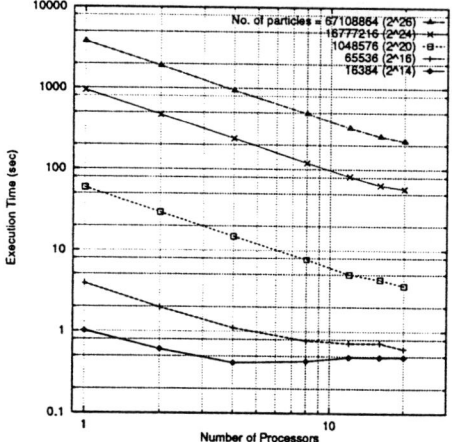

Fig. 3. Execution time on SPARCcenter 2000E

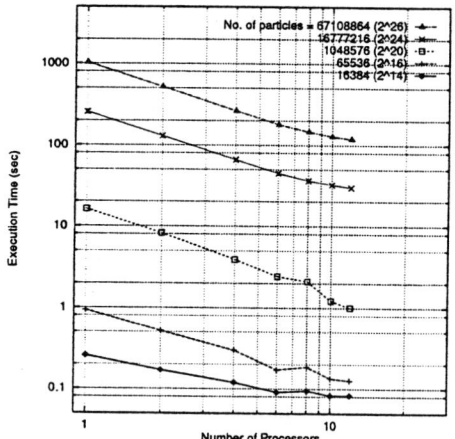

Fig. 4. Execution time on Power Challenge

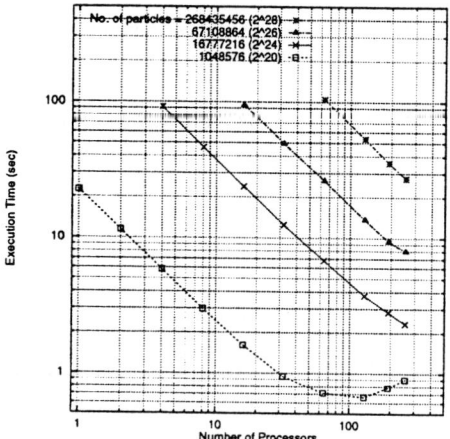

Fig. 5. Execution time on Hitachi SR2201

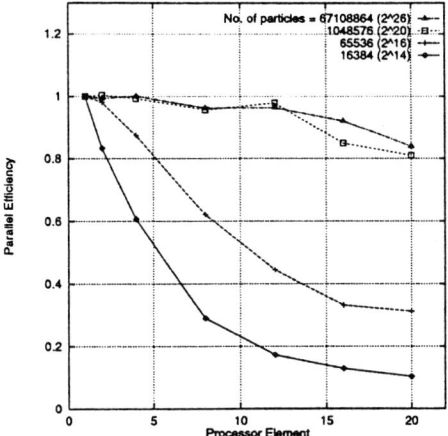

Fig. 6. Parallel efficiency on SPARCcenter 2000E

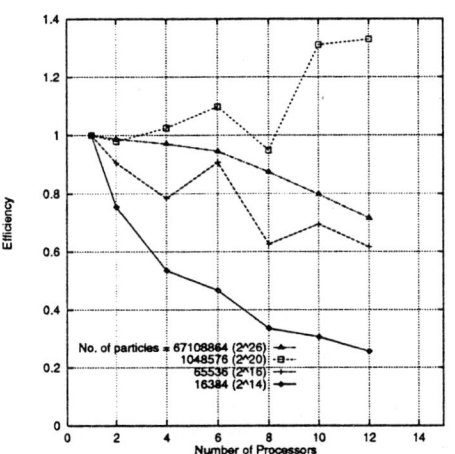

Fig. 7. Parallel efficiency on Power Challenge

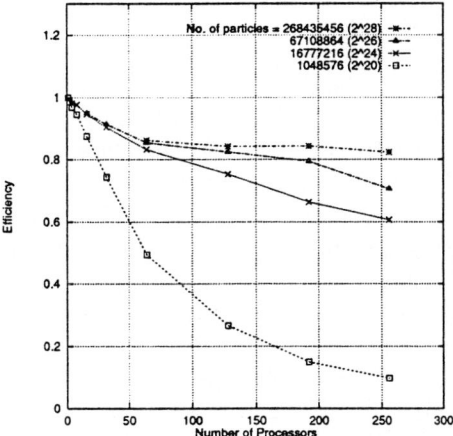

Fig. 8. Parallel efficiency on Hitachi SR2201

5 Discussion

5.1 Parallel random number generation

In a large-scale simulation on KEMPO1, generation of random numbers (in `inital` subroutine) takes a considerable amount of time (1,700 sec. for 2^{26} particles, and 7,000 sec. for 2^{28} particles on a single processor of SR2201), so that its parallelization is strongly desired. In order to keep the result reproductivity, independent from the number of processors used, random number sequences generated on distributed processors should exactly correspond to the sequentially generated random numbers.

For this purpose, parallel random number generator libraries have been developed, based on multiplicative congruential method [15, 16, 17], lagged Fibonacci method [16, 17], or generalized shift register method [17].

For initialization of a single particle, 19 uniformly-distributed random numbers are used (one for position x, three velocity v elements × six additions for making approximated Gauss normal distribution) in KEMPO1. To simulate 100 million particles, we need 1.9 billion random numbers which is almost equal to 2^{31} (2,147,483,648), the period of old random number generator routines. We need longer period sequences for a billion particles simulation.

As of this writing, we have implemented a sequential random number generator and a parallel version without full-compatibility to the sequential version. A fully-compatible parallel version will be developed in our future work.

5.2 Array padding

In KEMPO1 we should choose a power-of-2 number of grids (M) since Poisson equation in `ecrrct` (electric field correction) subroutine is solved by FFT technique. We have also been, without any strong necessity, using a power-of-2 number of particles N.

Each data array size, in the original KEMPO1 code, thus tends to be a power-of-2. It is well known that the use of multiple power-of-2-sized arrays have risks for cache line conflicts. On our SGI Power Challenge, for example, the L2 cache is 2MB 2-way set-associative (1MB/way). Therefore, the same cache line is used for every 2^{18}=(1MB/4B) FLOAT-type elements.

Suppose one-dimensional arrays x, v_x, v_y, and v_z are assigned contiguously on the main-memory space. $x(i), v_x(i), v_y(i)$, and $v_z(i)$ conflict each other on the same cache line, if the size (N/P) of each array is a multiple of $2^{18}/2$. Arrays are padded in order to prevent these cache conflicts.

Fig.9 shows an example of the effect of array padding which solves abnormal speed-down observed in all cases of power-of-2 (including 2^0) number of processors. The solid line is the performance before padding, and the dotted line is after padding, showing the execution time for 16 steps (Power Challenge). The results shown in section 4 use padding.

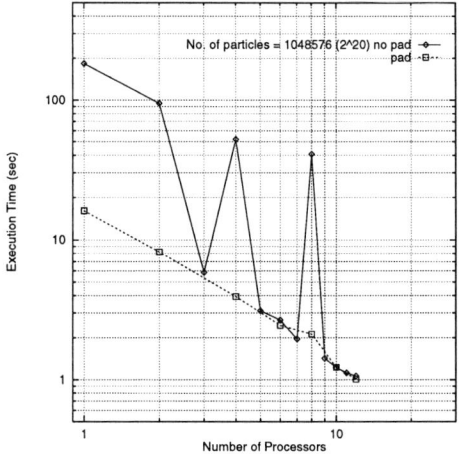

Fig. 9. Effect of Cache line conflict and array padding

5.3 Rearrangement of data array

In KEMPO1, the four one-dimensional data arrays, for particles' position x and velocity vector elements v_x, v_y, v_z, respectively, occupy the most part of memory usage. With respect to a single particle, the four array elements can be localized on memory address, when using a two-dimensional array like array(4,N), in stead of four independent one-dimensional arrays.

Our experiments showed a negative effect in total and we abandoned the two-dimensional representation. On currnt (current) subroutine, memory localization introduced by the two-dimensional representation brings major speed-up. However, on subroutines like positn (position) and charge (charge), which do not use v_y and v_z values, the execution time increased 3 to 4 times in the worst cases, because of low efficiency in cache utilization. We have used one-dimensional arrays to produce the results in the section 4.

6 Conclusion

We have parallelized a space plasma particle simulator KEMPO1, providing both PVM-version and MPI-version. The parallel version allows us to use billions of particles, as opposed to previously possible tens of millions. This enabled a high-resolution experiment, for example with 1,024 to 8,096 grids and 10^5 particles per Debye length, which is required for a realistic simulation of the ESW formation mechanism. About 200-folds speed-up of execution has been achieved (on Hitachi SR2201, 256 processors) and thus the simulation which would take one month on a single processor can be performed in three hours.

The particle decomposition method that we employed is suitable when the number of particles is sufficiently larger than the number of spatial grids, thus is greatly effective in the field of high-resolution plasma particle simulations.

Acknowledgment

The authors would like to thank Dr. Taisuke Boku at University of Tsukuba, Dr. Mitsuhisa Sato at Real World Computing Partnership, and Mr. Hideki Saito at CSRD, University of Illinois at Urbana-Champaign for their valuable comments.

References

1. H. Matsumoto and Y. Omura (Eds.), "Computer Space Plasma Physics: Simulation Techniques and Software", Terra Scientific Pub. Co., Tokyo (1993).
2. "PVM", http://www.epm.ornl.gov/pvm/
3. "MPICH", http://www.mcs.anl.gov/mpi/mpich/
4. Y. Omura and H. Matsumoto, "KEMPO1 Technical Guide to One-dimensional Electromagnetic Particle Code" in [1] (1993).
5. P. C. Liewer and V. K. Decyk, "A general concurrent algorithm for plasma particle-in-cell simulation codes", *Journal of Computer Physics*, **85**, 2, pp.302-322 (1989).
6. G. C. Fox, R. D. Williams and P. C. Messina, "Parallel Computing Works!" (Section 9.3: Plasma Particle-in-Cell Simulation of an Electron Beam Plasma Instability), Morgan Kaufmann (1994).
7. O. Buneman, "TRISTAN: The 3-D, E-M Particle Code" in [1] (1993).
8. NASA Inhouse Team Software Exchange,
   http://sdcd.gsfc.nasa.gov/ESS/inhouse-sw.html
9. Y. Omura, H. Kojima and H. Matsumoto, "Computer simulation of electrostatic solitary waves: A nonlinear model of broadband electrostatic noise", *Geophys. Res. Lett.*, **21**, pp.2923-2926 (1994).
10. H. Matsumoto, H. Kojima, S. Miyatake, Y. Omura, M. Okada, I. Nagano and M. Tsutsui, "Electrostatic solitary waves (ESW) in the magnetotail: BEN wave forms observed by GEOTAIL", *Geophys. Res. Lett.*, **21**, pp.2915-2918 (1994).
11. Y. Omura, H. Matsumoto, T. Miyake and H. Kojima, "Electron beam instabilities as generation mechanism of electrostatic solitary waves in the magnetotail", *Journal of Geophysical Research*, **101**, pp.2685-2697 (1996).
12. T. Miyake, Y. Omura, H. Matsumoto and H. Kojima, "Computer Experiments of Electrostatic Solitary Waves Observed by GEOTAIL Spacecraft", *Proc. of the 5th Int'l Symposium/School for Space Simulations*, (Mar. 13-19, Kyoto), pp.51-54 (1997)
13. H. Ueda, Y. Omura, H. Matsumoto, and T. Okuzawa, "A study of the numerical heating in electrostatic particle simulations", *Computer Physics Communication*, **79**, pp.249-259 (1994).
14. Y. Ueda, Y. Omura and H. Usui, "Electromagnetic Particle Simulations via Parallel Virtual Machines", *Proc. of 5th Int'l Symposium/School for Space Simulations*, (Mar. 13-19, Kyoto), pp.399-402 (1997)
15. NAS Parallel Benchmarks, NASA Ames Research Center,
    http://science.nas.nasa.gov/Software/NPB/Specs/RNR-94-007/node18.html
16. "Scalable Pseudorandom Number Generators Library for Parallel Monte Carlo Computations", http://141.142.3.70/Apps/CMP/RNG/RNG-home.html
17. "PRNGlib: A Parallel Random Number Generators library",
    http://www.cscs.ch/Official/SoftwareTech/CSCS-NEC/pubs_abs.html#tec:9608

Implementing Iterative Solvers for Irregular Sparse Matrix Problems in High Performance Fortran

E. de Sturler[1] and D. Loher[1]

Swiss Center for Scientific Computing (SCSC-ETHZ), Swiss Federal Institute of Technology Zurich, ETH Zentrum (RZ F-11), CH-8092, Zurich, Switzerland, email: sturler@scsc.ethz.ch, phone: +41-1-632 5566, fax: +41-1-632 1104

Abstract. Writing efficient iterative solvers for irregular, sparse matrices in HPF is hard. The locality in the computations is unclear, and for efficiency we use storage schemes that obscure any structure in the matrix. Moreover, the limited capabilities of HPF to distribute and align data structures make it hard to implement the desired distributions, or to indicate these such that the compiler recognizes the efficient implementation.

We propose techniques to handle these problems. We combine strategies that have become popular in message-passing parallel programming, like mesh partitioning and splitting the matrix in local submatrices, with the functionality of HPF and HPF compilers, like the implicit handling of communication and distribution. The implementation of these techniques in HPF is not trivial, and we describe in detail how we propose to solve the problems. Our results demonstrate that efficient implementations are possible. We indicate how some of the 'approved extensions' of HPF-2.0 can be used, but they do not solve all problems. For comparison we show results for regular, sparse matrices.

Keywords: High Performance Fortran, Irregular Sparse Matrices, Iterative Solvers.

1 Introduction

For large, sparse linear systems we often use iterative methods because of their efficiency in both memory requirements and work. On parallel computers, iterative methods have the additional advantage that they do not change the structure of the problem. Iterative methods use four major kernels: matrix-vector product, preconditioner, inner product (ddot), and vector update (daxpy). The choice of a preconditioner is very important for the efficient solution of a linear system, but we will not discuss preconditioning here because it is often problem-dependent. However, very effective parallel preconditioners have been derived that have essentially the same communication requirements as the matrix-vector product [2, 3, 6]. For regular sparse, linear systems, like those derived from regular grids, using High Performance Fortran (HPF) for iterative solvers is straightforward.

However, for irregular sparse matrices the efficient implementation of solvers in HPF becomes much harder.

First, the locality in the computations (a good partitioning) is unclear. Second, for efficiency we often use storage schemes that obscure even the simplest structure in the matrix (like rows and columns). Third, the limited capabilities of HPF to distribute data structures make it hard to implement the desired distribution. Fourth, data structures often have very different sizes and shapes, and matching the distributions for efficient implementation (locality) is a problem. Fifth, after implementing the distributions, we still must write the program in such a way that the compiler recognizes the efficient implementation, and leaves out unnecessary communication, synchronization, etc.

We propose techniques for handling these problems, and our results demonstrate that efficient implementations are possible. In [4, 5] we showed that, unless special implementations are chosen, on large parallel computers the global communication in the inner products dominates the parallel performance of iterative solvers. This is an architectural feature; it is independent of whether we use HPF or, say, MPI. The cost of an inner product is about the same in both implementations. Clearly, if we can find implementations for the sparse matrix-vector product for irregular matrices that are more efficient than the inner product, the sparse matrix-vector product will not be the bottleneck. We will show that this is indeed possible, even for relatively small problems. For comparison, we show results for regular, sparse matrices. The generalized block distribution GEN_BLOCK in the 'approved extensions'[1] of the HPF Language Specification version 2.0 (HPF-2.0) [7] solves the distribution problems (problem three in the discussion above), but not the other problems. We will discuss this at the end of the paper.

All our experiments are carried out using the Portland Group (PGI) HPF compiler (version 2.1) on the Intel Paragon at the Swiss Federal Institute of Technology (ETH Zurich).

In the next section we discuss the parallel performance of iterative methods for regular, sparse matrices for comparison. For our experiments we use the GMRES method [9] (see Fig. 1), one of the most widely used iterative methods. In Section 3, we outline how we address the problems mentioned above, and we discuss the performance results. In Section 4, we indicate in how far the approved extensions of HPF-2.0 improve the situation, and what is still needed. In the last section we provide some future directions.

2 Regular Sparse Matrix Problems

For regular problems, such as k-diagonal matrices resulting from discretizations over regular grids, the parallelization with HPF is straightforward. Our test problem comes from a convection-diffusion problem, discretized on a regular, two-dimensional grid (501×501), with Dirichlet boundary conditions. The matrix

---

[1] It will probably take more than a year before such extensions become available.

GMRES(m):

start:
  $x_0 =$ initial guess; $r_0 = b - Ax_0$;
  $v_1 = r_0/\|r_0\|_2$;

iterate:
  for $j = 1, 2, \ldots, m$ do
    $\hat{v}_{j+1} = Av_j$;
    for $i = 1, 2, \ldots, j$ do
      $h_{i,j} = (\hat{v}_{j+1}, v_i)$;
      $\hat{v}_{j+1} = \hat{v}_{j+1} - h_{i,j} v_i$;
    end;
    $h_{j+1,j} = \|\hat{v}_{j+1}\|_2$;
    $v_{j+1} = \hat{v}_{j+1}/h_{j+1,j}$;
  end
form the approximate solution:
  $y_m = \arg\min_{y \in \mathbb{R}} \left\| \|r_0\|_2 e_1 - \bar{H}_m y \right\|_2$;
  $x_m = x_0 + V_m y_m$;

restart:
  $r_m = b - Ax_m$;
  if $\|r_m\|_2 < tol$ then
    stop
  else
    $x_0 = x_m$; $v_1 = r_m/\|r_m\|_2$;
    goto *iterate*
  end

Fig. 1. The GMRES(m) algorithm.

and vectors are stored by grid point (reflecting the underlying two-dimensional structure), which leads to an efficient implementation of the communication in the matrix-vector product.

Tables 1 and 2 show that for one to sixteen processors the efficiency for the matrix-vector product is eighty percent or higher, and this is not significantly lower than the efficiency for the other routines. More important, these tables show that for larger numbers of processors (32 and more), the speed-up for GMRES(40) is dominated by the inner products (ddot), not by the matrix-vector product. Especially, notice the similarity between the speed-up figures for GMRES(40) and the inner product (ddot) for eight processors or more. Note that increasing the restart cycle (here 40) of GMRES leads to a quadratic increase in the number of inner products compared with a linear increase in the number of matrix-vector products. Hence, for larger numbers of processors a higher efficiency for the matrix-vector product than for the inner products has no influence

Table 1. Runtimes in seconds for sequential and parallel GMRES(40), matrix-vector product (matvec), inner product (ddot), and vector update (daxpy). The large run-times between brackets for one and two processors are caused by swapping.

#proc	1	2	4	8	16	32	64
gmres	(2.75E+03)	(1.18E+03)	9.17E+01	4.79E+01	2.68E+01	1.62E+01	1.28E+01
matvec	1.29E+00	6.44E-01	3.79E-01	1.96E-01	1.02E-01	5.50E-02	3.61E-02
ddot	1.91E-01	9.60E-02	4.95E-02	2.56E-02	1.46E-02	8.57E-03	6.72E-03
daxpy	1.12E-01	5.29E-02	2.66E-02	1.34E-02	7.02E-03	3.62E-03	1.88E-03

on the performance of the algorithm as a whole.

The large difference in the run-time of one GMRES(40) iteration for the sequential program and the parallel program on two processors compared with the parallel program on four and more processors is caused by swapping in the former case; the data does not fit in the memory of two processors, but it fits in the memory of four processors. This, of course, makes the timings incomparable, and therefore we normalized the speed-up for four and more processors against the run-time on four processors. So the speed-up on four processors is equal to four. The speed-up values for the separate routines, which have been timed without swapping, show that this is not far from the speed-up that would have been measured on a machine that can run the program on one processor without swapping.

3 Irregular Sparse Matrix Problems

For algorithms like GMRES(m) the only difference between structured and non-structured problems is in the implementation of the matrix-vector product and of the preconditioner. As argued above, for many popular parallel preconditioners the communication requirements and the implementation of the communication resemble those of the distributed matrix-vector product, and hence we will concentrate on the latter.

Table 2. Speed-ups for GMRES(40), the matrix-vector product (matvec), the inner product (ddot), and the vector update (daxpy). For GMRES(40), the speed-ups for four and more processors have been derived from the run-time on four processors.

#proc	1	2	4	8	16	32	64
gmres(40)	(1.00)	(2.33)	4.00	7.66	13.7	22.6	28.7
matvec	1.00	2.00	3.40	6.58	12.7	23.5	35.7
ddot	1.00	1.99	3.86	7.46	13.1	22.3	28.4
daxpy	1.00	2.12	4.21	8.36	16.0	28.4	59.6

3.1 Sparse Matrix Storage Schemes

We will first describe two often-used sparse matrix storage schemes, the so-called compressed sparse column (CSC) and compressed sparse row (CSR) scheme. The CSR scheme uses three arrays to describe the matrix. Let nnz be the number of non-zero coefficients in the matrix and let n be the number of rows (and columns) in the matrix.

- *val(1:nnz)* contains the values of the non-zero coefficients of the matrix in row-wise order.
- *col_idx(1:nnz)* contains the column indices for the corresponding elements of *val*: *col_idx(i)* gives the column in which coefficient *val(i)* appears.
- *row_ptr(1:n+1)* contains pointers to the start of each row in the arrays *val* and *col_idx*. The last pointer points one past the last element of *val* and *col_idx*.

We can implement the matrix-vector product $y = Ax$ (in Fortran77) as follows.

```
CSR matrix-vector product:
do row = 1, n
   y(row) = 0.0
   do j = row_ptr(row), row_ptr(row+1)-1
      y(row) = y(row) + val(j)*x(col_idx(j))
   end do
end do
```

The CSC scheme is equivalent to the CSR scheme except that the array *val* contains the non-zero coefficients in column-wise order, and therefore we have an array with column pointers and an array with row indices instead of the other way around. Historically the CSC scheme has received a certain preference because it often leads to superior performance on vector computers. However, on parallel machines this is not the case and the disadvantages (especially for implementation of the preconditioner) dominate. We will assume the CSR scheme in this paper; however, our approach is easily converted for the CSC scheme.

3.2 Distribution of the Matrix

Usually, we prefer a row-wise distribution (partitioning) of the matrix. With an appropriate ordering (see below) this amounts to a domain decomposition, which facilitates several effective preconditioning techniques. Also more generally, row-wise distribution makes preconditioning easier to implement (block-ILU type preconditioning e.g.). However, it is straightforward to adapt our approach to a column-wise distribution.

The distribution of the matrix assigns to each processor a 'local matrix' consisting of the set of rows in the partition assigned to that processor. For an efficient implementation of the matrix-vector product the distribution of the arrays describing the matrix must be as follows. For each partition the nonzero coefficients of the local matrix are stored in the local part of the array *val*, the column indices of the local matrix are stored in the local part of the array *col_idx*, and the pointers to the rows of the local matrix are stored in the local parts of the arrays *start_row* and *end_row*, which replace *row_ptr* (see below). We distribute the vectors in the same way as the matrix: we store the i-th coefficient of a vector on the same processor as the i-th row of the matrix.

We use graph partitioning techniques (currently in a separate off-line phase) on the underlying computational grid or directly on the matrix to find a partitioning of the rows of the matrix such that the distributed, sparse matrix-vector product yields low communication cost and a good load balance. Low communication cost means that the total number of non-local references in the matrix-vector product is minimized. Preferably, also the number of processor-pairs that need to exchange data, which is equal to the number of messages, should be minimized. For the matrix-vector product, load balancing means that the number of non-zero matrix coefficients in each local matrix is about the same. However, for the vector operations load balancing means that the number of unknowns on each processor, which is equal to the number of rows in each local matrix, is about the same. Currently we use the package by Simon and Barnard [8, 1] to compute partitionings. This package allows a trade-off between the two different load balancing requirements in computing the partitioning.

From the output of the graph partitioning routine we know which rows of the matrix should be grouped together in a partition (i.e., on a processor) to form the local matrix. For each partition the local matrix should consist of the local parts of the arrays *val*, *col_idx*, and *row_ptr*. This gives a problem for the array *row_ptr*. Since each pointer serves a double purpose, to indicate the start of a row and the end of the previous row, the distribution of this array means we cannot for all rows have the necessary pointers locally available. Therefore, we replace this array by two new arrays; *start_row(1:n)*, with pointers to the start of each row; and *end_row(1:n)*, with pointers to the end of each row. Now we must implement the distribution of the arrays describing the matrix. To achieve this we will use a block-wise distribution and renumber the rows and columns explicitly. However, renumbering by itself is not enough. In general, we do not have the same number of rows in each partition, and we do not have the same number of non-zero coefficients in each partition. So the regular distribution indicated by the HPF DISTRIBUTE (BLOCK) directive does not give the desired distribution. Moreover, there is no fixed ratio between the number of rows and the number of non-zero coefficients, because the number of non-zero coefficients per row may vary strongly between rows. So, the arrays of row pointers differ in size from the arrays with matrix coefficients and column indices, and the ratio between these sizes differs per partition. In short, we cannot use the HPF ALIGNMENT directive to enforce that the pointers to the rows in a partition are stored on the same processor as the coefficients and column indices of that partition.

We solve the distribution and alignment problems as follows. First we introduce dummy (empty) rows such that each partition has the same number of rows. Then we create dummy coefficients such that each partition has the same number of non-zero coefficients. These coefficients will be outside any row, so they will never be used in computations. The dummy rows can be masked, so also here no additional computation is introduced. Moreover, the mesh partitioning algorithm always generates a partitioning with a good load balance, so that the numbers of additional rows and coefficients are negligible and no overhead in memory results. Reordering the padded matrix and distributing the arrays regularly through the HPF block distribution directive now leads to the desired distributions and alignments. We have all information about the local matrices locally available.

However, one problems remains: the compiler has no way of knowing that the arrays with row pointers are actually 'aligned' with the arrays with column indices and coefficients. So a straightforward implementation of the matrix-vector product leads to a large overhead in unnecessary checks and synchronizations, or worse in unnecessary duplication and communication of data, and even in unnecessary computations. Unless we make additional changes to the sizes of the arrays we still cannot use the HPF alignment directive to align the arrays in their rank-one form, or indicate this alignment to the compiler. We have two ways of solving this problem. The first way is to make the necessary communication of the non-local vector values explicit (by a copy) and then use an HPF_LOCAL routine for the matrix-vector product. This way the local availability of all the data is explicitly given, and we avoid problems with the HPF compiler creating unnecessary overhead. The second way is to reshape the arrays describing the matrix into rank-two arrays, because then we can use the HPF alignment directive to align the arrays and make this locality in the matrix-vector product explicit. We will get the following arrays with a partition index and a local index: *val(part_index,loc_idx)*, *col_idx(part_idx,loc_idx)*, *start_row(part_idx,loc_row)*, *end_row(part_idx,loc_row)*. For the vectors we still have several options. This implementation is certainly the most elegant. However, it leads to further complexities in the actual implementation of the matrix-vector product that we cannot go into here. We will discuss this in a future paper. The HPF_LOCAL version has the additional advantage that the local part of the matrix-vector product resembles the sequential version. We will use the HPF_LOCAL version for our tests below.

Finally, we like to point out that the implementation is actually not as complicated as it may seem. Assuming that we read in the matrix in one of the standard storage schemes for 'sequential' matrices, the restructuring is accomplished in a relatively cheap preprocessing phase before the actual iterative solver.

3.3 Tests and Results

After distributing and reordering the data structures the matrix-vector product is quite efficient, and the scheduler creates an efficient communication scheme. In fact, the scheduler itself becomes the most costly part. In general, this is not important because scheduling needs to be done only once for many iterations, and hence the cost becomes negligible if the schedule is reused. In general, the

PGI compiler does move the computation of the communication schedule out of loops; however, in more complex routines like the GMRES algorithm, apparently, this no longer works (insufficient analysis capability), and the schedule is recomputed unnecessarily for every matrix-vector product. We assume that in the future such features will improve. If this is the case, irregular sparse matrix computations become quite feasible with HPF. In order to show that we can achieve a sufficiently efficient matrix-vector product provided the communication schedule is reused, we use three subroutines provided by L. Meadows of PGI to explicitly reuse the schedule.

In the previous section we showed that for larger numbers of processors the efficiency of the inner product tends to dominate the overall efficiency of the iterative solver. So, if the efficiency of the matrix-vector product is higher than that of the inner product, the efficiency of the matrix-vector product has no influence on the overall performance. Therefore, we will only discuss the results for the matrix-vector product and the inner product (ddot) here. For the purpose of analysis we have split the matrix-vector product in the part that fetches all non-local data (gather) and the actual computation (comp). Unfortunately, large, irregular, sparse test matrices are not so easily available, and hence we can only provide results for a problem that is much smaller than what we used for the regular case. The largest matrices in the Harwell-Boeing test collection are of the order of 35000 unknowns. Our test problem (bcsstk31) has 35588 unknowns and 1181416 nonzero coefficients in the matrix. The run-time for the sequential matrix-vector product is 0.268 s. For the inner product (ddot) the sequential run-time is given in Table 3. The parallel run-time of the matrix-vector product is the sum of the time for 'gather' and for 'comp'. Because of the implementation there is a non-zero run-time for 'gather' on one processor (basically a copy). Of course, this could have been masked, but we feel that this information provides useful insight for the performance on multiple processors. For example, we see that the gather (including communication) scales almost linearly from one to eight processors, and only for more processors the efficiency decreases sharply.

Table 3. Run-times in seconds for the irregular, sparse matrix-vector product (matvec), its gather part (gather) and its computation part (comp), and for the inner product (ddot).

#proc	1	2	4	8	16	32
gather	2.09E-01	1.09E-01	5.67E-02	3.16E-02	2.32E-02	2.26E-02
comp	3.75E-01	1.94E-01	9.66E-02	4.44E-02	2.30E-02	1.26E-02
matvec	5.84E-01	3.03E-01	1.53E-01	7.60E-02	4.62E-02	3.52E-02
ddot	1.29E-02	6.78E-03	4.22E-03	2.75E-03	2.25E-03	2.17E-03

We see that also in this case for larger numbers of processors the efficiency of the inner product drops below the efficiency of the irregular, sparse matrix-vector

Table 4. Speed-up and efficiency for the irregular, sparse matrix-vector product (matvec) and for the inner product (ddot).

#proc	1	2	4	8	16	32
matvec speed-up	0.459	0.884	1.75	3.53	5.80	7.61
efficiency (%)	45.9	44.2	43.8	41.1	36.3	23.8
ddot speed-up	1.00	1.90	3.06	4.69	5.73	5.94
efficiency (%)	100	95.1	76.4	58.6	35.8	18.6

product. So, for larger numbers of processors the irregular, sparse matrix-vector product will not be a bottleneck. Since the inner products are about as fast in HPF as they are in message-passing code (MPI/PVM), these results also show that a message-passing code cannot be much faster on large numbers of processors. We see on the other hand that for smaller numbers of processors the efficiency of the matrix-vector product is between forty and fifty percent. So, for our current implementation we will not see performance much above that level for irregular problems. Clearly, for small numbers of processors a message-passing code will do better. However, we think that the efficiency we achieve is high enough to be interesting for many applications. Especially since it seems to be fairly constant over a range of numbers of processors. This indicates that we can expect this level of performance on larger numbers of processors for larger problem sizes, because for many irregular, sparse problems the connectivity of the matrix or mesh is independent of the number of unknowns.

Furthermore, several improvements are still possible. We see that the actual computation (comp) is not so efficient; we do not have the same efficiency for the local computations (without communication) as we had for the original sequential program. We have to improve this. Also the gather part of the matrix-vector product seems too expensive. Probably, this is due to the work involved in masking (basically if-statements). We can adapt the implementation to prevent any references to dummy rows and coefficients.

In the new version of our program, we will split the matrix-vector product even further, so that each local matrix-vector product consists of two parts. One part refers only to the unknowns that are locally available, and the other part refers to the unknowns that are not locally available. The non-local references are stored in a data structure that resembles the one for the matrix as a whole. The implementation of the local part of the matrix-vector product can be exactly the same as the sequential version; this should bring a major improvement for the cost of both the computational part and the gather part. The implementation of the non-local part of the matrix-vector product will be the same as for the previously discussed program as a whole. This too should reduce the cost of the matrix-vector product. With these improvements we expect to significantly raise the efficiency and speed-up of the matrix-vector product for (relatively) small numbers of processors.

4 HPF-2.0 and Extensions

The new High Performance Fortran Language Specification version 2.0 [7] (HPF-2.0) includes a separate part on 'Approved Extensions'. These are *advanced features that meet specific needs, but are not likely to be supported in initial compiler implementations.* Given the fact the the standard HPF-2.0 features are expected to be implementable within a year, it is unlikely that the approved extensions will be commonly available soon. So, for portable programs we will have to continue for the moment on the way described in the previous section. However, it is important, to anticipate these new, and important, features. We will show that they solve some of our problems with irregular matrices, but that more is still needed.

The generalized block distribution in its executable form, HPF REDISTRIBUTE(GEN_BLOCK(*block_sizes*)) allows to compute or read in a desired partitioning and implement this in run time. The array *block_sizes* gives the sizes of the blocks. Using this directive we can distribute the arrays *val* and *col_idx* with appropriate block sizes to give the number of non-zero coefficients in each partition, and the arrays *start_row* and *end_row* with appropriate block sizes to give the number of rows in each partition. Note that a corresponding alignment directive ALIGN(GEN_BLOCK()) does not exist. This would allow us to align the arrays *start_row* and *end_row* with the arrays *val* and *col_idx* so that we have the description of a local matrix locally available. As mentioned before, the arrays with the row pointers differ in size from the arrays with the values and column indices, and there is no regular (linear) relation that matches the row pointers with the values or column indices in the row. The number of non-zero coefficients per row may vary strongly. For our purposes we would like an alignment that matches two compatible generalized block distributions. That is, two block distributions with different sizes but for the same number of partitions. We could map a block of row pointers to the processor that contains the block of coefficients that belong to that row. Such a directive could have the form (RE)ALIGN(GEN_BLOCK(*row_partition,coefficient_partition*)), where *row_partition* and *coefficient_partition* are two rank-one arrays that have the same size.

The REDISTRIBUTE(GEN_BLOCK()) directive can be used to get the desired distributions without padding the arrays with dummy coefficients or dummy pointers. However, since we cannot use an alignment directive, the fact that all information about the local matrix is locally available on each processor is still not clear to the compiler. So, as long as no generalized, block-wise alignment is available, the associated problems have to be handled in the same way as in the previous section.

5 Conclusions

We have outlined an approach to implement irregular, sparse matrix solvers in HPF. Our results clearly show that reasonable speed-ups are attainable, and that the performance of the sparse matrix-vector product does not play a significant role in the scalability. Indeed, for larger numbers of processors (relative to the problem size) the global communication in the inner products dominates the performance, and the efficiency of the inner product drops below the efficiency of the irregular, sparse matrix-vector product.

The current implementation is by no means optimal yet. We mainly concentrated on low communication costs, which we seem to have achieved. Moreover, we probably can improve the communication costs significantly by better partitioning algorithms. Some tests we performed indicated that the current partitionings are not so good. Also the computational cost is too high, and we have indicated several improvements.

Finally, we have indicated how the generalized block distribution in the 'approved extensions' of HPF-2.0 helps with some, but not all, of our problems. We propose a block-wise alignment to improve HPF programs for irregular, sparse matrix algorithms.

Acknowledgements

We would like to thank Larry Meadows (PGI) and Horst Simon (NERSC) for making their codes available to us.

References

1. S. T. Barnard and H. D. Simon. A fast multilevel implementation of recursive spectral bisection for partitioning unstructured problems. Technical Report RNR-92-033, NASA Ames Research Center, Mail Stop T045-1, Moffet Field, CA 94035, USA, 1992.
2. E. De Sturler. *Iterative Methods on Distributed Memory Computers.* PhD thesis, Delft University of Technology, Delft, The Netherlands, October 1994.
3. E. De Sturler. Incomplete Block LU preconditioners on slightly overlapping subdomains for a massively parallel computer. *Applied Numerical Mathematics (IMACS)*, 19:129–146, 1995.
4. E. De Sturler and H. A. Van der Vorst. Communication cost reduction for Krylov methods on parallel computers. In W. Gentzsch and U. Harms, editors, *High-Performance Computing and Networking*, Lecture Notes in Computer Science 797, pages 190–195, Berlin, Heidelberg, Germany, 1994. Springer-Verlag.
5. E. De Sturler and H. A. Van der Vorst. Reducing the effect of global communication in GMRES(m) and CG on parallel distributed memory computers. *Applied Numerical Mathematics (IMACS)*, 18:441–459, 1995.
6. F. Nataf, F. Rogier, and E. De Sturler. Domain decomposition methods for fluid dynamics. In A. Sequeira, editor, *Navier-Stokes Equations and Related Nonlinear Problems*, New York, 1995. Plenum Press.
7. High Performance Fortran Forum. High Performance Fortran Language Specification, version 2.0 Rice University, 1997
8. A. Pothen, H. D. Simon, and K.-P. Liou. Partitioning sparse matrices with eigenvectors of graphs. *SIAM J. Matrix Anal. Appl.*, 11:430–452, 1990.

9. Y. Saad and M. Schultz. GMRES: A generalized minimal residual algorithm for solving nonsymmetric linear systems. *SIAM J. Sci. Statist. Comput.*, 7:856–869, 1986.

Parallel Navigation in an A-NETL Based Parallel OODBMS

Lawrence Mutenda, Manabu Hiyama, Tsutomu Yoshinaga and Takanobu Baba

Department of Information Science, Utsunomiya University,
Utsunomiya 321, JAPAN.
Phone/fax: 028-689-6262,
mutenda@lynx.infor.utsunomiya-u.ac.jp

Abstract. A parallel OODBMS has been proposed based on a parallel object-oriented language, A-NETL. The OODBMS is designed for a shared-nothing environment. An overview of the database system is described. Accessing object-data in a parallel OODBMS is based on navigation. A parallel navigation algorithm being implemented for use in the system is presented including its features. The algorithm is based on the need to balance the load across all nodes in a parallel OODB accessing objects with set-valued attributes. An analytical evaluation of the features of the algorithm is presented.
Keywords: Parallel Processing, A-NETL, Hash, OODBMS, Navigation, Load Balancing

1 Introduction

Object-Oriented Database (OODB) systems have been a very active research area for the past several years [10]. A number of commercial systems have come on the market and many more have been designed as research prototypes, for example O_2 [2].

In the Relational Database field, parallel processing has long been a major research area and a number of successful projects have been carried out such as the Super Database Computer [11] at the University of Tokyo , GAMMA[7] at the University of Wisconsin and XPRS [8] at the University of California. Although parallel processing has been effectively employed in relational database systems, only a few projects are reported in the literature for OODBs. A parallel OODB is described in [5]. This is based on a shared nothing paradigm and introduces the concept of multi-wavefront query processing. DeWitt et. al.[6] also describes the Shore OODBMS based on the notion of ParSets. Parallel pointer-based joins for OODBMSs are described in [9]. However, it should be noted that research into aspects of OODB parallel processing is still limited and compared to relational systems, it is still much less clear whether parallelism can be effectively applied to OODBMS[6].

An important process in joining object-oriented data is navigation. In this process pointers stored within objects are followed to locate related objects. In a shared-nothing parallel system there is a need to follow pointers to data stored in other nodes. It is necessary to design an algorithm that allows these pointers to be efficiently followed and at the same avoid potential load imbalance.

This research seeks to apply parallel processing to an object oriented database system. The parallel OODBMS is being implemented in A-NETL, a new object-oriented parallel programming language designed by Baba *et. al.*[1] and will run on a cluster of workstations. We are implementing a parallel navigation algorithm for this new OODBMS.

The rest of this paper is organized as follows. Section 2 discusses the OODBMS system overview. Section 3 describes the parallel navigation algorithm proposed for the database system. Section 4 presents an evaluation model for the algorithm and section 5 describes future work.

2 System Overview

A-NETL assumes the existence of a number of processing nodes each of which can participate in the execution of a program. An A-NETL program is basically a network of objects co-operating to solve a programming problem. Program objects are allocated among processing nodes. Objects communicate by message passing. Our OODBMS design consists of system modules designed as objects distributed across processor nodes, in a shared-nothing environment.

2.1 Software

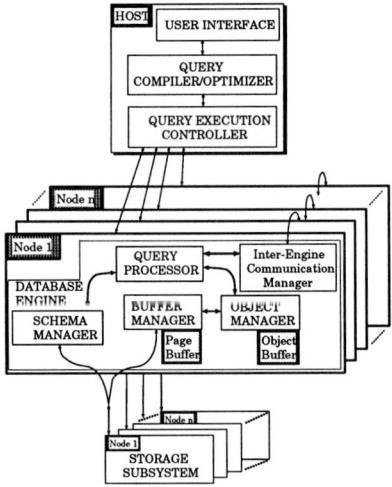

Fig. 1. The Parallel A-NETL OODB System Overview

Fig.1 shows the outline of the A-NETL OODBMS design . As shown, the system consists of the following sub-systems: the user interface, the query compiler/optimizer and the Query Execution Controller to be written in the C language; the Database Engine being implemented in A-NETL; the storage subsystem also being implemented in A-NETL.

The Query execution controller will be responsible for receiving an optimized and compiled query and initiating execution on system nodes. It will itself reside on the Host.

Each node in the system consists of a database engine. The engine contains 5 main modules implemented as A-NETL objects, namely the query processor, the schema manager, the buffer manager, the object manager and the inter-engine communication manager. The query processor executes system algorithms to solve compiled user queries. The schema manager maintains the class descriptions of all the classes defined within a database. The buffer manager manages the page buffers in each processor node memory. When the query processor requires an on-disk object to be loaded into memory, the buffer manager loads the whole page containing that particular object. The object manager manages all objects loaded in from buffer memory and controls the conversion of objects from disk form to memory form. The inter-engine communication manager facilitates communication between database engines executing at different nodes.

2.2 Hardware

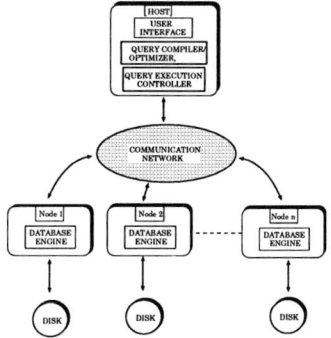

Fig. 2. The Parallel A-NETL OODB Hardware Overview

Fig 2 shows the outline of the hardware on which the A-NETL OODBMS software is being implemented. This is basically a shared-nothing system. The configuration consists of a host machine, and an arbitrary number of processing nodes. Each processing node manages its own disk and communicates with others via the network by message passing. A Database Engine software executes on each node.

System hardware consists of a workstation cluster running the PVM message passing system. Our laboratory has designed an A-NETL compiler that runs on the PVM message passing library.

3 Parallel Navigation Algorithm

The A-NETL OODBMS design includes algorithms for implementing parallel query processing as well as standard functions like select, project and object method invocation [12]. The system includes parallel navigation with dynamic load balancing for processing, in parallel, path expressions found in object-oriented queries. In this section we describe the parallel navigation algorithm.

3.1 Navigating Objects in a Parallel OODBMS

Every object has a set of attributes which describe its composition. Some of these attributes are set valued, with either absolute values or object-identifier (OID) values . For example an object such as a three-dimensional prism may have an attribute ,*sides*, which may be a collection of OIDs dereferencing the sides of the object. Set-valued objects whose elements are OIDS are quite common in OODBMSs.

It is important to consider the parallel navigation of object pointers of such objects. Consider the object whose structure is shown in fig.3.

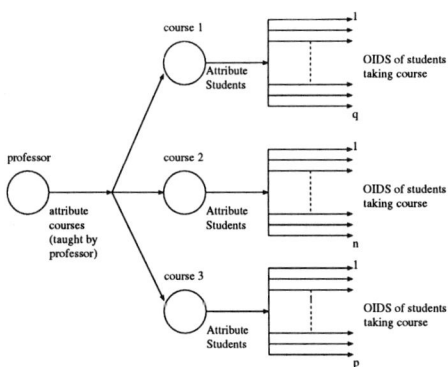

Fig. 3. A Complex Object with Set-valued Attributes

The *professor* object has a set-valued attribute, *courses* which is a set of OIDs representing the courses that professor teaches. In turn each course has a set-valued attribute whose value is the set of *students* (OIDs) taking that course.

Assume that the OID of an object includes information about its node location. Operating on shared-nothing hardware, the simplest way to navigate the complex object of fig.3, would be as shown in fig.4. Here we start with professor object OID1 on node 1 which points to course object OID2 on node 2, which in turn points to object OID3 on node 3. A query searching for these objects would move from node 1 to node 2, examine the contents of OID2, which directs it to OID3 on node 3. This strategy is simple but immediately presents the following problems:

- For a given query, some nodes may contain more query relevant objects than others and therefore have more pointers from other objects referring to its resident objects leading to load imbalance.
- The network will be overloaded with data being transferred between nodes.

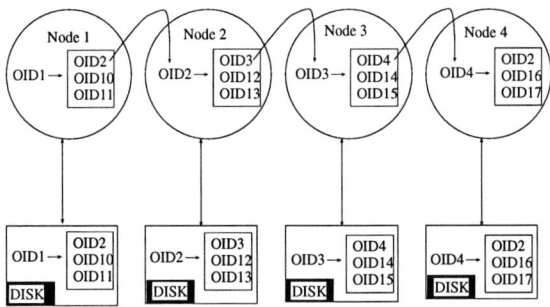

Fig. 4. Navigation in a Parallel OODBMS

Inter-node navigation can be reduced by a good data-placement strategy. However, this is a whole research area by itself and since our main focus is parallel processing in OODBMSs we decided that we will consider efficient data placement strategies later on in the research. At this point we will use simple data placement strategies such as round-robin, hash and range partitioning. We assume that each class is declustered evenly across system nodes.

3.2 Parallel Navigation with Dynamic Load Balancing

In this research we propose dynamic object data redistribution as a load balancing technique. In the proposed technique, if a node determines that it has to process too many data objects compared to other nodes it can initiate data redistribution to those other nodes.

```
result  =  SELECT  S.name
           FROM    F in faculty ,
                   S in F.course.students
           WHERE   40 ≤ F.age ≤ 50 and F.rank = 'professor'
```

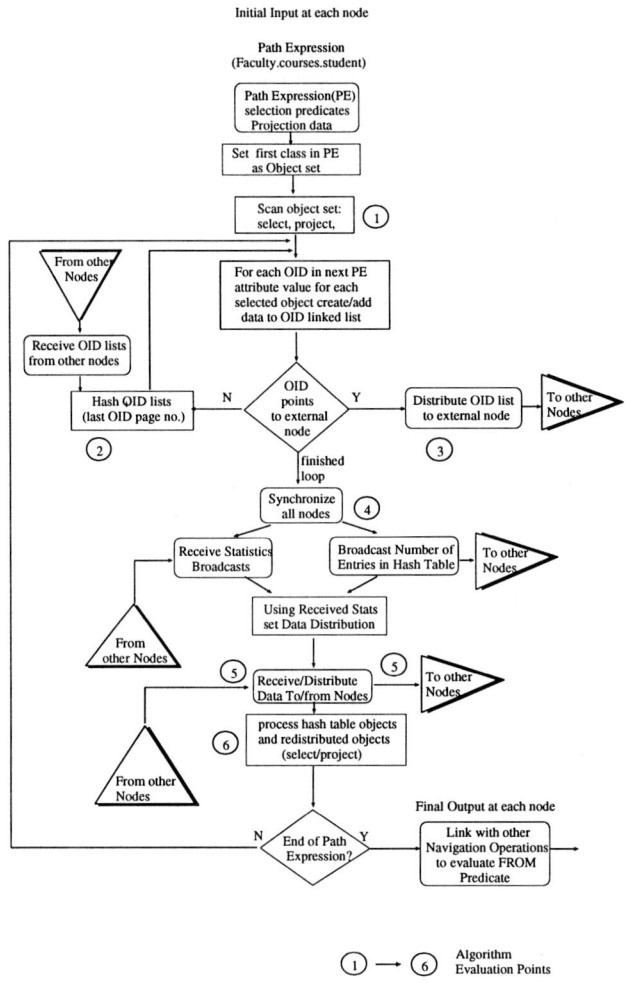

Fig. 5. Parallel Navigation with Dynamic Load Balancing Algorithm

Fig. 5 is a flow chart of the algorithm proposed to implement parallel navigation as described. We assume that the an object OID is physical and identifies the node and page on which an object is located. The input to the algorithm is a path expression (PE) of arbitrary length. We will briefly explain the algorithm below.

We will consider the query given above to find those students taking a course from a professor who is between 40 and 50 years of age.

Parallel Navigation Take the *faculty.courses.student* as an example PE. The algorithm executing at each system node will take the faculty class, scan the portion of the faculty at that node, selecting and projecting as required by the query. The query relevant data from these objects is used to create a stripped down version of the object. The value of the next attribute in the PE, in this case *courses*, is then loaded. for each object, every OID in this multi-attribute value is combined with the stripped down version of the object to form an OID linked list. This linked list, as shown in fig.6, consists of the source object data and an OID from a multivalued attribute as the last element.

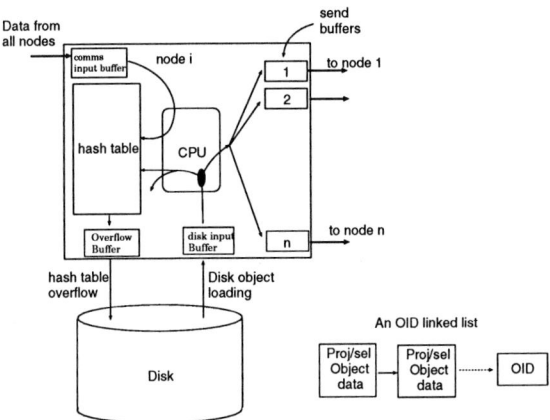

Fig. 6. Data Distribution Diagram

Each OID linked list is then transmitted to the node pointed to by its last element (a multi-valued attribute OID). Lists bound for the node executing the algorithm are hashed on the page number of the last element. Lists going to other nodes are transmitted accordingly and each node receives lists sent from other nodes and inserts theminto its hash table. The process is illustrated in fig.6. When all OID lists have been processed a synchronisation step then is executed. Dynamic load balancing, explained in the next subsection, is then initiated.

After load balancing the second level of the path expression is executed and with the algorithm looping until all PE elements are processed.

Dynamic Load Balancing Data in the hash table is such that all OID linked lists with the terminal element (an OID) pointing to one page are stored in one hash line.

The need for load balancing was explained in section 3.1. In brief, the load at a node is considered to be the number of OID linked lists in its hash table. Load imbalance is said to occur if the number of OID linked lists at any load is greater or less than $\pm\epsilon\%$, of the average number of OID linked lists per node O_{avg}.

To ensure that the system load is balanced across all nodes, data is redistributed in the system as follows. Each node broadcasts the total number of OID linked lists in its hash table. Upon receiving broadcasts of this data from all nodes, a node uses that data to determine a data redistribution that balances the system load. Each node finds the average number of OID linked lists per node O_{avg}. A range from this average, $\pm\epsilon\%$, is defined such that a node with more than $O_{avg} + \epsilon\%$ objects or less than $O_{avg} - \epsilon\%$ objects is considered to be unbalanced. The value for $\epsilon\%$ will be derived experimentally to produce a value that gives good load characteristics.

The unit of data redistribution is the hash line and its corresponding page. We assume that the number of OID linked lists per hash line in a node is approximately equal to simply the redistribution algorithm.

Data is redistributed using the following algorithm. Each node creates two lists of nodes, list 1, a list of overloaded nodes and list 2, a list of under-loaded nodes. List 1 is arranged in order of decreasing load with the heaviest loaded node at the head of the list. List 2, is arranged in order of increasing load with the least loaded node at the head of the list. Each node on the system runs the same algorithm and produces a redistribution plan before any data is actually transferred. Data is set for transfer as follows:

The node at the top of list 1 assigns some of its hash lines table to the node at the top of list 2 until either it own load is within the balanced load range or until the node at the top of list 2 is within balanced range. If the list 1 node is now balanced but the list 2 node is not then the list 1 node is removed from the list and the next node in that list is put at the head. The new head then sets data for transfer to the node at the head of the list 2 until balance is achieved in either. The basic algorithm is then that hash lines are assigned for transfer until one of the nodes participating is balanced and any balanced node is removed from its corresponding list. The algorithm terminates when there are no more nodes in either list 1 or list 2 or both.

When a node has produced a complete redistribution plan, actual data transfer is initiated. In A-NETL data is transferred as parameter of a message.

The granularity of redistribution has been chosen as a page, since it was felt that objects are too fine a granularity and would increase message passing overhead unnecessarily. Since OID lists are hashed on a page value, a page and all the OIDs pointing to it are transferred to the appropriate node.

4 Analytical Evaluation of the Navigation Algorithm

In this section we discuss an analytical evaluation of the navigation algorithm. Table 1 lists the definitions of the terms used in the evaluation. We focus on three aspects of processing namely reading and writing to disk in pages , communication time for an object of size s , and CPU time for selecting and projecting and hashing an object [13].

4.1 Analysis Assumptions

We assume that sending an object od size s takes a fixed time at the sending node t_{send} and a fixed time at the receiving node t_{recv}. Inputing or outputting a page of disk data takes a constant time t_{disk}. Projecting and selecting an object takes t_{proc} and hashing an object takes t_{hash}. Finding the amount of time spent in each part the algorithm then becomes a matter of estimating the number of I/O, communication and CPU operations and multiplying by the relevant time constant.

We also assume that that all query relevant attributes are multi-valued and that each object has only one multi-valued attribute, the query relevant attribute. We also assume that we are processing the following PE: **A.b** where **A** is a class and **b** is a multi-valued attribute with domain class **B**.

4.2 Cost Equations

Name	Description
P	size of a disk page
n	number of nodes
s	size of an object
Φ_{oi}	total number of class O objects at node i
sel_{oi}	selectivity for predicate for class O at node i
M	total available main memory per node
$y_{o,ij}$	average fraction of pointers from node i to node j for class O
x_{oi}	average number of multi-attribute pointers per object for node i for class O

Table 1. Parameter Definition List

Referring to fig.5 we can identify six points in the algorithm that are critical to determining execution time and algorithm flow. These are numbered 1-6 in the figure. We develop equations for each of these points below.

Processing first class A in PE (1) Node i scans and selects objects from class **A** and the number of pages input is given as:

$$PageIO_{ai,scan} = \frac{\Phi_{ai} \cdot s}{P}$$

where Φ_{ai} is the number of class A objects at node i, the where A is the first class in path expression **A.b**.

The number of selection/projection operations are given as:

$$SelPro_{ai,op} = \Phi_{ai}$$

since each object in class A is examined.

Hashing OID Linked Lists (2) At this point each of the selected objects is inserted into a hash table after creating an OID linked list. Since $y_{a,ji}$ is, for node i, the average fraction of multi-valued attribute OIDs pointing to node i from node j for class A, the total number of OID linked lists received by node i is equal to the total number of OID linked lists sent to this node by all the other nodes for class A. This value, termed $A_{i,recv}$ for class A at node i, can be calculated as follows:

$$A_{i,recv} = \sum_{j=1}^{n,i \neq j} y_{a,ji} \cdot x_{aj} \cdot \Phi_{aj} \cdot sel_{aj}$$

This figure is represents the receive communication time for node i.

The total number of OID linked lists, one for each multi-valued attribute OID at node i, added to the total number of received OID lists equals the total number of hashing operations:

Number of hashing operations =

$$Hash_{ai} = A_{i,recv} + (1 - \sum_{j=1}^{n,i \neq j} y_{a,ij}) \cdot x_{ai} \cdot \Phi_{ai} \cdot sel_{ai}$$

where $1 - \sum_{j=1}^{n,i \neq j} y_{a,ij}$ is, for class A, the fraction of multi-valued OIDs at node i pointing to node i itself (the rest point to external nodes).

If the size of an OID linked at the first level is s_{OID}, the total memory used to store this hash table will be $s_{oid} \cdot Hash_{ai}$. If this figure exceeds the amount of main memory available then some of the hash table overflows to disk. The corresponding disk operations so generated are given as:

$$I/O_{ai,dsk-hsh} = \frac{s_{oid} \cdot (Hash_{ai}) - M}{P}$$

Distributing OID Linked Lists (3) Point 3 consists only of operations for distributing data to other nodes. The number of OID lists distributed here is given below, calculated from the fraction of pointers pointing to external nodes:

$$A_{i,send} = \Phi_{ai} \cdot sel_{ai} \cdot x_{ai} \cdot \sum_{j=1}^{n,i \neq j} y_{a,ij}$$

If we multiply $A_{i,send}$ by the time for a a send operation for an OID linked list, we can estimate the time for this phase of the algorithm. Note that network time is ignored in this analysis since it is assumed that such is overlapped with the other times. Only the in case of a slow network will the network time probably become a factor. In our analysis, only the time a node spends in injecting the message into the network is considered. The same is true for receiver operation. However it will be important to experimentally verify the validity of this assumption and make any allowance that might be necessary to correctly represent the real world.

Synchronisation (4) The slowest node to reach synchronisation will determine the amount of time stages 1,2 and 3 will take. The amount of time to reach synchronisation will be as follows:

$$T_{synch,1,2,3} = max\{(PageIO_{ai,scan} + SelPro_{ai,op} \\ + A_{i,recv} + Hash_{ai} \\ + I/O_{ai,dsk-hsh} + A_{i,send}) : i = 1 \text{ to } n\}$$

where where i is the node and each term is multiplied by the corresponding time unit.

Load Balancing (5) Once data has been distributed statistics are gathered for load balancing. Generating a data redistribution plan should take negligible time since it involves simple CPU operations.

At the beginning of the load balancing phase node i contains $Hash_{ai}$ OID linked lists. Any node that deviates from the value by more than $\pm\epsilon\%$ is considered a candidate for redistribution. However at this point to simplify analysis we assume that only one node i is overloaded and only one node j is underloaded. if we assume that the OID linked lists in its hash table are distributed evenly across hash lines then the overloaded node will only need to transfer OID linked lists and the data pages they point to until the underloaded node has OID linked list within the acceptable range.

Processing Next Class B in PE (6) This point is now basically a repetition of the algorithm, and can be analysed in the same way as for point 1.

4.3 Discussion

In the analysis we defined Φ_{oi}, the number of Class O objects at node i. This figure can be calculated from experimental data. The parameter sel_{oi} can also be estimated from available data. The parameters x_{oi} and $y_{o,ij}$ can be estimated from experimental data as well but in practice may vary considerably from the average for each query on the same class. Experiment should determine how serious this variation is.

We have also chosen to ignore network communication time assuming that this time is overlapped with node operations. This will need to be validated experimentally as well.

5 Conclusion and Further Work

We have described the design outline of a parallel OODBMS based on the A-NETL language to be implemented in a shared-nothing environment on a cluster of workstations. We have described in detail a parallel navigation algorithm and its analytical evaluation. presently we are implementing the algorithm in a simulated environment to enable us to experimentally evaluate the algorithm and tune the analytical model to more accurately reflect the real world environment. In tandem we are developing the overall system which includes other necessary functionality. Our ultimate aim is to produce a fully fledged parallel OODB that can be evaluated using standard benchmarks like such as OO1 [4] and OO7 [3] adapted for a parallel environment.

References

1. Baba T., Yoshinaga T., Furuta T.,: *Programming and Debugging for Massive Parallelism: The Case for a Parallel Object-Oriented Language A-NETL*, Proc. Wkshp. OBPDC 95, June (1995).
2. Bancilhon F. et. al.: *Building an Object-Oriented Database System: The Story of 0-2*, Morgan Kauffman, (1992).
3. Carey M.J., DeWitt D.J., Naughton J.F.: *The OO7 Benchmark*, Univ. Wisconsin January (1994).
4. Cattell R.G.G., Skeen J.: *Object Operations Benchmark*, ACM TODS Vol 17, No. 1, March (1992).
5. Chen Y.H., Su S.: *Identification and Elimination-based Parallel Query Processing Techniques for Object-Oriented Databases*, Journal of Parallel and Distributed Computing, Vol. 18, No. 2, August (1995).
6. DeWitt, D.J., Naughton J.F., Shafer J.C., Venkataraman S.: *ParSets for Parallelizing OODBMS Traversals: Implementation and Performance*, Univ. Wisconsin Tech Report (1995).
7. DeWitt D. J., Ghandiharizadah S., Schenider D, Bricker A, Hsiao H., Rasmussen R,: *The Gamma Database Machine Project*, Univ. Wisconsin Tech Report (1990).
8. Hong W.: *Parallel Query Processing Using Shared Memory Multiprocessors and Disk Arrays*, PhD Thesis, Univ. Southern California (1992).
9. Lieuwen, D.F., DeWitt, D.J., Mehta M. *Parallel Pointer-based Join Techniques for Object-oriented Databases*, Univ. Wisconsin Tech Report (1993)
10. Kim W.: *Modern Database Systems*, Addison Wesley Co. , (1995).
11. Kitsuregawa Masaru, Yang W., et. al.: *Overview of the Super Database Computer* , IEICE Trans. Electron. Vol. E77-C, No. 77 July (1994).
12. Mutenda L., Yoshinaga T. and Baba T.: *A-NETL Based Parallel Object-Oriented Database Management System* 8th Data Engineering Workshop 97, March 1997 pp233-238.
13. Walton C.B., Dale A.G., Jenevein R.M.: *A Taxonomy and Performance Model of Data Skew Effects in Parallel Joins*, Proc. of the Int. Conf. on VLDB, September 1991 pp537-548.

High Performance Parallel FFT on Distributed Memory Parallel Computers

Naohiko Shimizu[1] and Takehiko Watanabe[2]

[1] School of Engineering, Tokai University
1117 Kitakaname Hiratsuka-shi, Kanagawa 259-12, Japan
email: nshimizu@et.u-tokai.ac.jp
[2] NTT corp.
email: watanabe@system.tsh.cae.ntt.jp

Abstract. In this paper, a high performance parallelizing method of FFT is presented. Well known four or six step parallel algorithm with standard index map is not suitable for highly parallel computers, because it requires all-to-all communications between two phases of sub-FFTs which can not be overlap the computation of the each sub-FFT over the communication. We introduce another index map and algorithm which is intended to overcome the problem, and our results shows that our method out-perform the four step method in the 26 case out of 32 experiments. The results was obtained with up to 128 processors NEC Cenju-3 using the mini-MPI library.

1 Introduction

Fast Fourier Transforms (FFTs) have a variety of applications, and they are very important algorithms of scientific computing. FFTs need $O(Nlog_2 N)$ floating point operations and there are many proposals for the parallel or vectorized execution of FFTs.[1][2][3] Recently, distributed memory parallel processors are very attractive for high performance computing, if there are certainly algorithms to exploit the parallelism.

Parallelizing of FFTs on the distributed memory parallel processors can be carried out by dividing the problems and execute each parts in parallel. As shown in the reference[1], there are two well known frameworks for distributed memory parallel processors, 1) distributed Cooley-Tukey framework based on the distributed butterflies, 2) distributed four-step and six-step frameworks based on distributed transposition. There are not significantly difference on the communication overhead between the two frameworks. But the four step algorithm has more attractive features as a distributed memory parallel algorithm[1][2]. This algorithm is consist of two phases of sub-FFTs, and between these two phases there is a transposition step and a twiddle-factor scaling. For distributed memory parallel computers, this transposition requires all-to-all communication, and the volume of the communication is in proportion to the volume of the problem. Then if one want to do a larger FFT, the communication cost will be larger. And it will restrict the performance even with massively parallel multi-processors.

In this paper we introduce a new algorithm for distributed memory parallel processors. Our method is an extension of the *four step algorithm* and it introduces another sub-FFT step and another twiddle-factor scaling step to make overlapping of the computation over the communications. Also in this paper we describe the performance results of our method comparing with the *four step algorithm*.

Our results are obtained on a NEC Cenju3 parallel computer with 128 processing elements. And we used C language with mini MPI library which is a variant of the MPI (Message-Passing Interface) standard.

2 The Index Mapping Method for Distributed Memory Parallel Processors

In this section we describe the index mapping method for distributed memory parallel processors. The index mapping method is known as a *twiddle-factor* method because of the *twiddle-factor scaling* in the method[1]. First we show the famous index mappiing method *four-step* algorithm briefly. Then we will introduce a new index mapping for distributed memory parallel processors which is an extension of the *four-step* algorithm.

2.1 The Four-Step Algorithm

Given sequence x_n, we have the definition of the discrete Fourier transform

$$g_k = \sum_{n=0}^{N-1} x_n \omega_N^{nk} \quad (1)$$

where, $\omega_N = e^{\frac{-2\pi jnk}{N}}$ and $j^2 = -1$. If N has factors $N = N_1 N_2$, then by using a index maps

$$\begin{aligned} n &= n_1 N_2 + n_2 \\ k &= k_1 + k_2 N_1 \end{aligned} \quad (2)$$

we can define the two-dimensional arrays

$$\begin{aligned} x_{n_1,n_2} &= x_n, \; n_1 = 0, \ldots, N_1 - 1, n_2 = 0, \ldots, N_2 - 1 \\ g_{k_1,k_2} &= g_k, \; k_1 = 0, \ldots, N_1 - 1, k_2 = 0, \ldots, N_2 - 1 \end{aligned} \quad (3)$$

Substituting (2) and (3) into (1) we obtain

$$g_{k_1 k_2} = \sum_{n_2=0}^{N_2-1} \sum_{n_1=0}^{N_1-1} x_{n_1 n_2} \omega_{N_1}^{n_1 k_1} \omega_{N_1 N_2}^{n_2 k_1} \omega_{N_2}^{n_2 k_2} \quad (4)$$

Therefore $g_{k_1 k_2}$, and hence g_k, can be computed as two multiple transforms. This DFT problem can be computed as follows:[1]

$$\begin{aligned} x_{n_1 \times n_2} &\leftarrow x_{n_2 \times n_1}^T \\ x_{n_1 \times n_2} &\leftarrow F_{n_1} x_{n_1 \times n_2} \\ x_{n_1 \times n_2} &\leftarrow F_n(0 : n_1 - 1, 0 : n_2 - 1) . * x_{n_1 \times n_2} \\ x_{n_1 \times n_2} &\leftarrow x_{n_1 \times n_2} F_{n_2} \end{aligned} \quad (5)$$

2.2 The New Index Mapping

If N has factors $N = N_1 N_2 N_3$, then by using a index maps

$$n = n_1 N_2 N_3 + n_2 N_3 + n_3$$
$$k = k_1 + k_2 N_1 + k_3 N_1 N_2 \qquad (6)$$

we can define the three-dimensional arrays

$$x_{n_1,n_2,n_3} = x_n, \; n_1 = 0, \ldots, N_1 - 1, n_2 = 0, \ldots, N_2 - 1, n_3 = 0, \ldots, N_3 - 1$$
$$g_{k_1,k_2,k_3} = g_k, \; k_1 = 0, \ldots, N_1 - 1, k_2 = 0, \ldots, N_2 - 1, k_3 = 0, \ldots, N_3 - 1 \qquad (7)$$

Substituting (6) and (7) into (1) we obtain

$$g_{k_1 k_2 k_3} = \sum_{n_3=0}^{N_3-1} \sum_{n_2=0}^{N_2-1} \sum_{n_1=0}^{N_1-1} x_{n_1 n_2 n_3} \omega_{N_1}^{n_1 k_1} \omega_{N_1 N_2}^{n_2 k_1} \omega_{N_2}^{n_2 k_2} \omega_{N}^{n_3 k_1} \omega_{N_2 N_3}^{n_3 k_2} \omega_{N_3}^{n_3 k_3} \qquad (8)$$

Therefore $g_{k_1 k_2 k_3}$, and hence g_k, can be computed as three multiple transforms. The three dimensional arrays $g_{k_1 k_2 k_3}, x_{n_1 n_2 n_3}$ are presented in the Figure 1.

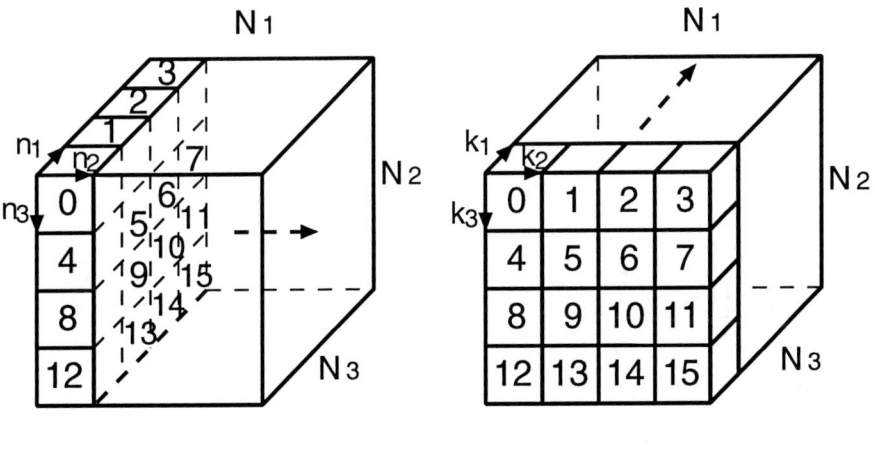

Fig. 1. The three dimensional arrays derived from index mapping

The equation (8) can be decomposed into tree transforms, and we name the transforms as XFFT, YFFT, ZFFT.

$$XFFT \Rightarrow X_{k_1 n_2 n_3} = \sum_{n_1=0}^{N_1-1} x_{n_1 n_2 n_3} \omega_{N_1}^{n_1 k_1} \omega_{N_1 N_2}^{n_2 k_1} \qquad (9)$$

$$YFFT \Rightarrow Y_{k_1 k_2 n_3} = \sum_{n_2=0}^{N_2-1} X_{k_1 n_2 n_3} \omega_{N_2}^{n_2 k_2} \omega_{N}^{n_3 k_1} \omega_{N_2 N_3}^{n_3 k_2} \qquad (10)$$

$$ZFFT \Rightarrow Z_{k_1 k_2 k_3} = \sum_{n_3=0}^{N_3-1} Y_{k_1 k_2 n_3} \omega_{N_3}^{n_3 k_3} \qquad (11)$$

$$g_{k_1 k_2 k_3} = Z_{k_1 k_2 k_3} \qquad (12)$$

Each transforms are corresponding to the X, Y, Z axes of the figure 1, respectively. The result of the XFFT will directly be the input of the YFFT, and the result of the YFFT will directly be the input of the ZFFT. The output of the ZFFT will be the FFT of the input vector x_n. These transforms can be carried out as a conventional three dimensional Fourier transform which is designated in figure 2.

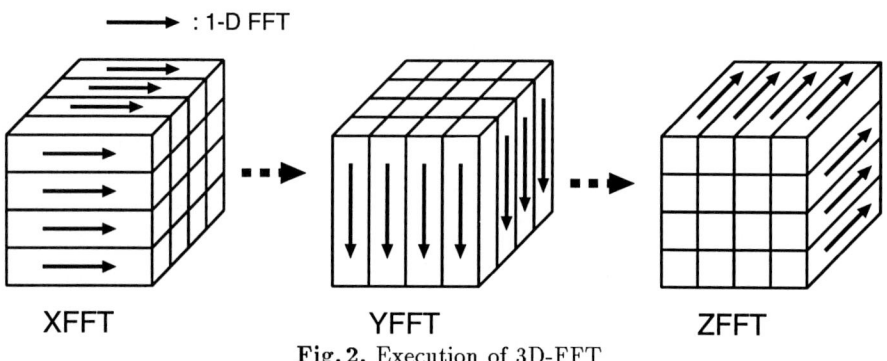

Fig. 2. Execution of 3D-FFT

We can derive our algorithm as followings:

1. Permutate the input vector x_n as a three dimensional array with index mapping of $n = n_1 N_2 N_3 + n_2 N_3 + n_3$.
2. For each line with length N_1 orthogonal to the y-z plane, make DFTs.
3. For each transformed data, multiply the twiddle factor $\omega_{N_1 N_2}^{n_2 k_1}$.
4. For each line with length N_2 orthogonal to the z-x plane, make DFTs.
5. For each transformed data, multiply the twiddle factor $\omega_N^{n_3 k_1} \omega_{N_2 N_3}^{n_3 k_2}$.
6. For each line with length N_3 orthogonal to the x-y plane, make DFTs.

3 Parallelization of the Index Mapped FFT

In general, it is required to evaluate the amount of data communication and the distribution strategy to exploit maximum performance on a parallel algorithm. These parameters are normally a function of the problem size and the number of processing unit and the network performance and topology. In this section we describe the data distribution of our method and the algorithm with data communication.

The figure 3 is the communication diagram for the execution of the 3D FFT which is described in the following algorithm. This algorithm will not overlap

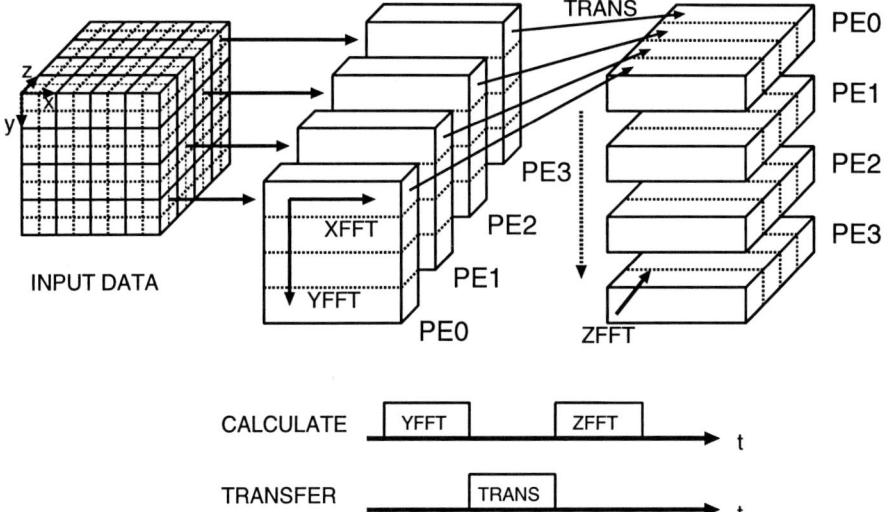

Fig. 3. Data distribution and transform execution of the 3D FFT

the communications over the computations. And the performance will not be optimal.

1. Permutate the input vector x_n as a three dimensional array with index mapping of $n = n_1 N_2 N_3 + n_2 N_3 + n_3$.
2. Distribute the data as y-x plane will fit on a processor.
3. For each line with length N_1 orthogonal to the y-z plane, make DFTs.
4. For each transformed data, multiply the twiddle factor $\omega_{N_1 N_2}^{n_2 k_1}$.
5. For each line with length N_2 orthogonal to the z-x plane, make DFTs.
6. For each transformed data, multiply the twiddle factor $\omega_N^{n_3 k_1} \omega_{N_2 N_3}^{n_3 k_2}$.
7. Execute all-to-all data communication.
8. For each line with length N_3 orthogonal to the x-y plane, make DFTs.

4 Overlapping the Communication over the Computation

The algorithm described in the previous section does not overlap the communication over the computation. In this section we discuss about the overlapping to exploit more parallelism. Unlike the four step algorithm, our method can start to transfer the 2nd transform result before the all 2nd transforms have completed.

4.1 Grouping YFFT Transforms for the Overlapping

The YFFT transforms can be grouped so as to start data transmission before the completion of the transforms. The figure 4 shows the grouping of the YFFT to 4 groups and the way of the overlapping transforms computation over the transmission. The parallel algorithm of the figure4 is shown below:

1. Permutate the input vector x_n as a three dimensional array with index mapping of $n = n_1 N_2 N_3 + n_2 N_3 + n_3$.
2. Distribute the data as y-x plane will fit on a processor.
3. For each line with length N_1 orthogonal to the y-z plane, make DFTs.
4. For each transformed data, multiply the twiddle factor $\omega_{N_1 N_2}^{n_2 k_1}$.
5. For each line within a group with length N_2 orthogonal to the z-x plane, make DFTs.
6. For each transformed data within the group, multiply the twiddle factor $\omega_N^{n_3 k_1} \omega_{N_2 N_3}^{n_3 k_2}$.
7. For each transformed data within the group, start the non-blocking transmission to the appropriate processors.
8. If there are any groups which does not processed yet, goto step 5.
9. For each line with length N_3 orthogonal to the x-y plane, make DFTs.

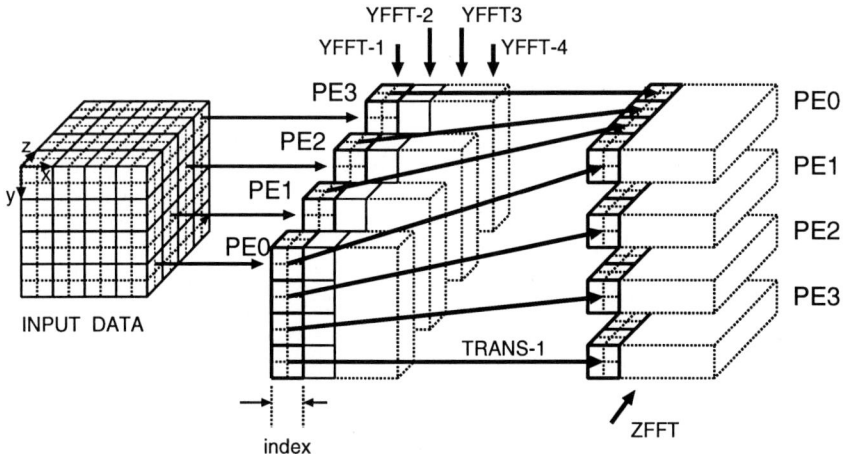

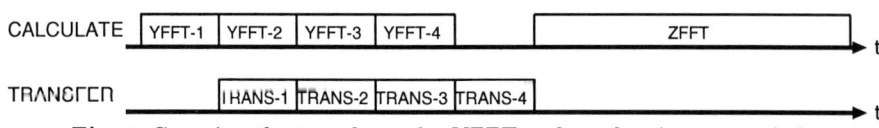

Fig. 4. Grouping the transforms for YFFT and overlapping transmission

Even with this algorithm, the last transmission data(TRANS-4) will not be able to overlap over the computation.

4.2 Scheduling of YFFT and ZFFT for Overlapping of the Transmission over the Computation

To avoid the problem that is he last transmission can not be overlapped, we can start some part of the transform computation of the ZFFT before the YFFT transform is completed. Then we can reduce the waiting time for receive data. And as we have more things to do during the transmission of the YFFT data, the overall performance will be improved even for the machines with slower transmission performance. The algorithm will be shown below:

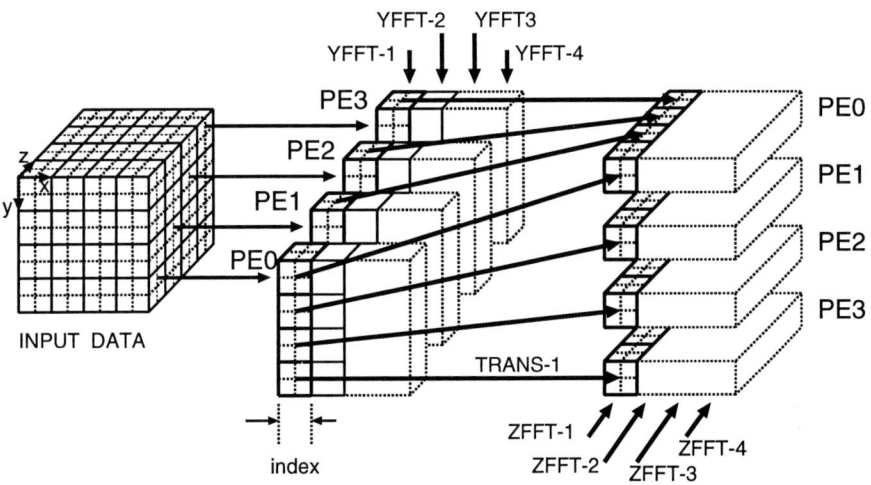

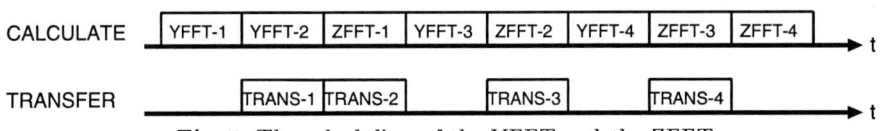

Fig. 5. The scheduling of the YFFT and the ZFFT

1. Permutate the input vector x_n as a three dimensional array with index mapping of $n = n_1 N_2 N_3 + n_2 N_3 + n_3$.
2. Distribute the data as y-x plane will fit on a processor.
3. For each line with length N_1 orthogonal to the y-z plane, make DFTs.
4. For each transformed data, multiply the twiddle factor $\omega_{N_1 N_2}^{n_2 k_1}$.
5. For each line within a group with length N_2 orthogonal to the z-x plane, make DFTs.
6. For each transformed data within the group, multiply the twiddle factor $\omega_N^{n_3 k_1} \omega_{N_2 N_3}^{n_3 k_2}$.
7. For each transformed data within the group, start the non-blocking transmission to the appropriate processors.
8. Do the ZFFT for the receiving data.
9. If there are any groups which does not processed yet, goto step 5.
10. For each line with length N_3 orthogonal to the x-y plane, make DFTs.

The figure 5 shows the scheduling of the YFFT and ZFFT.

4.3 Granularity of the Grouping

To exploit the better performance, it is important that the transmission time is fully overlapped by the meaningful computation. But it depends on the processor performance, the amount of the computation and/or the amount of the communication data, the communication performance, and the number of the peer processors. Then the granularity of the communication and the transmission will be very important for higher performance. To control the granularity of the algorithm, we introduce a parameter named the *index*. The *index* designates the number of lines within a $x - y$ plane to be computed at a time. The figure 5 shows the case of $index = 2$. The more *index* values, we will get coarser granularity.

5 The Experiments and the Evaluation

In this section, all data are obtained with a NEC Cenju3 parallel computer. This machine has 128 processors and each processor has 64MB of local memory and peak 50MFlops performance. The time data shown in the tables and figures are the average of the three measurements.

5.1 The Result of Four Step Algorithm

The execution time of FFTs with four step algorithm are shown in the table 1. The **M** in the table means that the specified size of the problems could not work due to the memory restriction.

Table 1. The execution time of the four step algorithm[sec]

#PE	2^{14}	2^{16}	2^{18}	2^{20}	2^{22}	2^{24}	2^{26}
1	0.485	2.461	11.566	M	M	M	M
2	0.266	1.253	6.180	27.273	M	M	M
4	0.138	0.629	3.042	13.915	62.854	M	M
8	0.073	0.319	1.469	7.026	31.579	M	M
16	0.043	0.171	0.740	3.519	16.217	70.906	M
32	0.034	0.096	0.378	1.673	8.024	35.192	M
64	0.043	0.075	0.214	0.840	3.927	17.605	M
128	0.078	0.095	0.167	0.474	1.956	8.834	38.473

5.2 Our Method

The execution time of our method is shown in the table 2. As in the previous results, the term **M** in the table means that corresponding problems could not work due to the memory restriction. Our method is not fully tuned for utilize the processor memory yet, and is restricted earlier than the four step algorithm. And the asterisk in the table mean that there are too many processors compared to the problem size. The figure 6 shows the performance of our method based on the MFLOPS.

Table 2. The execution time of FFT with our method

#PE	2^{14}	2^{16}	2^{18}	2^{20}	2^{22}	2^{24}	2^{26}
1	0.541	2.579	11.485	M	M	M	M
2	0.279	1.263	5.856	25.830	M	M	M
4	0.140	0.627	2.896	12.916	M	M	M
8	0.072	0.353	1.426	6.508	28.702	M	M
16	0.037	0.164	0.710	3.211	14.404	M	M
32	*	0.086	0.348	1.597	7.242	31.448	M
64	*	*	0.188	0.801	3.645	15.868	M
128	*	*	*	*	1.854	8.146	M

5.3 Comparing Our Method with the Four Step Algorithm

The table 3 shows the performance rating of our method comparintg with the four step algorithm.

Table 3. The performance rating of our method to the four step algorithm

#PE	2^{14}	2^{16}	2^{18}	2^{20}	2^{22}	2^{24}	2^{26}
1	0.896	0.953	1.011	M	M	M	M
2	0.953	0.991	1.055	1.055	M	M	M
4	0.985	1.003	1.051	1.084	M	M	M
8	1.013	0.903	1.030	1.077	1.096	M	M
16	1.162	1.042	1.042	1.095	1.131	M	M
32	*	1.116	1.086	1.048	1.112	1.107	M
64	*	*	1.138	1.048	1.078	1.125	M
128	*	*	*	*	1.054	1.079	M

We obtained a better performance for relatively large problems and/or for more processors. With these results we can say the overlapping computation over the communication and divided sized FFT will be worth for consideration.

6 Conclusion

In this paper we describe the algorithm and experimental result of a parallel Fast Fourier Transforms with new index mapping method. We obtain about 5

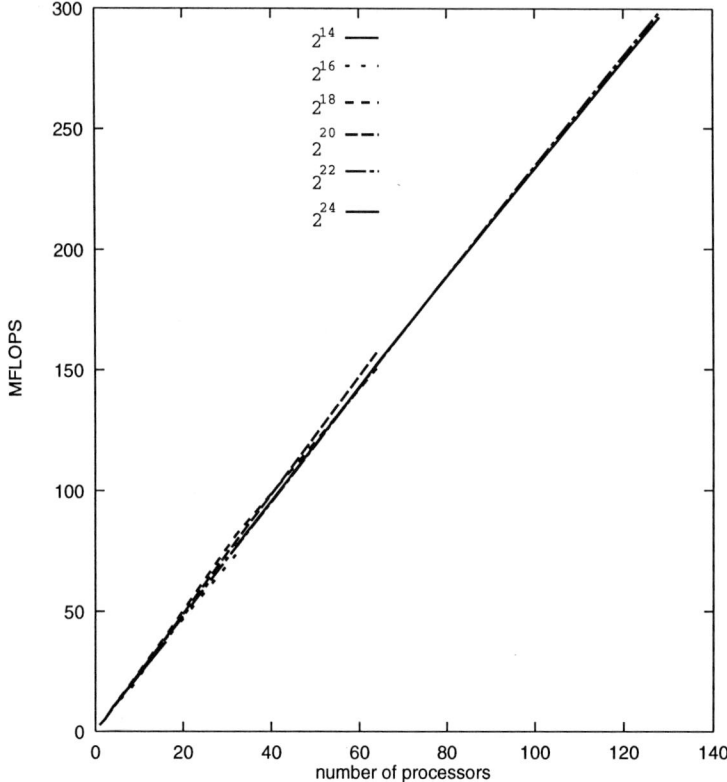

Fig. 6. The performance of our method on Cenju3

to 13 percent improved performance compared to the four step algorithm on the problems for 2^{18} to 2^{24}. And we got better performance with the 26 case out of 32 experiments.

For the future work we should,

1. Improve memory management for more size of FFT,
2. Improve the performance.

References

1. Van Loan, C: Computational Frameworks for the Fast Fourier Transform, SIAM, 1992
2. Swartztrauber, P.N.: Multiprocessor FFTs.Parallel Computing,no.5, (1987)197-210.
3. Hegland, M: Block Algorithms for FFTs on Vector and Parallel Computers, Parallel Computing:Trends and Applications, Elsevier Science, 1994
4. Takahashi, D., Kaneda, Y.: Implementation and Evaluation of 1-D FFT with External Memory on Parallel Computers, IPSJ SIG Notes, Vol.97, No.22, pp.7-12, 1997

Parallel Computation Model LogPQ *

Takayoshi TOUYAMA *and* Susumu HORIGUCHI

Graduate School of Information Science,
Japan Advanced Institute of Science and Technology
1-1 Asahidai, Tatsunokuchi, Ishikawa 923-12, Japan

Abstract We propose a parallel computation model LogPQ by taking account of the communication queue into the LogP model. We assume that the message length is fixed and the three parameters SQ, RQ and TQ indicate the-capacities of the sending queue, the receiving queue, and the transport channel queue, respectively. Since real parallel machines avoid the overflow of communication queues by using hardware, the message queue management takes up a lot of time in every communication. A parallel computation model should take into account the flow of processor communications. The LogPQ model considers the message length, and the communication queue management depended on its physical restriction in contrast with the LogP model. First, the LogP model is analyzed to show the requirements of the processor communication. We propose the LogPQ model by taking account of the communication requirement in parallel computers. To evaluate the accuracy of the LogPQ model, we compare with the LogP model for the behavior of parallel matrix multiplication algorithm on CM5.

keywords: parallel computers, parallel computation model, LogP model

1 Introduction

Many parallel computers have been developed, and they represent the alternative to vector computers. However, parallel programming on parallel computers is difficult and the efficiency of parallel programming depends on real parallel computers.

Recently, many researchers have reported realistic parallel computation models that reflect the hardware constraints of parallel machines. Most of the commercial parallel machines adopt message passing for processor communications. nCUBE2[1] and CM-5[2] use the Send-Receive model which achieves processor communication by a pair of SEND and RECEIVE commands. J-Machine[3] uses the Message Driven Model in which the receive-side processor executes the message command. Eicken et.al.[4] proposed the Active Message Model using a handler to interrupt receiving messages. Horie et.al.[5] proposed a method to overlap communications and computations. Culler et.al.[6] proposed the LogP model which takes account of the latency, the overhead, the gap and the number of processors to realize processor communications.

---

* This research is partly supported by Grant-in-Aide for Scientific Research (B) 09480051.

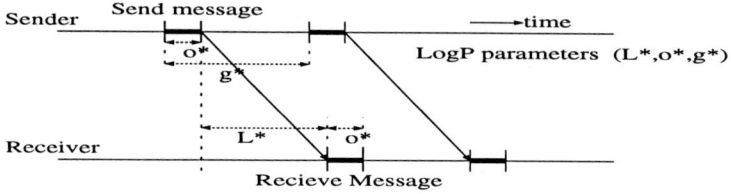

Fig. 1. Communication behavior in the LogP model.

Our main purpose in the parallel model is to make parallel programs efficient on many machines, so that the program uses the model parameters as constraint variables. We investigate the parallel model from this viewpoint.

In this paper, we investigate the LogP model on real commercial machines and show that it is not sufficient for processor communications. And we propose the LogPQ model which includes the message length and message queue on practical parallel machines. To evaluate the LogPQ model, we compared it with the LogP model regarding the behavior of the parallel matrix multiplication algorithm on CM5.

2 LogP model

The LogP model[6] considers four parameters: the communication latency L, the communication overhead o, the communication bandwidth g and the number of processors P. Figure 1 shows the behavior of message passing in the LogP model. We indicate L^*, o^* and g^* instead of L, o and g of the LogP parameters because L, o and g are used in the LogPQ model. L^* is the upper-bound of the latency of message passing from a source processor to a destination processor. o^* is the overhead of sending and receiving a message. g^* is the minimum time of the interval between two sending messages and two receiving messages at each processor. g^* corresponds to the reciprocal of the bandwidth of the interconnection network. P is the number of processors. Each processor executes a local command in one clock time. Parameters L^*, o^* and g^* are expressed in clock units.

The LogP model can describe the communication restriction between processors using three parameters; L^*, o^*, and g^*. So, the LogP model can analyze parallel algorithms practically. The LogP was investigated by showing the correspondence between four parameters and practical machines. Several algorithms were investigated on the LogP model, and the usefulness of the LogP model was discussed in the reference [6].

Although the LogP model is useful, it has not practical enough for our purpose for the following reasons.

(1)Behavior of buffering

The LogP model assumes discontinuous execution of message passings. However, most practical machines have buffers on communication lines and can execute message passing continuously.

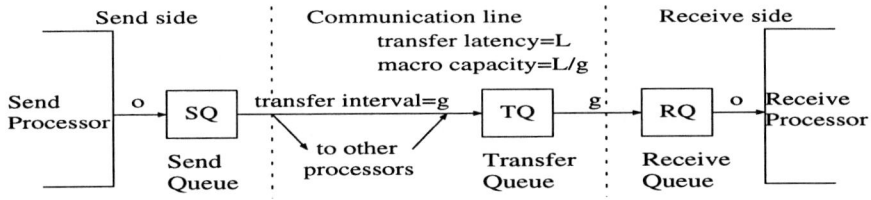

Fig. 2. A structure of LogPQ model.

(2) Concentration of communication

A contention occurs on the communication lines when the multiple processors send messages to the same processor at the same time. The contention in the LogP model prohibits sending messages to the same processor at the same time if the communication lines don't have enough buffering. Therefore, the asynchronous message passings are not managed easily in the LogP model.

(3) Message length

The LogP model has to define L, o, and g parameters for each message length of communications. The LogP model requires a number of parameters to describe real parallel algorithms. The relationship between message length and parallel algorithm becomes complex. Therefore, the LogP model does not analyze the behavior of parallel algorithms in detail.

3 LogPQ model

3.1 Structure of the LogPQ model

The LogPQ model is a parallel computation model that takes account of message queue parameters on communication lines to remedy the problem in the LogP model. Figure 2 shows the structure of the LogPQ model. The LogPQ model has a send queue on the send channel of each processor, a receive queue on the receive channel, and a transfer queue on the concentration point at the receiving channel. The LogPQ model introduces the capacity of queues: the sending queue SQ, the receiving queue RQ, and the transferring queue TQ. In figure 2, $transfer\ interval = g$ indicates the bandwidth of communication lines. And the $transfer\ latency\ L$ and the $macro\ capacity\ (L/g)$ corresponds to the latency L and the finite capacity (L/g) of communication lines between processors without contention.

3.2 Message passing

The relationship between the LogPQ parameter and the message passing on the communication line is discussed. Figure 3 shows the message passing in the LogPQ model. The LogPQ model uses a fixed word-length for communication, since commercial computers prepare communication lines with fixed word-length, which depends on the word-length of the CPU. Define L_0, o_0, and

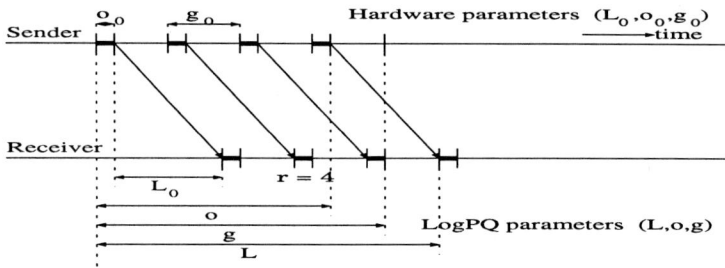

Fig. 3. Message passing in LogPQ model.

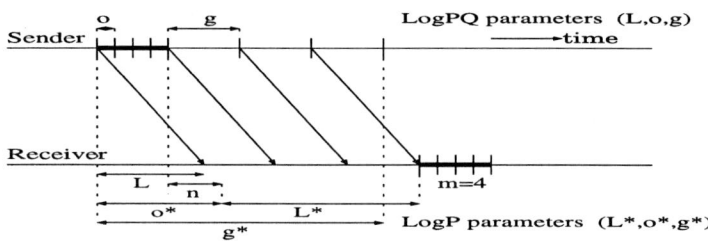

Fig. 4. Relation between LogP and LogPQ.

g_0 as the physical parameters in the LogP model based on the one-word communication. We derive the parameters L, o, and g in the LogPQ model as follows:

$$L = L_0 + o_0 + (r-1) \cdot g_0$$
$$o = o_0 + (r-1) \cdot g_0 \qquad (1)$$
$$g = r \cdot g_0$$

where r is the ratio of processor word-length w_p to the communication word-length w_c. L is the time units between sender and receiver. o is the overhead time required to send commands. g is the time interval between continuous message passings. The LogPQ model executes the message passing of tied m words data by m iterations of 1 word messages. P is the number of processors on the LogP model.

3.3 Relation between LogP and LogPQ

Figure 4 shows the relationship between the message passing of m words on the LogP model and on the LogPQ model. The LogPQ executes 1 word message passing m times. Let L, o, and g be the parameters in the LogPQ model in figure 4. The overhead of message passing in the LogP model requires n clocks for wait. The n-clock wait is applied to parallel programs because the LogPQ communicates by the one word-length message. The relationship between the LogP parameters (L^*, o^*, g^*) and the LogPQ parameters (L, o, g) are as follows:

$$L = L^* + o^*$$
$$o = (o^* - n)/m \qquad (2)$$
$$g = g^*/m.$$

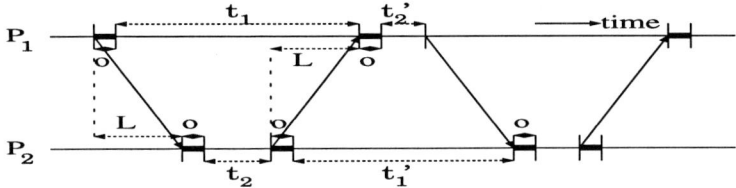

Fig. 5. Behavior of iterative communication on LogPQ.

When the message length of communication is fixed at a constant, the overhead between the sender and the receiver to communicate messages are distributed to each word of its message, that is, $n = 0$.

3.4 Communication parameters

Let's explain communication behavior of the LogPQ model using L, o, and g parameters. Figure 5 shows the iterative communication method between two processors in the LogPQ model. The processors P_1 and P_2 execute iterative processings by communicating between two processors. The overlap between communication and computation is optimum when $t_1 = t_2 + 2L$, where t_1 and t_2 describes the computation time on P_1 and P_2, respectively. The behavior between t'_1 and t'_2 is similar to the above. $t_1 + t'_2$ and $t_2 + t'_1$ describe the computation time in an iteration on P_1 and P_2, respectively. The overhead time o does not affect the mutual communication between the two processors. The total times required to communicate between PEs is equal to o times over message length if there is no wait time to receive messages. The parameter g corresponds to the bandwidth of communication lines.

3.5 Queue length

The parameters SQ, TQ and RQ are the maximum capacities of sending queue, transferring queue and receiving queue on communication lines, respectively. Although some parallel machines don't need to consider this facter, they have a large latency due to inspection and processing for the state of the receiver. The LogP model has the finite capacity of communication lines $\lfloor L^*/g^* \rfloor$. The communication lines are saturated when the latency significantly increases for a large number of messages. The communication capacity $\lfloor L/g \rfloor$ determines the average communication message, and the limited queue capacity SQ and RQ determine the maximum permissible load. The LogP model is a special case of the LogPQ model where $SQ = RQ = TQ = 1$.

When the communication load is low, SQ and RQ correspond to the number of continuous messages of the sender and the receiver. A processor can send/receive m words data at a time if m is less than SQ or RQ. RQ is the maximum delay in receiving the messages at the receive-side processor.

The LogPQ model expresses the constraints in many-to-many communication by the queue length of communication. Each processor can receive messages from

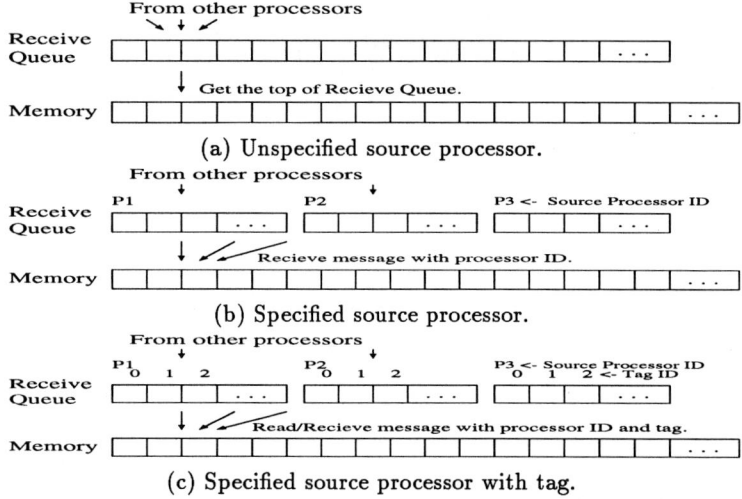

Fig. 6. Three types of the receive handler for the LogPQ.

multiple processors simultaneously. The communication latency depends on the queue length of the communication queues; SQ, RQ and TQ.

3.6 Receive handler

The communication handler greatly influences the performance of parallel algorithms on real machines. The LogPQ model can analyze the processing cost of the receive handler more detail than the LogP model.

This paper investigates three types of receive handlers as follows: the unspecified source processor, the specified source processor and the specified source processor with tags.

(1) Unspecified source processor

Figure 6(a) shows the unified receive queue which does not specify the source processor. Messages from many processors are stacked into the unified receive queue. The processor reads the message from the unified receive queue in the FIFO order. The processing cost of receive handler is included in the execution time of applications and it is not included in the parameters: L, o, and g.

(2) Specified source processor

Figure 6(b) shows the partitioned receive queue which can specify the source processor. The messages are stacked into the partitioned receive queue to specify the source processor. The processing cost of the receive handler is included in the parameters: L, o, and g. Therefore, we do not investigate the complicated communications so that each processor receives multiple-length messages. However, this type of communication model makes the analysis of a parallel algorithm on the LogPQ model much simpler.

(3) Specified source processor with tags

Figure 6(c) shows the partitioned receive queue with tags which is able to specify both the source processor and the order of the message in the partitioned queue. Using this receive handler, we easily analyze parallel algorithms which are programmed by a high-level communication interface on parallel machines.

By assigning a tag-value to each received message, the message has the absolute address of messages in parallel programs. This type of handler can exchange the order of messages.

We consider two receive commands: Receive and Read. When we use the Receive command, it reads and removes the data from the receive queue. When we use the Read command, it only reads the data from the receive queue. Since the receive queue is described as a received buffer, its constraint becomes clear.

When the length of the receive queue is unlimited, we do not need to use the Receive command. In this case, the receive queue is the same as ROM (Read Only Memory). Since the tag corresponds to the address of ROM, we can easily implement parallel algorithms, which are the set of jobs with partial order, by locating them on PEs. The RQ shows the maximum width between received data and eliminated data in the receive queue. A parallel algorithm has to be designed within this restriction of RQ.

4 Evaluation of LogPQ

The LogPQ model has been evaluated by comparing it with the LogP model on a commercial machine CM5[2]. We implemented Cannon's matrix multiplication algorithm[7] using C language on CM5. This algorithm multiplies two matrices with size $N \times N$ using partial matrices with size $l \times l$ in each processor, where $l = N/\sqrt{P}$ and P is processor count. All parameters of the LogP and LogPQ are measured by the experiment of size 32×32 on 64 PEs. Analytical execution times of the LogP and LogPQ model are estimated for any size of matrix by using these parameters.

Figure 7 shows the execution times and the analytical execution times of the LogP and LogPQ models for Cannon's algorithm on 64 PEs of CM5. The figure shows that the LogPQ model can expect a more exact execution time than the LogP model for another matrix size which makes another message length transported among PEs in each iteration.

5 Conclusion

The LogPQ model was proposed by taking account of the message length and the three queue parameters to the LogP model.

We investigated the execution times of the LogP and LogPQ models for matrix multiplications on a parallel machine CM5. Comparison among the experimental result and the analytical execution times on both models shows that the LogPQ model has more accuracy than the LogP model.

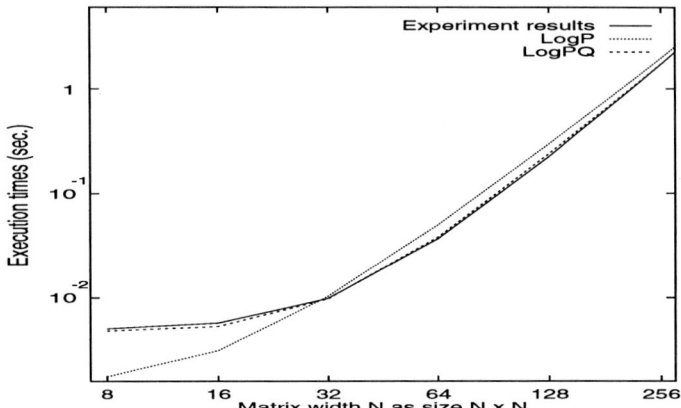

Fig. 7. Execution times and analytical times of the LogP and LogPQ models for Cannon's matrix multiplication as a function of matrix size on 64 PEs of CM5.

References

1. "nCUBE2: Technical Overview", nCUBE Corporation (1992).
2. "Connection Machine CM-5 Technical Summary", Thinking Machines Corporation (1992).
3. William J.Dally, Andrew Chien, Stuart Fiske, Waldemar Horwat, John Keen, Michael Larivee, Rich Lethin, Peter Nuth and Scott Wills, "The J-Machine: A Fine-Grain Concurrent Computer", Information Processing 89, pp.1147-1153 (1989).
4. Thorsten von Eicken, David E.Culler, Seth Copen Goldstein and Klaus Erik Schauseret, "Active Messages: a Mechanism for Integrated Communication and Computation", Proceeding of the 19th Annual International Symposium on Computer Architecture pp.256-266 (1992).
5. Takeshi Horie, Youichi Koyanagi, Nobutaka Imamura, Kenichi Hayashi, Toshiyuki Shimizu and Hiroaki Ishihata, "Effect of Message Communication on Distributed-Memory Parallel Computer Performance (in Japanese)", Transactions of IPSJ, Vol.35 No.4, pp.609-618 (1994).
6. David Culler, Richard Karp, David Pattersom, Abhijit Sahay, Klaus Erik Schauser, Eunice Santos, Ramesh Subramonian and Thorsten von Eicken, "LogP: Towards a Realistic Model of Parallel Computation", Proceeding of the 4th ACM SIGPLAN Symposium on Principles and Parallel Programming (1993).
7. Vipin Kumar, Ananth Grama, Anshul Gupta and George Karypis, "Introduction to Parallel Computing", Benjamin/Cummings Publishing (1994).

Cost Estimation of Coherence Protocols of Software Managed Cache on Distributed Shared Memory System

Takeshi Nanri[1] and Hiroyuki Sato[1] and Masaaki Shimasaki[2]

[1] Kyushu University
[2] Kyoto University

Abstract. In recent years, software managed cache systems are becoming widely used on parallel computing environments, because of its portability and applicability. However, cache managing costs on those systems are higher than hardware managed cache systems.
Therefore, we estimate the cache managing cost of the cache system we have implemented on distributed shared memory system. As a coherence protocol, programmers of our system can select invalidating or updating. We estimate the cache managing costs for both protocols. As an example program, we use two application to evaluate our system, Gaussian elimination and sum total calculation. From the result, we can conclude that the cost of cache information management cannot be ignored for estimating software managed cache system.

1 Introduction

To improve performance of global memory access on parallel computers, private cache is becoming widely used. The software managed cache has advantages over hardware managed cache for some reasons, such as the easiness of implementation and scalability. However, cache managing costs on those systems are higher than hardware managed cache systems. Therefore, it is important to estimate the cache management cost to evaluate the software managed cache systems.

In this paper, we propose a cost estimation method of coherence protocols for fully software managed cache on a distributed shared memory system. We have implemented a fully software managed cache system. The coherence protocol and the cache line size of our cache system can be selected at compile time. Thus programmers can select the most suitable parameter for each application program. We estimate cache management costs of our cache system.

This paper is organized as follows. The second section surveys related works. The third section introduces the organization of our software cache. The fourth section shows the result of our experiments. In the fifth section, we discuss about our result.

2 Related Works

In recent years, selective cache memory systems which support two or more coherence protocols have been proposed. Carter *et al.* [CBZ95] propose techniques for reducing consistency-related communication in Munin. In this system, a special programming environment is provided so that the programmer can annotate the declaration of shared variables to specify what protocol to use to keep shared data consistent. Mounes-Toussi et al. [MTL95] use the predictive capability of the compiler to select updating or invalidating for each write reference. From their simulation result of miss ratio and total network traffic using memory traces, they conclude that there is a trade-off between updating and invalidating cache coherence enforcement schemes.

There have been many evaluation of these cache memory system on distributed shared memory system(DSM). Some of them, such as Prete [Pre95], evaluate cache systems by a trace-driven simulator. Veenstra et al. [VF92] evaluate cache system with hybrid cache coherence protocols. As for performance models of DSM, Archibald [JJL93] discusses modeling methods of cache coherence overhead. However, these evaluation methods are mainly based on miss-ratio and communication cost.

There have also been efforts to develop techniques for reducing coherence-related costs on DSM. By using selective cache systems [PRAH96], hardware supports [BKP+96] or compiler supports [SS96]. In our previous work [NSS97], we evaluate applying a RISC-oriented cache optimizing technique to software-managed cache.

3 Organization and Management of Our Cache System

3.1 Platform

As an experimental platform, we use Split-C, a parallel SPMD C language. Split-C provides SPMD programming environment and a virtual shared memory on a distributed memory parallel computer CM-5. Each processor directly accesses to the virtual shared memory. This means that any cache system for global memory access is supported by neither CM-5 nor Split-C.

Each global memory access of Split-C program is implemented as an interprocessor communication. therefore, the overhead of the access is heavy. This means that some kind of cache system can significantly improve performance of Split-C.

3.2 Organization of Our Cache

Cache Memory We implement the private cache of our system on a part of local memory of each processor. With this system, we can provide a hierarchical memory model of fast small cache memory, and slow large global memory.

As ordinary cache system, our system does block based cache management. Each memory block is represented by using global block address. A global address of Split-C specifies the processor number and the local block address of memory on the processor.

Currently, our cache system does not change the owner of memory blocks, dynamically. Therefore, our cache does not maintain the status of cache blocks. This kind of improvement will be a subject to the next version of this system.

Cache Coherence Protocol Our system provides both write invalidate protocol and write update protocol for cache coherence enforcement. When a global write access occurs, the write invalidate protocol sends the invalidation message to all processors that have cached the modified block, and invalidate the block. The write update protocol, on the other hand, sends the new data directly to all processors that have cached the data, and updates all the data.

For each write access function, cache coherence protocol is specified by a programmer. Therefore, the programmer can select a suitable coherence protocol.

Cache Consistency Model The memory consistency model supported by our system is the weak consistency model[Lil93]. This model only guarantees the orders of the synchronization points defined by a programmer. Therefore, all memory references by different processors to global data variables between accesses to synchronization points can occur in any arbitrary order.

Cache Placement We use *4-way set associative* cache placement. Therefore, in our cache, a copied block can be placed in a restricted set of places in the cache. The number of blocks in a set, *way*, can be specified at compile time. The set is chosen by bit selection; that is,
$(local\_block\_address) modulo (number\_of\_sets\_in\_cache)$.

Cache Information Structure Our cache management system uses two cache information structures, the directory and the cache table. The directory keeps the summary information of memory blocks which is cached to other processor. Each element of a directory specifies its states and processors which have a copy of the block in private cache. The cache table, on the other hand, keeps the information of private cache blocks, such as the original processor number, the original address, and the head address of the block.

On the cache table, to provide 4-way set associative cache placement, information of blocks in the same set are linked as one list structure, and each list consists of up to 4 elements. Each set of the cache table is accessed via a *set array*.

The directory of our cache has almost the same structure as the cache table. The only difference is that each list of directory can consists of arbitrary number of elements.

3.3 Cache Management

The cache management is done by manipulating the cache information structure, and by communicating with other processors.

Each phase of cache information management, such as adding information, deleting information, searching in a table and sorting elements in a table, manipulates lists of directories and cache tables. Sorting is used to move the information of the last accessed block to top of the list. This means that elements on the tail of lists are the least recently accessed elements in the list. This element will be swapped out when a set of the cache becomes full.

Communication at cache management consists of sending requests and transferring cache block. Cache related requests, such as cache copy or coherence enforcement, are sent as active messages. Furthermore, transferring cache blocks are also done by sending block-transferring active messages.

Let us support that, a processor P1 fails to search a block in the cache table, and that a cache miss occur. P1 first prepares for a port for receiving a cache block, then sends a cache copy request to the target processor P2. The request message consists of the address of the handler function for copying cache block, the address of the required block, the processor number of P1 and the port number for receiving a cache block. When the message is received by P2, P2 suspends its computation and does the computation of the handler function for copying cache block. In the handler, P2 first adds the information of the required block to the directory, then replies a block transfer message to P1 and resumes its computation. P1 receives the message and adds the information of this cache block to the cache table.

3.4 Programming Model

Currently, our cache system is used via extended global-access functions, listed below.

cache_d_read (*dest, src*)
cache_i_read (*dest, src*)
cache_d_write (*dest, src, protocol*)
cache_i_write (*dest, src, protocol*)

`cache_d_read` and `cache_i_read` are global read functions using our cache system for double precision floating point number and integer number, respectively. `cache_d_write` and `cache_i_write`, on the other hand, are global write functions. *dest* specifies the address to where data is written, while *src* specifies address from where data is read. Furthermore *protocol* specifies which coherence protocol to be used for this write operation. We are considering to use optimizing compiler to specify optimal coherence protocol automatically. In the following sections, we experiment and discuss the performance of our system for cost estimation of such compiler.

4 Experiment

4.1 Experimental Environment

In this paper, we use the environment shown in Table 1 to evaluate our software managed cache system. We have not used any optimization in the compiler.

In this environment, we have implemented our system with 512KBytes of software managed cache memory. Each block of our system consists of 2KBytes.

Table 1. Experimental environment

Machine	CM-5(Thinking Machines Co.)
Processors	16
Memory	32MBytes/proc
OS	CMOST 7.4
Compiler	Split-C 1.2(based on gcc 2.6.3)

Table 2. Basic parameters

Parameters	Costs
cache miss	107.0
searching in list	7.10
sorting list	11.1
invalidate protocol	41.4
update protocol	45.1

(μsec)

4.2 Parameters

Table 2 shows the basic parameters of our cache system. It is to be noted that costs of searching in a list and sorting elements of a list is about $\frac{1}{10}$ of the cost of cache miss. Therefore, the cost for managing cache information cannot be ignored in programs in which searches in information tables occur more than ten times as frequently as cache misses.

4.3 Example programs

```
for (k = 0; k < N - 1; k++){
  find_pivot();
  swap_row();
/* decompose pivot block */
  for (i = k + 1; i < N; i++){
    t = a[row[i] + k * N];
    for (j = ; j < local_N; j++)
      a[row[i] + j * N] -=
        t * a[row[k] + j * N];
  }
}
```

Fig. 1. Gaussian elimination program

```
for (p = 1; p < PP; p++){
  dest = (MYPROC + p) % PP;
  for (i = 0; i < LOCAL_N; i++){
    ack_lock(dest);
    cache_d_read(&val,
      &(g_total[dest][0]));
    val += g_a[MYPROC][i];
    cache_d_write(
      &(g_total[dest][0]),
      val, PROTO);
    cache_rel_lock(dest);
  }
}
```

Fig. 2. Sum total program

We evaluate our system with two kernel routines, the Gaussian elimination with partial pivoting and the sum total.

Gaussian elimination with partial pivoting

Gaussian elimination decomposes a square matrix into upper and lower triangular submatrices by repeatedly eliminating the elements of the matrix under the diagonal, one column at a time(Fig. 1). By partial pivoting, the numerical stability of the program is improved.

In our experiment, we use parallelized version of this program. The parallelization is done as follows. The computation is decomposed by column so that the pivoting phase, which can be a synchronization bottleneck, can be performed on a single processor. Each processor accesses the target matrix indirectly by using *pivoting row* vector. Therefore, swapping row after pivoting phase is done by only swapping elements of this vector.

Sum total

The sum total program computes over all the elements of distributed array(Fig. 2). In this program, each processor adds elements of its part to the variables on all the other processors. Therefore, all processors have the sum total of the distributed array. Each add operation to a variable on a processor is mutually excluded, by using *locks*.

4.4 Result

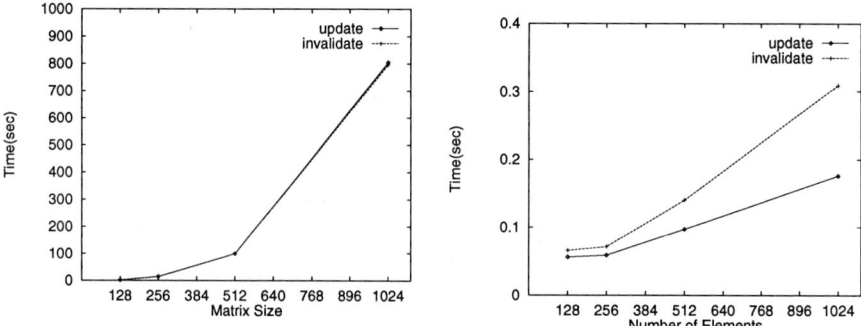

Fig. 3. Execution time of Gaussian elimination

Fig. 4. Execution time of sum total

Gaussian elimination As shown in Figure 3, the execution time of the Gaussian elimination is not affected by changing cache coherence protocol. This is because the cache managing cost of each protocol does not differ on this program(Table 3). The reason of this will be discussed in the next section.

Sum total Figure 4 shows the execution time of the sum total program. For all sizes, the cost of write update protocol is lower than that curve of write invalidate protocol.

Table 3. Cache managing cost (Gaussian elimination)

	Coherence Protocol		
	(a)inv	(b)upd	(a) - (b)
Miss	6.58	6.62	−0.04
Search	159.4	161.1	−1.70
Sort	0.315	0.315	0.0
Total	804.3	796.0	8.3

(sec)

Table 4. Cache managing cost (Sum-total)

	Coherence Protocol		
	(a) inv	(b) upd	(a) - (b)
Miss	213.0	336.0	−123.0
Search	38.6	57.1	−18.5
Sort	0.0	0.112	−0.112
Total	309.0	176.0	133.0

(msec)

Table 4 shows the differences of cache management cost between the case with write invalidate protocol and the case with write update protocol. The cache miss cost of write invalidate protocol is 2.1×10^{-1}(sec) higher than write update protocol. However, the difference of total execution time is only 1.33×10^{-1}(sec).

The main reason is the difference of search cost. On write update protocol, every write access causes searching directory. Moreover, because all information remains on the cache until the cache system becomes full, the directory can be a long list. Therefore, the searching cost of write update protocol for each write is higher than that cost of write invalidate protocol.

5 Discussion

The cache managing cost does not differ between write invalidate protocol and write update protocol on Gaussian elimination. This is because, ideally no coherence enforcement is needed on this program, since a row of the matrix is copied to private caches of other processors after all computations for the row is finished. Actually, coherence enforcement are done only when the false sharing occurs.

On the other hand, performance of the sum total program on different protocols differ significantly. This is because this program writes and reads the same address repeatedly. Therefore, the data of the global address which once copied into a private cache may be changed by the time when the address is next accessed. If the cache system uses write invalidation protocol, almost all global read may cause cache miss, and the block transfer occurs frequently. With the write update protocol, on the other hand, changes of the global address are propagated to all cache copies. Therefore, only the first read from the address causes the cache miss. Thus, on this program, the write update protocol is more suitable than the write invalidate protocol.

6 Conclusions

In this paper, we have evaluated a fully software managed cache system on distributed shared memory. Our system supports selective coherence protocol. To evaluate our system, we measured not only cache miss cost but also costs for cache information management.

References

[BKP+96] R. Bianchni, L. Kontothanassis, R. Pint, M. De Maria, M. Abud, and C.L. Amorin. Hiding communication latency and coherence overhead in software dsms. In *Proc. of the 7th Intl. Conf. on Architectural Support for Programming Languages and Operating Systems*, pages 198–209. ACM, 1996.

[CBZ95] J.B. Carter, J.K. Bennet, and W. Zwaenepoel. Techniques for reducing consistency-related communication in distributed shared-memory systems. *ACM Trans. on Computer Systems*, 13(3):205–243, 1995.

[JJL93] Archibald J. and Baer J.-L. Cache coherence protocols: Evaluation using a multiprocessor simulation model. *ACM Trans.Comp. Syst.*, 4(4):273–298, 1993.

[Lil93] D.J. Lilja. Cache coherence in large-scale shared-memory multiprocessors: Issues and comparisons. *ACM Computing Surveys*, 25(3), 1993.

[MTL95] F. Mounes-Toussi and D.J. Lilja. The potential of compile-time analysis to adapt the cache coherence enforcement strategy to the data sharing characteristics. *IEEE Trans. on Parallel and Distributed Systems*, 6(5):470–481, 1995.

[NSS97] T. Nanri, H. Sato, and M. Shimasaki. Using cache optimizing compiler for managing software cache on distributed shared memory system. In *Proc. HPC Asia'97*, 1997.

[PRAH96] V.S. Pai, P. Ranganathan, S.V. Adve, and T. Harton. An evaluation of memory consistency models for shared-memory systems with ilp processors. *SIGPLAN Not.*, 31(9):12–23, 1996.

[Pre95] C.A. Prete. A trace-driven simulator for performance evaluation of cache-based multiprocessor systems. *IEEE Trans. on Parallel and Distributed Systems*, 6(9):915–929, 1995.

[SS96] J. Skeppstedt and P. Stenstrom. Using dataflow analysis techniques to reduce ownership overhead in cache coherence protocols. *ACM Trans. on Programming Languages and Systems*, 18(6):659–682, 1996.

[VF92] J.E. Veenstra and R.J. Fowler. A performance evaluation of optimal hybrid cache coherency protocols. In *Proc. of the 5th Intl. Conf. on Architectural Support for Programming Languages and Operating Systems*, pages 149–157. ACM, 1992.

A Portable Distributed Shared Memory System on the Cluster Environment: Design and Implementation Fully in Software

SATO Hiroyuki[1], NANRI Takeshi[2], and SHIMASAKI Masaaki[3]

[1] Dept. Computer Science and Comm. Engr., Kyushu University, Japan,
[2] Computer Center, Kyushu University, Japan,
[3] Dept. Electrical Engr., Kyoto University, Japan

Abstract. Cluster of workstation or personal computers connected with high speed network has become one of major architectures of distributed memory parallel computers. However, software on the cluster environment is still not improved in performance. The distributed shared memory can be a solution of programming style on distributed memory parallel system including clusters because we know from experiences that shared memory model ease programming. However, it must be implemented with care for the performance problem where there is no hardware support for shared memory system. Another, but serious problem is the portability. In this paper, we discuss the design and implementation of portable distributed shared memory system. Our shared memory system is based on PVM in consideration of portability. Our contributions in this paper is the design and implementation of portable shared memory system on the cluster environment using faithful implementation of active messages fully in software, together with an enhancement of PVM to support active messages.

1 Introduction

Cluster of workstation or personal computers connected with high speed network has become one of major architectures of distributed memory parallel computers. Its advantage over parallel computers provided with special hardware is the cost effectiveness. On the other hand, on performance, there has not been put stress. However, as the performance of general purpose network such as Ethernet or FDDI has been improved rapidly, the parallel performance of clusters has also been improved. However, software on the cluster environment is still not improved in performance. Specifically, porting of software designed and implemented for parallel computers which have hardware support for parallel processing is a serious problem. In particular, the porting of software originally for shared memory parallel computers almost resulted in failure in performance. This is mainly because there has been not proposed a standard distributed shared memory mechanism on the cluster environment, and the rewriting of programs from that for shared memory to that for distributed memory is not so easy.

In this paper, we examine the design issues of portable distributed shared memory system. Moreover, we implement our distributed shared memory system and show its performance problem. We choose Split-C originally designed for CM-5 as a language with shared memory, and PVM as the message passing layer. Our contributions in this paper is the design and implementation of portable shared memory system on the cluster environment using faithful implementation of active messages fully in software, together with an enhancement of PVM to support active messages.

Related Work

Today, we can say that the cluster environment is a major class of distributed memory parallel architectures. Projects on the cluster environment includes NOW of UCB[8], RWC of RWCP[9], and SWCP of U. Penn, together with the well-tuned message passing libraries such as active message of NOW and the PM library of RWCP. All of these projects aim at proposing alternatives of "real" parallel machines by building parallel system by connecting high performance workstations with high speed network, which is a significant difference from conventional clusters. Our project shares the problem with those projects. However, we also put stress on portability, and we do not use high performance network such as Myrinet[2], but restricting the network interface within 10BaseT or 100BaseT and connecting switches. TreadMark[1] is a project of distributed shared memory system. Memory allocation and access are implemented as library calls.

Porting of Split-C[4] is reduced to the implementation of active messages on the target environment. In [10], Culler implements active message on Myrinet. In the UCB implementation of Split-C on PVM, however, the active message is not faithfully implemented. [5] is an implementation of Split-C on PVM. However, the implementation of active messages is not safe. HPF is another major distributed shared memory programming language. [7] reports the implementation of HPF using Split-C as its intermediate language.

2 Design of Portable DSM

2.1 Functions to be implemented in software

In implementing the distributed shared memory system on the cluster environment, the fact that we cannot expect any hardware support nor any operating system support is a severe handicap. As for the hardware support, the function which handles the request of memory access and the transfer of data is strongly required. In other words, we need a kind of DMA method to a remote processor for the performance improvement. As for the operating system support, the function which resolves the address of the data possessed by a remote processor is indispensable.

In real distributed memory parallel machines, those functions are fully provided, or at least they are given a consideration. However, in the cluster environment, there is no support of such functions. Therefore, we must implement them fully in software.

Major troubles in implementing them are summarized as: we do not have any DMA to remote memory. This means that we must implement the memory access method using software interrupt and user-level event queue, and

2.2 Portability

The next major problem is portability. Today, portability is one of major concerns in programming. To keep portability, we write programs in

- a standard programming language which is widely used, and
- a standard message passing library functions.

We choose Split-C as the target of the programming language which supports distributed shared memory. We implement its distributed shared memory system on PVM. The protocol stack of our implementation is, therefore, figured in Fig. 1.

Split-C	↔	Language Layer
Active Message		
PVM	↔	Message Passing Layer
UDP, (Domain Socket)	↔	Transport Layer
IP		

Fig. 1. Protocol Stack

MPI vs. PVM PVM and MPI are two major portable message passing libraries. In choosing one of them, we consider their expressive power in implementing the distributed shared memory system. As discussed in Section 2.1, we must design our distributed shared memory system by virtually simulating hardware functions for parallel machines. From this viewpoint, PVM has advantage over MPI in that PVM has the inter-processor signal handling functions by which we simulate the hardware interrupt of each processing element of parallel machines, and implement a parallel *virtual* machine.

In the implementation of the demand-driven data transfer functions of MPI-2[11], windows which are defined to be the pair of memory block and its owner process for the remote memory access must be declared first. Moreover, access *phase* and open *phase* are defined. RMA(remote memory access) can be allowed only when the origin is in the access phase, and the target is in the open phase. These phases *must* be controlled by a programmer. The principles of RMA of MPI-2 are different from that of shared memory in which remote memory is accessed in the same way as the local memory without phase control nor domain restriction. Active messages do not have the above phase problem because in any phase, they interrupt the target process with the handler specified by the message.

3 Our Choice in Design and Implementation

Our design problems of distributed shared memory are summarized as the address resolution of the remote data and the safe and faithful implementation of active messages.

We design and implement a distributed shared memory system on the cluster environment by porting Split-C on PVM. The porting is divided into two stages: implementation of active messages on PVM and address resolution mechanism on the cluster environment.

3.1 Problems in Unix

Design of the shared address space of Split-C depends on the implementation of functions of 2.1 which are supported by hardware in real parallel machines. As our implementation of these functions, we use software signals as the substitute of hardware interrupt, and we make the SPMD assumption: every object on each processor has the same image. However, these decisions can cause other kind of problems. In Unix, software signals are not guaranteed to be raised. at least, sending a signal while handling a signal and having a waiting signal results in the loss of the signal last sent, and there is no method of knowing whether a signal is lost or not.

SPMD assumption makes the address resolution easy. We omit the details, but we have enhanced the address resolution mechanism so that the address can be resolved at the runtime.

3.2 Active Message

We use PVM signal handling functions in implementing handler invocation of active messages. In PVM, this is implemented as the daemon-daemon communication and daemon-task communication as in Fig. 2.

First, we discuss the phase reduction in active messages. The principle in phase reduction in active messages is that by using active message mechanism, we can avoid the unnecessary acknowledgment between communication request and the data send/receive action. In our implementation, our version of pvm_sendsig is sent with the signal number and the data to be sent/received on UDP. Unlike TCP, the implicit acknowledgment is not sent in UDP. Therefore, if we ignore the Unix domain socket protocol used in communication between intra-processor tasks, we can conclude that we can avoid three-phase protocols in our implementation.

Second, we discuss the handler invocation in active messages. The handler invocation is implemented using PVM signal invocation. because PVM signal invocation depends on the Unix signal invocation. We solve this problem by enhancing the daemon around signals.

We add two entries of library functions of PVM to send signal safely as shown in Table 1.

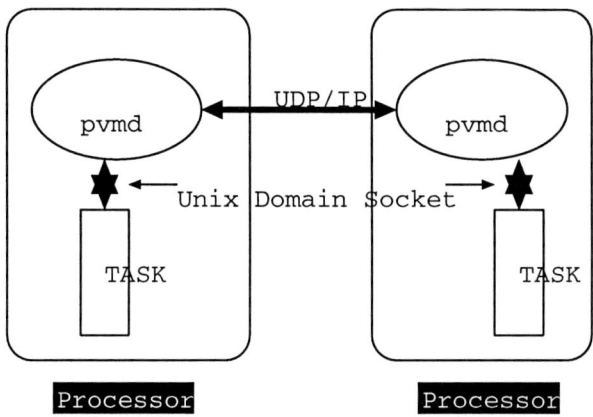

Fig. 2. Communication in PVM

Table 1. Signal Handling Functions of PVM(* our enhancement)

Function (Description)
pvm_sendsig(tid, signum) Raise Signal(conventional semantics)
pvm_safesigsend(tid, signum, mid)* Raise signal safely, and send data to the handler. Signal is guaranteed to be raised.
pvm_safesigsendrecv(tid, signum, mid, bufid)* Raise signal safely, and send data to the handler, and receive the data from the handler. Signal is guaranteed to be raised.

In pvmd, we have made the pending queue of signal request for each PVM task. It is guaranteed that no two signals are not raised at the same time by a pvmd. Changes around dm_sendsig(the entry for signal handling in pvmd) are figured in Fig. 3

Just before the invocation of dm_sendsig, signal handling functions in the daemon, a flag is tested whether a signal is raised or not. If a flag is down, dm_sendsig is invoked as usual. If a signal is raised, however, the request of dm_sendsig is put into the message queue, and waits in the queue for the completion of the signal handler.

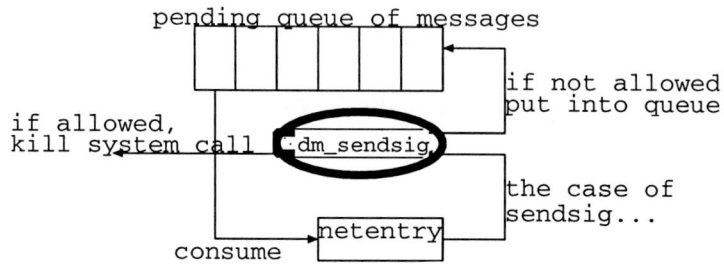

Fig. 3. Changes around signal handling routines

3.3 Memory Access Sequence on PVM

With thus enhanced PVM, we can implement the access to remote memory as illustrated in Figure 4:

1. Issue a request of memory access with its address offset calculated as (corresponding address of the local processor) - (bss) to the PVM layer,
2. send the request to the task of the remote processor using pvm_safesigsendrecv,
3. receive the request, and invoke the signal handler,
4. calculate the address as (the offset sent in the request) + (bss), and access the data of the address in the signal handler, and
5. return the value of the data.

As the consequence, we can conclude that we faithfully implement active messages, and that we implement the address resolution mechanism with SPMD assumption.

4 Experimental Results

We experimentally implement the distributed shared memory on the cluster environment of Table 2 fully in software. Table 2 shows the latency and throughput of our distributed shared memory system. We find that the bottleneck of the performance is on the latency, rather than the throughput.

It is to be noted that the overhead of signal invocation is not yet a problem of the latency in our system. Let us consider the transfer of 8 bytes. In BSDI BSD/OS, the overhead of signal invocation is measured as 29 μsec. If we estimate the ideal communication cost as the throughput performance, it is approximately 16 μmsec(8 / ((0.613 + 0.428)/2)). Therefore, we can conclude that most of the latency is the overhead incurred by PVM and UDP/IP handling. To improve the latency, we must first improve the performance of PVM and UDP/IP.

There is another approach by which we can improve the performance. We can restructure the communication pattern by overlapping the communication and computation, by collecting the small size communication(by using bulk data transfer)[7], and by utilizing data locality (by using software cache)[6].

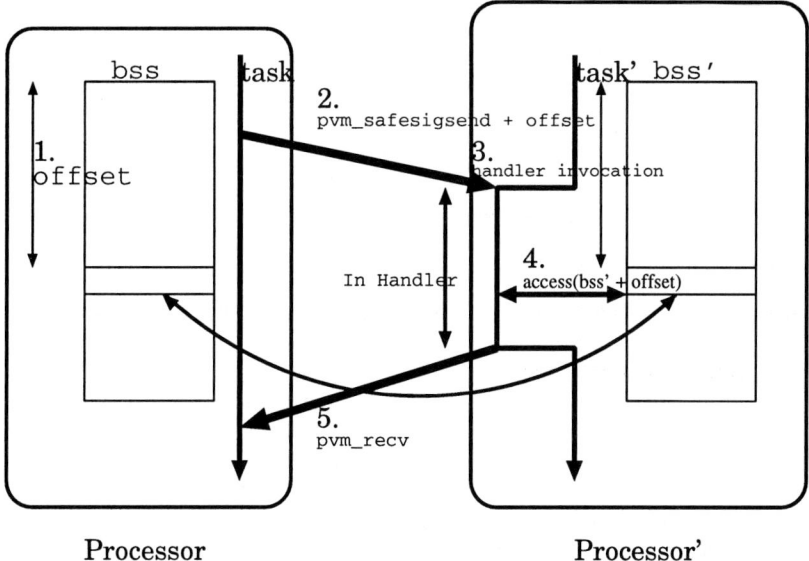

Fig. 4. Sequence of Remote Memory Access

5 Concluding Remarks

In this paper, we discussed the design problems of distributed shared memory system on the cluster environment. Faithful implementation of active messages was discussed, and an enhancement of PVM to support active messages was also discussed. Finally, the performance of our implementation was discussed and some approaches for the performance improvement were investigated.

Acknowledement

The authors have been partially supported by The Japan Society for The Promotion of Science Research for the Future Program (JSPS-RFTF96P00505: Software for Distributed and Parallel Supercomputing) and by The Ministry of Education, Science and Culture(No. 08458071).

References

1. Amza, C., Cox, A.L., Dwarkadas, S., Keleher, P., Lu, H., Rajamony, R., Yu, W., Zwaenepoel, W.: "TreadMarks: Shared memory computing on networks of workstations," IEEE Computer, 1996, pp. 18–28.
2. Boden, N.J., Cohe, D., Felderman, R.E., Kulawik, A.E., Seitz, C.L., Seizovic, J.N., Wen-King Su: "Myrinet – A Gigabit-per-Second Local-Area Network," IEEE Micro, VOl. 15, No. 1, 1995, pp. 29–36.

Table 2. Latency and Throughput of the Remote Memory Access

Environment:

Processor	Pentium 166MHz
Network	10BaseT with XYLAN Switching Hub
OS	BSDI BSD/OS 2.1

Performance:

(get)

# bytes	latency	thruput
8	3.34	—
1024	5.94	0.172
8192	18.79	0.436
65536	114.54	0.572
644640	1050.93	0.613
	(msec.)	(MB/sec.)

(put)

# bytes	latency	thruput
8	3.26	—
1024	5.58	0.183
8192	17.24	0.475
65536	114.45	0.573
644640	1529.97	0.428
	(msec.)	(MB/sec.)

3. von Eicken, T., Culler, E., Goldstein, S., Schauser K.: "Active Messages: a Mechanism for Integrated Communication and Computation," Proc. 1992 Int. Sympo. Computer Architecture, 1992, pp. 256–266.
4. Krishnamurthy, A., Culler, E., Dusseau, A., Goldstein, S., Lumetta, S., von Eicken, T., Yelick, K.: "Parallel Programming in Split-C," Proc. Supercomputing'93, 1993, pp. 262–273.
5. Nanri, T., Sato, H., Shimasaki, M.: "Implementing a Portable SPMD Shared Memory Model Parallel Language in a Distributed Computing Environment," Proc. Int. Symp. Parallel and Distributed SuperComputing, 1995, pp. 243–252.
6. Nanri, T., Sato, H., Shimasaki, M.: "Using Cache Optimizing Compiler for Managing Software Cache on Distributed Shared Memory System," Proc. HPC Asia 97, 1997, pp. 312–318.
7. Sato, H., Nanri, T., Shimasaki, M.: "Using Asynchronous and Bulk Communication to Construct an Optimizing Compiler for Distributed-Memory Machines with Consideration Given to Communication Costs," Proc. 1995 ACM ICS, 1995, pp. 185–189.
8. http://now.cs.berkeley.edu
9. http://www.rwcp.or.jp
10. http://www.cs.berkeley.edu/AM/lam_release.html
11. http://ftp.cs.wisc.edu/pub/lederman/mpi2

An Object-Oriented Framework for Loop Parallelization

Yoichi Omori[1], Kazuki Joe[2] and Akira Fukuda[1]

[1] Graduate School of Information Science
Nara Institute of Science and Technology
8916-5 Takayama, Ikoma, Nara, 630-01, JAPAN
[2] Faculty of Systems Engineering
Wakayama University
930 Sakaedani, Wakayama City, 640, JAPAN

Abstract. The main goal of parallelizing compiler research is the ability to produce efficient parallel programs and the portability because of wide choices of current and future available parallel computer architectures. Since the design of parallelizing compilers tends to be more complicated than conventional compilers, it is extremely difficult to achieve both the efficiency and the portability. To meet this problem, we have investigated an application of object oriented design to parallelizing compilers. Our parallelizing compiler design is based on abstractions of intermediate representations of loops and class definitions for them. In this paper, we focus on loop parallelization and propose a framework where loop parallelization process is divided into three phases and the optimization of loops is performed via cyclic use of those three phases. The class of each phase is hierarchically derived from intermediate representations of loops. This increases the portability of resultant parallelizing compilers. Furthermore, one of the phases uses a reservation table of hardware resource to obtain practical optimization of parallel programs for given hardware resource.

1 Introduction

Distributed memory multiprocessor computers are a recent trend of high performance computer architectures from the viewpoint of scalability. Parallelizing compilers, which translate sequential programs into parallel forms, have been investigated for centralized memory architectures, and it is not straightforward to apply them to distributed memory because of various optimization problems on distributed memory environments such as task partitioning, scheduling or process synchronization. Furthermore, the large number of processors, which is a characteristic of distributed memory systems, makes communication paths among the processors complicated, and causes optimization phases of parallelizing compilers extremely difficult. Thus the design of parallelizing compilers for new parallel computers tends to be complicated year by year.

The conventional design of parallelizing compilers consists of two parts: A front-end to exploit high level parallelism[11] and a back-end to gain instruction level parallelism[9]. Given a new parallel computer such as distributed memory architectures, conventional parallelizing compilers can not be applied, since 1) the range of high level parallelism is different, 2) the instructions for low level parallelism are not valid for new ones and 3) cooperation between the front-end and the back-end has not been achieved yet[3].

We have worked for introducing an object oriented methodology to design of a parallelizing compiler[10]. In the object oriented method, the design of parallelizing compilers can employ a framework, which has finer granularity of program modules than conventional front-end/back-end style parallelizing compilers. It is evident that the finer granularity of program modules has better re-usability if they are modularized in an appropriate manner. We believe the re-usability of modules is the indispensable characteristic of a parallelizing compiler for current and future parallel computers.

In [10], presenting and analyzing a prototype, we have shown that the parallelizing compiler designed with Object Modeling Technique (OMT) [12] is flexible (to be expanded and maintained), portable (in the sense of re-usability) and theoretically clear (by the nature of OMT). But we did not present the loop parallelization design of the prototype compiler in detail, which is the kernel of parallelizing compilers. Therefore we present an object oriented framework for loop parallelization of parallelizing compilers in this paper.

In general, the following forms must be clearly defined to modularize programs for improving re-usability[4, 7].

1. External data structure
2. Internal data structure
3. Interface to access internal data

External data of parallelizing compilers are programs to be parallelized (input data) and have been parallelized (output data). In this paper, they are an intermediate representation of a parallelizing compiler front-end since we focus on loop parallelization. The intermediate representation for output data includes a description model, which does not assume any specific target architectures, such as MPI[3]. It enhances re-usability of the external data.

Internal data of parallelizing compilers includes reservation table, abstract syntax tree and symbol table, which have been investigated as intermediate representations of conventional parallelizing compilers. Since those conventional representations heavily depend on parallel computing models or architectures, an inappropriate choice of intermediate representations causes the poor execution of the resulting parallel program. To obtain good performance of the execution, two techniques must be employed: 1) Quantitative evaluation of given hardware and 2) Feedback of the evaluation result to program optimization phases. If we adopt the above techniques in the conventional procedure oriented design of parallelizing compilers, frequent data sharing would be observed and it leads to inferior maintenancibility and portability[10]. In this paper, we employ an internal data structure which does not only contain the conventional intermediate representations but also keep internal data from being shared among the representations to

---

[3] Recently, the integration of a front-end and a back-end is reported in [2]

improve their re-usability. The re-usability enhances the maintenancibility and portability of parallelizing compilers.

The interface to access the internal data should be defined so that access methods are unchangeable even if the internal data are revised because of the change of hardware configurations. We take an advantage of encapsulation mechanism of object oriented method to define the above three forms. The encapsulation increases safety and consistency of the data.

In the rest of this paper, we analyze a conventional method to parallelize a loop of sequential programs, divide the method into finer phases to obtain required intermediate representations, give definitions of hierarchical interfaces, and propose a framework for easy implementations of loop parallelization algorithms.

In this paper, for the sake of simplicity, the target of loop parallelization is limited to each procedure level of loops which can be normalized, and pointer analysis is not investigated.

2 Loop Parallelization Process

Object oriented design stands on problem formalization[8]. Hens we describe formalization of conventional and typical loop parallelization in this section.

There have been so many parallelization methods proposed of which formalization is not easy even if input programs are limited to in FORTRAN[13]. Given an intermediate representation after abstract analysis, a process of loop parallelization is known to the following steps[5, 14].

1. Loop Detection
2. Loop Normalization
3. Analysis of Scalar Variable
4. Transformation of Scalar Variable
5. Analysis of Vector Variable
6. Transformation of Vector Variable

A loop is detected in a phase of control flow analysis by depth-first search of executable paths and marking them[6]. The detected loop can be normalized so that the following conditions are satisfied.

- The normalized loop is a single nest or a perfectly nested loop.
- Each normalized loop has just one index variable and the variable is only accessed during the execution of the loop.
- The initial and incremental value of the index variable is 1.
- There is no assignment to the index variable during the execution of the loop body.

The normalized loop is uniquely represented by the first statement of the loop body and the number of loop iteration[14]. The program, of which loops have been normalized, can be represented as a directed graph. The directed graph is hierarchically constructed so that loops as well as basic blocks are presented as a node of the graph.

The next step of loop parallelization is data dependence analysis for scalar variables and vector variables.

The data dependence analysis for scalar variables investigates inter and intra basic blocks. For each variable, the existence and the scope of data dependence, life time and use-definition chain are obtained here.

The data dependence analysis for vector variables usually employs an approximation based on their index variables. In typical programs, the array index expressions are most likely defined as a linear form of loop index variables, so the analysis can make use of Diophantine equation. Through the approximation, distance vectors and dependence vectors are found to obtain detailed information such as maximum parallelism or maximum and minimum length of distance vectors. The detailed information is managed by each loop.

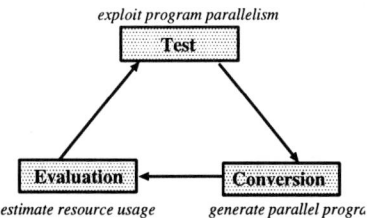

Fig. 1. Loop parallelization phases

Since conventional loop parallelization methods have been heavily based on the above data dependence analysis, we take a careful investigation on the common processes among the above analysis to succeed to classify them into three phases as shown in Fig. 1. In the figure, the *test* phase exploits parallelism of given programs with consideration of the relation between parallelization and data dependence constraint when control dependence and resource constrains are not taken in account. In the same way, the *conversion* phase generates parallel programs with considering control dependence constraint, and the *evaluation* phase takes care of resource constraint. Namely, the three phases correspond to data dependence, control dependence and resource constraints respectively, which are the key points for efficient parallelization.

Of course, it is true that results of the three phases may be inconsistent. But the cycle of the three phases absorbs the inconsistency to produce more effective and consistent results. Therefore, we can say the three phases model proposed here is very flexible.

3 The Framework for Loop Parallelization Process

Since parallelizing compilers contain a lot of parallelization and optimization functions and those functions sometimes are associated each other, the design of intermediate representations, which are commonly accessed by the functions, of the parallelizing compilers is very critical.

In this section, we derive an intermediate representation for loop parallelization process from the analysis of conventional process. The intermediate representation is designed as a set of external classes of which derivation is used to design each of three phases described in section 2. This design is a framework to make various parallelizing and optimizing functions access the intermediate representation with keeping its consistency.

3.1 Intermediate Representation of Loops

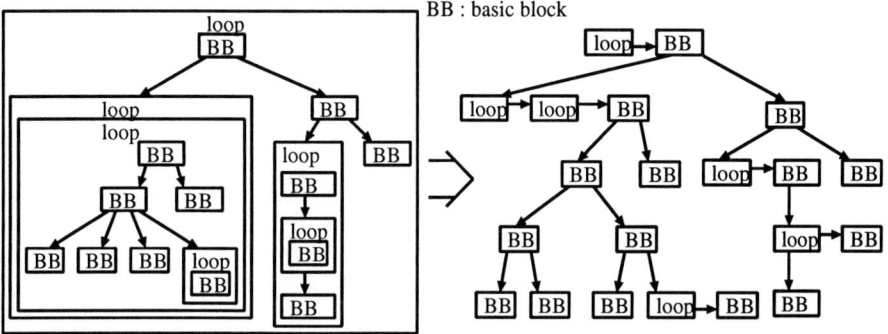

Fig. 2. Structure of a normalized loop

Since normalized loop bodies are a set of nodes of basic blocks and loops, it can be hierarchically transformed into a binary tree as shown in Fig.2. Therefore, loops are described by an intermediate representation constructed from classes shown in Fig.3. Remark that loops are distinguished from basic blocks by the iteration number *times*.

```
class loop_IM {              class source_line {        struct {                          class parallel_loop_IM {
    int head;         refer      int line;       array      strings filename;                 int initialize_head;
    int times;       ----->      strings image;  ----->     int file;                         int initialize_tail;
    int tail;                    source_line* child;        class source_line[N];             int closing_head;
    int child;                   bool valid;            }                                     int closing_tail;
    int next_left;               bool complex;          //source file structure               int distance;
    int next_right;          };                                                               int sync[];
};                           //source line image                                          }
//Class for loop inetmediate representatin                                                // Class for parallelism
```

Fig. 3. Classes of loop parallelization phases

To produce output of loop parallelization as external data, a set of information, initialization processes, termination processes, dependence vectors and synchronizing points, are required. The information is given in Fig.3 as a class for parallelism. While input of loop parallelization as external data is the program to be parallelized. The input program codes are kept as another data structure and referred via line number for the sake of consistent accesses as shown in Fig.3.

We describe the three phases, test, conversion and evaluation, in the rest of this section.

3.2 Test Phase

Input to test phase, namely sequential program codes, is analyzed so that their parallelism is exploited. Separation test or Banerjee test[1] is known to be a typical test phase.

```
struct {
    strings IV;        //index variable
    strings IVE[];     //expression images using IV
    strings UV[];      //refered only
    strings DV[];      //given a value in loop body
    strings LV[];      //local variables
    struct VV VV[];    //vector variables
};
//loop variable

struct VV {
    strings name;
    void* head;
    int size;
    int num;
};
//operation object

struct branch{
    int true;          //distination
    int false;         //distination
    int size;          //jump distance forward
    strings cond;      //conditional expression
    strings dep[];     //depending variables
};
//branch structure
```

Fig. 4. Classification of variables related to loops and Branch structure

To construct a framework for this phase, variables related to loops are classified into five types: four scalar variables and a vector variable. This classification as well as loop detection can be done at structure analysis with recording them on a symbol table. Therefore, we use the combination classes shown in Fig.3 and Fig.4 as an intermediate representation of input loops to this phase. Some classes belong to variables while other classes belong to statements, and each of them is kept in its own object.

The purpose of the interface of this phase is to manipulate those variables. For scalar variables, two kinds of interfaces are given: the low level interface shown in Tab.1 and the middle level shown in Tab.2.

The low level interface is used for direct accesses of variables or type checking. It contains a special field *flags* where users can save the value of obtained maximum parallelism, for example.

The middle level interface works in a more functionable fashion. For example, *rename()* does not only replace the strings of a variable but also produce a unique name and give the produced variable a property so that the produced variable works as a local variable of the original variable. *what_branch* inserts a branch condition to a branch structure shown in Fig.4, and obtain the direction and the jump distance of the branch.

Table 1. Interface to operate scalar variables (low level)

function	explanation
head()	return head
tail()	return tail
child()	return child node
left()	return left node
right()	return right node
next_branch(line)	return next branch after line
next_call(line)	return next call after line
is_local(name)	true if local variable
is_global(name)	true if local variable
is_vec(name)	true if vector variable
is_live(name, line)	true if live variable
is_linear(name, line)	true if the expression is linear
is_use(name, line)	true if the variable used
is_define(name, line)	true if the variable defined
strings newname()	generate unique name
ctype(name)	return variable type for parallelism
type(name)	return variable type
eval(name)	return variable value
touch(line)	set flag to the line
clear(line)	unset flag to the line
try(line)	return flag
int dis_vec(name)	return depending node
int dep_vec(name)	return depended node
int flags	user flags

The interface for vector variables is defined in Tab.3.

Since data passing among phases is done by setting values, we prepare copy constructors by which checking of the linearity of index variables can be done to exploit parallelism.

3.3 Conversion Phase

In the conversion phase, inputed loops are transformed into a parallel form or a form which can be directly parallelized. Regular operations to each loop iteration can be represented as matrix transfers. The set of such transformations as loop expansion or loop interchange is the interface of the conversion phase and shown in Tab.4.

The loop transformation techniques which can be used in this phase have been investigated in a field of vector processing research[14].

Table 2. Interface to operate scalar variable (middle level)

function	explanation
induction(source, dist, size, last)	induce variable
reduction(source, dist)	reduce vector
search(name, line)	search next statement including name after line
search_all(name)	search all statements including name
rename(source, dist, head, tail)	rename variable in the range
copy(source, dist, head, tail)	copy variable in the range
next_line(line, name)	show next line in the dataflow
prev_line(line, name)	show previous line in the dataflow
what_branch(line, branch)	show branch property
what_io(line, io)	show input/output property
what_level(line)	loop level
max_level(loop)	max loop level
replace(source, dist)	replace line image
eliminate(line, name)	eliminate dead code from line
eliminate(node, line)	eliminate dead line from node
const(node, name, val)	substitute name for val in node

Table 3. Interface for vector variables in test phase

function	explanation
diophantine(name, head, tail)	calculate name's Diophantine equation extracted from head to tail
divide_node(node)	divide the node into two
divide_body(source, dist)	copy link from source to dist
copy_data(sourece, shead, stail, dist, dhead, dtail)	copy vector from source to dist
divide_body(source, dist)	copy link from source to dist
inline(line)	expand line

3.4 Evaluation Phase

In the evaluation phase, inputed parallel programs are evaluated based on the amount of use of hardware resource through resource reservation table shown in Tab.5. In the table, each entry represents an abstracted hardware resource. This can be a quantitative evaluation of parallel programs based on hardware resource.

The interface to access each entry of the table is just *show*(read) and *set*(write) operations.

Conventional models for the performance evaluation of parallel programs often assume the infinite number of processors or memory modules. Such models can easily obtain the optimal solution because of the unrealistic assumptions.

Table 4. Interface of conversion phase

function	explanation
is_relate(loop1, loop2)	return relation of loop1 and loop2
fpeel(loop, times)	expand loop for times from head
tpeel(loop, times)	expand loop for times from tail
expansion(loop)	expand loop
exchange(loop1, loop2)	exchange loop1 and loop2
exchange_all(loop)	reverse loop nest
fold(loop)	fold loop into half the length
serial(loop)	leave for serial execution
tail_recursion(name, loop)	remove name's tail recursion

Table 5. Resource reservation table for evaluation phase

attribution	explanation
int min	known minimum path
int max	known maximum path
times	repeat times
memory	memory size
io	io times

However, a lot of trials are needed to increase the execution performance of practical parallel programs since they frequently meet non-continuity and finiteness. Therefore, the quantitative evaluation of parallel programs based on hardware resource is inevitable.

4 Conclusion

In this paper, we proposed a framework for loop parallelization process, which had tended to require complicated design and implementation, based on object oriented design. We first extracted the information of intermediate representations of loops, and gave class definitions to them. Then we divided the loop parallelization process into three phases under the framework.

The class of each phase was derived from the intermediate representations of loops. Especially, we hierarchically divided the class which represents complicated descriptions of algorithms to increase the maintenancibility and the portability. This means descriptions of algorithms are separated from basic data structures, and a new parallelizing algorithm requires just defining new subclasses. Furthermore, the evaluation phase uses resource reservation table to achieve quantitative optimization.

Thus, the framework we proposed in this paper has resultant parallelizing compilers have the characteristic of portability, re-usability and expandability as well as providing efficient parallelization.

Our future work should be to generalize the framework. Since we focused on loop parallelization in this paper and the framework was not designed to consider the situation where other phases might share the classes of the framework. Now we are designing the integrated framework for a parallelizing compiler.

References

1. U.Banerjee. *Dependence Analysis for Supercomputing.* Kluwer Academic Pub., USA, 1988.
2. C.Brownhill, A.Nicolau, S.Novack and C.Polychronopoulos. Achieving Multi-level Parallelization. In *Proceedings of ISHPC*, will appear, 1997.
3. M.Snir, S. Otto, S.Huss-Lederman, D. Walker, and J.Dongara. *MPI:the complete reference.* MIT press, 1996.
4. S.Gossain and B.Anderson. An iterative-design model for reusable object-oriented software. In *Proceedings OOPSLA/ECOOP '90, ACM SIGPLAN Notices*, pages 12–27, October 1990. Published as Proceedings OOPSLA/ECOOP '90, ACM SIGPLAN Notices, volume 25, number 10.
5. K.Kennedy S.Hiranandani and C.Tseng. Compiling fortran d for mimd distributed-memory machines. *Communications of the ACM*, 35(8):66–80, Aug 1992.
6. J.Kam and J. Ullman. Global data flow analysis and iterative algorithms. *journal of ACM*, 23(1):158–171, 1976.
7. J. Lewis, S. Henry, D. Kafura, and . Schulman. An empirical study of the object-oriented paradigm and software reuse. In *Proceedings OOPSLA '91, ACM SIGPLAN Notices*, pages 184–196, November 1991. Published as Proceedings OOPSLA '91, ACM SIGPLAN Notices, volume 26, number 11.
8. M.Loomis, A.Shah and J. Rumbaugh. An object modelling technique for conceptual design. In P. Cointe J. Bézivin, J-M. Hullot and H. Lieberman, editors, *Proc. ECOOP '87*, LNCS 276, pages 192–202, Paris, France, jun 1987. Springer Verlag.
9. S.Novack. *The EVE Mutation Scheduling Compiler: Adaptive Code Generation for Advanced Microprocessors.* PhD thesis, UCI, 1997.
10. Y.Omori, K.Joe and A.Fukuda. A Parallelizing Compiler by Object Oriented Design. In *Proceedings of COMPSAC*, will appear, 1997.
11. C.Polychronopulos, et al. Parafrase-2: An environment for parallelizing, partitioning, synchronizing, and scheduling programs on multiprocessors. In *J. of High Speed Computing*, Vol.1, No.1, 1989.
12. J.Rumbaugh et al. *Object Oriented Modeling and Design.* Prentice-Hall, 1991.
13. M. Wolfe and M. Lam. A loop transformation theory and an algorithm to maximize parallelism. *IEEE Transactions on Parallel and Distributed Systems*, 2(4):452–470, Oct 1991.
14. H. Zima and B. Chapman. *Supercompilers for parallel and vector computers.* ACM Press, New York, NY, 1991.

A Method for Runtime Recognition of Collective Communication on Distributed-Memory Multiprocessors

Takeshi Ogasawara and Hideaki Komatsu

Tokyo Research Laboratory, IBM Japan Ltd., Shimotsuruma 1623-14, Yamato-shi, Kanagawa, Japan

Abstract. In this paper, we present a compiler optimization for recognizing patterns of collective communication at runtime in data-parallel languages that allow the dynamic data decomposition. It has a calculation time of the order $O(m)$, and is appropriate for large numerical applications and massively parallel machines. The previous approach took $O(n_0 + ... + n_{m-1})$ time, where m is the number of dimension of an array and n_i is the array size on the i-th dimension. The new method can be used for data redistribution and intrinsic procedures, as well as data pre-fetch in parallelized loops.

1 Introduction

Programs written in data-parallel languages cause many communications in which a group of processors communicate with each other simalteneously. Many such communications are organized into a class called *collective communication* [1]. In collective communication, it becomes more difficult to reduce the communication cost if by using only library routines for point-to-point communication as the number of processors increases on massively parallel machines, since the cummulative software overhead becomes non-negligible. We should therefore also exploit special hardware for communication on multiprocessors.

Library routines for collective communication are normally provided, to exploit the hardware architecture of multiprocessors. Recently, MPI [1] has become generally accepted as a de-facto standard for message-passing programming interfaces, and has been implemented for many distributed-memory multiprocessors. It defines an interface with routines for collective communication such as MPI_Allgather and MPI_Bcast. To reduce the overhead of collective communication and make generated codes to run efficiently on different machines, compilers of data-parallel languages should use library routines for collective communication that follow a standard such as MPI.

We focus on data-parallel languages that allow the dynamic data decomposition and the dynamic data realignment. In those languages, multiple data decompositions can reach a point of a program. For example, data realignmnet statements that are conditionally executed and actual arguments of subroutines of which data decompositions are inherited from different call sites. Thus, it is necessary for compilers to calculate communications at runtime.

Compilers for data-parallel languages can improve the performance of collective communication by generating codes that use library routines for it instead of point-to-point communication routines. However, since there are many cases in which communications should be calculated at runtime, compilers should recognize collective communication at runtime.

In this paper, we present a method for recognizing collective communication at runtime in data-parallel languages that allow the dynamic data decomposition and realignment. The method handles arrays of linear subscripts, or regular problems, and exploits the regularity of array decomposition and array reference. The regularity means that the same function is used for a decomposition of or a reference to an array of all processors. The order of the calculation time is $O(m)$ for communications and recognition of collective communication in the method, where m is the number of dimensions of an array. The time is independent of the size of the array and the number of processors. Thus, the method is practical for very large numerical applications and for massively parallel machines.

The rest of this paper is organized as follows. Section 2 describes previous work in the area of collective communication recognition. In section 3, first of all, we present the basic concepts of our method. Next, we explain basic data structures and operators on those data structures that make our algorithm work efficiently. We then present an algorithm for recognizing collective communication at runtime. Section 4 gives experimental results for estimating the runtime overhead of our method. Section 5 concludes the paper.

2 Related Work

Compile-time approaches for recognizing collective communication [2] do so for assignment statements in parallelized loops. Such approaches assume that decompositions of arrays are statically specified at compile-time: that is, that a dimension of an array must be distributed onto a dimension of a processor shape that consists of a constant number of processors, in a manner such as BLOCK. They also assume that the parameters of subscripts of array references are constant values, and that sections of processors that perform loops are not concerned [3]. For compilers that use compile-time approaches, there are two situations in which collective communication cannot be recognized or an extra communication occurs because of imprecise recognition. One is when the decomposition of an array is resolved at runtime, typically by dynamic array redistribution. The other is when sections of an array to be accessed are resolved at runtime, typically when the lower and upper bounds of the index space of the array are variables. Our runtime approach handles these two situations and recognizes collective communication precisely without extra communication.

For irregular problems, a runtime support method called *inspector/executor* method and a runtime compilation method [4] are proposed, to calculate communications and to schedule communications at runtime. The approaches expand all communications into a table, and schedule the communications on each processor without taking account of regularity [5]. Each processor schedules communications in which it does not participate, thus performing redundant computations. Agrawal [6] discusses exploiting the regularity in multi-block problems on a restricted condition and recognizes shift communication patterns. No existing studies of irregular problems offer general approaches for recognizing collective communication.

Communication sets are calculated at runtime in array redistribution [7]. The communication cost can be reduced in redistribution between specific decompositions on the same processor shape [8]. Ramaswamy [9] can handle redistribution between any decompositions on different processors. Kalns [10] discusses using MPI_Scatter,

MPI_Gather, and MPI_Alltoall; however, the algorithm takes $O(n_0 + ... + n_{m-1})$ time, where m is the number of dimensions of an array and n_i is the array size in the i-th dimension. It is not practical in very large numerical applications, where array sizes are very large.

3 Methodology

3.1 Overview

The collective communication is characterized by the same pattern of communication that is performed by multiple processors. We should therefore preserve information that which processors participate in the same communication during the construction of the communication.

Basic concepts are as follows. For each dimension of an array to be communicated, we construct a 1-D communicatin set that has information of a 1-D communication, that is, processor set A communicates 1-D array section R to processor set B (for example, processors p0, p2, p4 communicates 100th column of array A(*,*) to processors p1, p3, p5). Here, we represent such a 1-D communication set as a relational expression $P_d^{local}(i_d) - A_d(i_d) - P_d^{remote}(i_d)$. d means the d-th dimension of an array to be communicated. P_d^{local} and P_d^{remote} mean sending processors and receiving processors of the d-th dimension of the array if the communication set represents sending, and mean receiving processors and sending processors if the communication set represents receiving. A_d means an array section of the d-th dimension. P_d^{local}, P_d^{remote} and A_d are indexed by i_d. A processor ID is associated with i_d. A 1-D communication is calculated by specifying i_d in the expression and the same communication is calculated among processors of which ID are associated with the same i_d.

To construct a complete communication that has information of all array dimensions, we combine 1-D communications through all array dimensions. First, processors construct the 1-D communication set and each processor calculates a 1-D communication that is specific to the processor by specifying i_d in the 1-D communication set. Next, those 1-D communications are combined and the processor sets such as $P_d^{local}(i_d)$ are intersected through all dimensions. Each of intersected P^{local} and P^{remote} specifies a processor set. A_ds are combined and construct an array section.

Processors in a local processor set calculates the same communication during the combination. Thus, a processor can know other processors that participate in a communication that the processor is going to perform. The type of the collective communication depends on the relation of each expression $P_d^{local}(i_d) - A_d(i_d) - P_d^{remote}(i_d)$ (discussed in 3.4).

In Section 3.2, basic data structures for structuring 1-D communication sets are explained. Since those data structures exploit the regularity of array decomposition and array reference, those data structures are compact and have information for all processors. Such a compact 1-D communication set is called an *ITR list* and a group of those 1-D communication sets for all array dimensions is called an *In/Out Set (IOS)*. Section 3.3 describes the operation of combining ITR lists. The efficiency of our method is based on those data structures and operation. Section 3.4 presents an algorithm for recognizing collective communication patterns, using the data structures and the operation. Finally, we show that the time needed to calculate a complete communication set and recognize collective communication is $O(m)$.

3.2 Data Structures

We use a data structure called an *ITR* to structure IOSs by taking advantage of their regularity. An ITR specifies a dimension of an array section and processors with which the array section communicates. The processors specified in the ITR are expressed by a form called a *quadruplet*. The quadruplet is also used to locate ITRs in IOSs.

Quadruplets We define a compact form called a *quadruplet* for specifying processors in regular computations. A quadruplet consists of four integer values: the processor decomposition index dx, the processor stride ps, the number of processor decompositions nd, and the total number of processors np. A quadruplet $(dx : ps : nd : np)$ specifies more than a processor on a 1-D processor shape.

We shall introduce *processor decomposition*, PD, to specify a sequence of ps processors that refer to or own the same array section. In a quadruplet, a 1-D processor shape of np processors is split into groups of PDs, and a group of PDs consists of nd PDs. A group of PDs is repeated rf times if the number of processors, $ps*nd$, specified by groups of PDs is less than np, and rf is given by Equation 1.

$$rf = np/ps/nd \quad (1)$$

As an example, assume a quadruplet (1:2:4:32). The processor stride of a PD is 2 on a 1-D processor shape of 32 processors. The size of a PD group is 4, and the replication factor is 4 according to equation 1. The PD index is 1, and specifies the first PD in a group. As a result, the quadruplet specifies eight processors: p0, p2, p9, p10, p17, p18, p25, and p26.

ITRs, ITR Blocks, and ITR Lists The IOS for an operation on a m-dimensional array consists of m *ITR lists*, each for a dimension of the IOS. An ITR list consists of a list of *ITR blocks* and an *ITR master* that maintains the ITR block list. An ITR block consists of one or more *ITRs*.

An ITR is a pair of a triplet $R_{ITR} = (bg : ed : st)$ that specifies an array section and a quadruplet $Q_{ITR} = (dx : ps : nd : np)$. bg, ed, and st are the start, the end, and the stride of a dimension of an array section, respectively. We use the following expression for an ITR:

$$ITR = [R_{ITR}, Q_{ITR}] = [bg : ed : st, dx : ps : nd : np] \quad (2)$$

An ITR block consists of one or more ITRs and extra ITRs have information that cannot be expressed by the regularity. An ITR block in an ITR list is located by a quadruplet called a *remote quadruplet*. Quadruplets in ITRs are called *local quadruplets*, to distinguish them from remote quadruplets. Local quadruplets and remote quadruplets specify sending processors and receiving processors when IOSs represent sending, and specify receiving processors and sending processors when IOSs represent receiving. The ITR master M_{ITRL} for the ITR blocks has parts of a local quadruplet, mps, mnd, and mnp, corresponding to ps, nd, and np of remote quadruplet. Thus, an ITR list $ITRL$ is represented by the ITR expression:

$$ITRL = \langle mps : mnd : mnp \rangle \{[bg : ed : st, dx : ps : nd : np]\}. \quad (3)$$

In each processor and in each ITR list of IOSs, the processor ID is translated into a local quadruplet and an ITR block is specified by the local quadruplet. The PD index

mdx of a local quadruplet for processor ID pid is given by Equation 4, using mps and mnd of the Master ITR as an ITR list:

$$mdx = 1 + mod((pid - 1)/mps, mnd)). \qquad (4)$$

mdx specifies the position of an ITR block on the ITR list.

An ITR list of ITR blocks can be compacted by taking advantage of the regularity of ITR blocks such as the augmented Regular Section Descriptor [11]. For example, $[i*10-9:i*10:1, i+1:4:4:32]_{i=1..3}$ is a compact form of an ITR list of three ITR blocks, each of which consists of an ITR. The regularity expressed in the compact form allows each processor to know without traversing the ITR list that processors that specify the i-th ITR in the ITR list have an ITR $[i*10-9:i*10:1, i+1:4:4:32]$.

3.3 ITR Product

Information in the ITRs of the ITR blocks specified by Equation 4 are combined by *ITR product* through all dimensions of IOS. We use the symbol $\diamond$ for an ITR product. A combined ITR shows an array section and communication partner(s) for a communication.

For an operation on a d-dimensional array, assume that $\{ITR^i_{k_i}\}_{k_i=1..n_i}$ is an ITR block selected on the i-th dimension of the array, and that $ITR^i_{k_i}(1 \leq k_i \leq n_i)$ is one of n_i ITRs $(1 \leq i \leq d)$ in the ITR block. By using an ITR product, the communication set for the operation can be represented as

$$\{ITR^1_{k_1} \diamond ... \diamond ITR^d_{k_d}\}_{k_1=1..n_1,...,k_d=1..n_d},$$

where $ITR^1_{k_1} \diamond ... \diamond ITR^d_{k_d}$ specifies a communication.

An ITR product is performed as follows. Assume that the array section and the remote quadruplet in $ITR^i_{k_i}$ are $R^i_{k_i}$ and $Q^i_{k_i}$, and that the ITR product $ITR^1_{k_1} \diamond ... \diamond ITR^d_{k_d}$ is given by Equation 5:

$$ITR^i_{k_i} \diamond ... \diamond ITR^d_{k_d} = [R^1_{k_1}, Q^1_{k_1}] \diamond ... \diamond [R^d_{k_d}, Q^d_{k_d}]$$
$$= \|(R^1_{k_1}, ..., R^d_{k_d}), \bigcap_{i=1}^{d} Q^i_{k_i}\| \qquad (5)$$

The last form of Equation 5 is expressed as $\|R, Q\|$, and specifies a communication of array section R and remote processors Q. The array section part $R^i_{k_i}$ of an ITR is considered as a element of the set $\{R^i_{k_i}\}_{k_i=1..n_i}$, and R is an element of the Cartesian product of the sets on all dimensions $i = 1..d$. Each $Q^i_{k_i}$ is considered as the set of processors, and Q is the intersection, $\bigcap$ of $Q^i_{k_i}$ on all dimensions.

The operation $\bigcap$ is also applied to local quadruplet $mdx : M_{ITRL}$ on the ITR product. The operation results in a local quadruplet: all processors specified by the local quadruplet have the same communication set. Thus, each processor calculates its own communication set using IOS, and can find other processors that have the same communication set as that of the processor during the calculation.

Table 1 shows an example of IOS for an operation on a 2-D array. We demonstrate how a communication set is constructed from the IOS for processor p3. In the first array dimension, Equation 4 with the values of the ITR masters in Table 1 gives $1+mod((3-1)/1, 2) = 1$. In the second dimension, the equation gives $1+mod((3-1)/2, 2) = 2$. Thus two ITRs for the first dimension and an ITR for the second dimensions are specified

Array dimension	Local quadruplet			ITR block 1		ITR block 2	
	Processor stride	Number of PDs	Total number of processors	Array section	Remote quadruplet	Array section	Remote quadruplet
1	1	2	8	(1:49) (50:50)	(1:1:2:4) (2:1:2:4)	(51:99)	(2:1:2:4)
2	2	2	8	(0:49)	(1:2:2:4)	(50:99)	(2:2:2:4)

Table 1. Example of an IOS

in the IOS. The ITR product of ITR [50:50,2:1:2:4] in the first dimension and ITR [50:99,2:2:2:4] in the second dimension results in array section (50,50:99) and a remote quadruplet (4:1:4:4). This shows that processor p3 communicates the array section to p4. The ITR product is also applied to the local quadruplet, and the resulting local quadruplet of (1:1:2:8) ∩ (2:2:2:8) shows that processor p7 has the same communication set.

3.4 Algorithm

The algorithm of runtime recognition of collective communication proceeds as follows: (1) *array ownership set (AOS)* calculation, (2) *local index set (LIS)* calculation, (3) *in/out set (IOS)* calculation, and (4) *collective data movement (CDM)* recognition. AOS specifies how arrays are decomposed among processors. In HPF [12], block(n) or cyclic(n) can be specified with an alignment in each array dimension [12]. LIS is a collection of the loop index spaces that are local to each processor [13]. We use the term *collective data movement* to distinguish the operation to which it refers from collective reduction. Both are involved in collective communication [1], since reduction is not handled in our method.

As many as possible of procedures (1), (2) and (3) are performed at compile-time. Calculations that have to be performed at runtime may be cached to reduce the runtime overhead. As explained at th ebeginning of Section 3, every processor has the same IOS, because of SPMD, and each processor calculates communication sets from the IOS and recognizes collective communication at runtime. Thus procedure (4) is performed at runtime. The following sections explain the details of the procedures.

AOS Calculation *Array Ownership Set (AOS)* of an array is information that section R of the array is distributed to processor A. In data-parallel languages such as HPF, a processor shape $P(m_1,..,m_{pd})$ decomposes array $A(n_1,..,n_{ad})$. The i-th dimension of A, A_i, is decomposed by the j-th dimension of P, P_j, or is not decomposed by any processors. A decomposed array is replicated on dimensions of P that do not decompose the array.

Assume that $np = \prod_{l=1}^{pd} m_l$ is the number of processors and that $R_{mdx}^{A,i}$ is the array section of the mdx-th ITR. If A_i is decomposed by P_j, the i-th dimension of AOS of A is represented by an ITR list $ITRL_{AOS_i^A}$:

$$M^{P,j} \equiv \prod_{l=1}^{j-1} m_l : m_j : np$$

$$ITRL_{AOS_i^A} = \langle M^{P,j} \rangle \{[R_{mdx}^{A,i}, \emptyset]\}. \tag{6}$$

$M^{P,j}$ is the ITR master. How to calculate R_k depends on the type of distribution [12]. Quadruplets in ITRs are not used in AOS.

LIS Calculation A *Local Iteration Set (LIS)* is a set of iterations that are local to processors [13]. Each processor executes a loop by using its local index space in the LIS of the loop. The index space of a nested loop is called the *Global Iteration Space (GIS)*. The d-th dimension of a GIS, GIS_d, is decomposed by using the *owner computes* rule [11] into the LIS LIS_d, if subscript S_{il} of the il-th dimension of the left-hand side array A in the loop body contains the loop index iv_d of GIS_d. The array subscript is expressed as a function of iv_d: $S_{il}(iv_d)$. For a dimension of a GIS that does not have any corresponding array subscript, the LIS for the dimension is the dimension of the GIS.

To calculate the LIS, we need to calculate the array sections on the left-hand side to which each processor refers, called the Local Write Set (LWS). Calculating the intersection of $S_{il}(GIS_d)$ and each array section of $ITRL_{AOS_{il}^A}$ gives LWS:

$$RLWS_{mdx}^{il,d} \equiv \{S_{il}(iv_d)\}_{iv_d=lb_d:ub_d:by_d} \cap R_{mdx}^{A,il}$$
$$ITRL_{LWS_{il}^A} = \langle M^{P^l,j^l}\rangle\{[RLWS_{mdx}^{il,d},\emptyset]\}, \qquad (7)$$

where lb_d, ub_d, and by_d are the initial value, the limit, and the stride for GIS_d, respectively.

LIS_d is calculated by applying the inverse of the subscript function, S_{il}^{-1}, on array section $RLWS_{mdx}^{il,d}$ of each ITR in the LWS $ITRL_{LWS_{il}}$:

$$RLIS_{mdx}^{il,d} \equiv S_{il}^{-1}(RLWS_{mdx}^{il,d})$$
$$ITRL_{LIS_d} = \langle M^{P^l,j^l}\rangle\{[RLIS_{mdx}^{il,d},\emptyset]\}. \qquad (8)$$

IOS Calculation The IOS for the right-hand side array B is calculated from *Local Read Set (LRS)* and the AOS. The LRS for B is a set of array sections that each processor accesses by using the LIS.

The subscript of the ir-th dimension of B, B_{ir}, is expressed by the function of iv_d, $T_{ir}(iv_d)$, if the subscript contains the loop index. The LRS is calculated by $T_{ir}(LIS_d)$ and is given by the equations

$$RLRS_{mdx}^{ir,il,d} \equiv T_{ir}(RLIS_{mdx}^{il,d})$$
$$ITRL_{LRS_{ir}^B} = \langle M^{P^l,j^l}\rangle\{[RLRS_{mdx}^{ir,il,d},\emptyset]\}. \qquad (9)$$

We define an operation called *ITR division* of two ITR lists, for which we use the symbol /. In ITR division, every ITR in the dividend ITR list is divided by a corresponding ITR in the divisor ITR list. The array section part of the result ITR of the division is $R_s \cap R_d$, where R_s is the array section part of a dividend ITR and R_d is that of a divisor ITR. The quadruplet part of the result ITR is (PD index of divisor : ITR Master of divisor). The division is performed for every R_d where $R_s \cap R_d \neq \emptyset$.

Assume that P^r is the shape of the processor to which B is distributed, and that jr is the dimension of P^r by which B_{ir} is decomposed. The ir-th dimension of the AOS for B $ITRL_{AOS_{ir}^B}$ is given by the equation

$$ITRL_{AOS_{ir}^B} = \langle M^{P^r,jr}\rangle\{[R_{mdx}^{B,ir},\emptyset]\}. \qquad (10)$$

ITR division of the two ITR lists given by Equations 9 and 10 gives the ir-th dimension of the IOS for B. The equations

$$ITRL_{IS^B_{ir}} = ITRL_{LRS^B_{ir}} \tilde{/} ITRL_{AOS^B_{ir}} \tag{11}$$

$$ITRL_{OS^B_{ir}} = ITRL_{AOS^B_{ir}} \tilde{/} ITRL_{LRS^B_{ir}} \tag{12}$$

show the ir-th dimension of *In Set* and *Out Set*, respectively.

CDM Recognition As explained in Section 3.3, each processor locates ITR blocks on each ITR list of the IOS, and calculates their ITR product to create a communication set. Transmissions are created from Out Set and receptions are created from In Set. This section describes how CDM patterns are recognized during the creation of a communication set.

Shift. If all ITR lists of all dimensions of IOS are in compact form, every processor that participates in the communication can find that those processors communicate in the same way. In addition, assume that the remote quadruplet part of all ITRs is expressed by a form such as $mdx + c$, where mdx is the PD index of a dimension of an ITR list and c is constant. If all the conditions are satisfied, the method recognizes a communication set as *Shift*. In Shift, every processor communicates with its peer at a distance given by the vector $(.., c, ..)$ on the processor shape and stores remote data into the overlap area [14].

Broadcast. The method recognizes *Broadcast* in a communication set, if the local quadruplet part of ITR product contains multiple processors when calculating the communication set from In Set. That is, more than one processor receive the same array section in the communication. In this case, the remote quadruplet that specifies the same group of receivers is created from Out Set.

Concatenation. This is also called *Allgather* [1]. Processors that participate in *Concatenation* make a processor group and gather data from all other processors in the group. As a result, the processors have the same collection of data. The method recognizes Concatenation in a communication set, if the product of the number of ITR blocks through all ITR lists of In or Out Set and the number of processors specified by the remote quadruplet is equal to the number of processors specified by the local quadruplet.

Gather/Scatter. Assume that each ITR list of In Set has only a non-empty ITR block (empty ITR blocks are compacted) and that the remote quadruplet part of ITRs in the ITR block form a processor group. In addition, assume that each ITR list of Out Set has an ITR block whose remote quadruplet part specifies only a processor. In this case, the method recognizes *Gather*. *Scatter* is recognized if the reverse condition is satisfied.

4 Experimental Results

We have implemented our method in our HPF compiler [15, 16]. For our experiments, we used a RS/6000 SP with 16 POWER2-66MHz thin-2 nodes. We currently force the calculation of the communication set to be dynamically performed on our system in order to simulate the dynamic data decomposition.

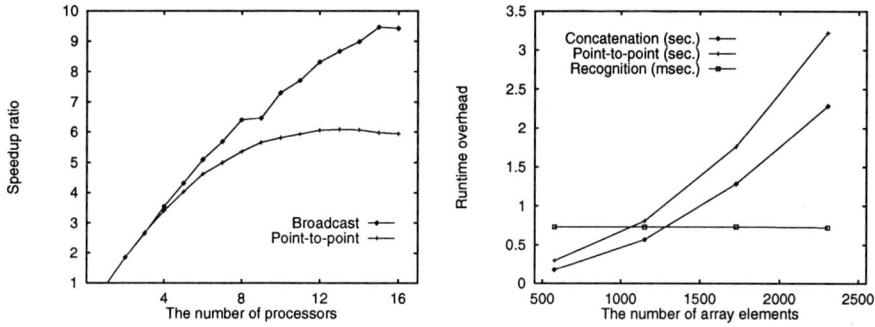

Fig. 1. Speedup ratio (left) and algorithm overhead (right)

We used two applications, LU factorization and matrix multiplication, to evaluate the efficiency of our method. These examples are simple, however, but enough to evaluating the efficiency and the applicability of our method.

We estimate the extent to which how our method efficiently works the speedup ratio in LU factorization with respect to the number of processors. The array is passed to the subroutine and its decomposition of (*,CYCLIC) is inherited. The compiler generates The left graph in Figure 1 shows the speedup ratio against the single-processor for two kinds of executions of the generated SPMD code for 1 to 16 processors. One is for calling only point-to-point routines, and the other is for using a broadcast routine that is recognized by our method and includes the runtime overhead of the method. In all cases, the array size is 2000. As the left graph in Figure 1 shows, a broadcast routine recognized by our method reduces the overhead of Broadcast communication even if our method runs at runtime, especially when the code runs on 16 processors.

We estimate that our algorithm works independent of the array size as regards the computation time by using a example of matrix multiplication. For matrix multiplication, in which three arrays $X(n,n)$, $Y(n,n)$ and $Z(n,n)$ of real*8 are operated by $Z = XY$, we assume that the array decomposition of (*,BLOCK) is dynamically specified for those arrays. Concatenation communication is recognized in this example. The right graph in Figure 1 shows the runtime overhead of our method against the cost of Concatenation communication on eight processors with respect to the array size. The runtime overhead is measured in mili-seconds and the total communication cost is measured in seconds. As the right graph in Figure 1 shows, the method works in a constant time though the array size increases and has the advantage of the computation time against the previous approach [10]. The communication cost is not dramatically improved by the collective communication because RS/6000 SP does not have a special hardware for Concatenation.

5 Conclusions

We have presented a method of $O(m)$ for recognizing collective communication at runtime. Compared with the previous approach, our method is appropriate for large numerical applications since the runtime overhead of our method does not depend on the array size. Our method can optimize the communication overhead by collective communication routines in data-parallel languages that allow the dynamic data decomposition and the dynamic data realignment. Experimental results from a HPF compiler show a practical efficiency of our method in the situation of the dynamic data decomposition. We can use the same modules for data pre-fetch, intrinsic procedures, and array redistribution that require the dynamic data decomposition.

To reduce the runtime overhead of our method, the hyblid approach [17] is applicable to our method in which calculations that use only information resolved as constant values at compile-time are performed at compile-time. The reuse of calculations is also effective to reduce the runtime overhead, as demonstrated in experimental results.

References

1. William Gropp, Ewing Lusk, and Anthony Skjellum. *Using MPI: portable parallel programming with the message-passing interface.* The MIT Press, 1994.
2. Jingke Li and Marina Chen. Compiling communication-efficient programs for massively parallel machines. *IEEE Trans. Parallel and Distributed Systems*, 2(3):361–376, July 1991.
3. Manish Gupta and Prithviraj Banerjee. A methodology for high-level synthesis of communication on multicomputers. In *Procs. 6th ACM International Conference on Supercomputing*, pages 357–367, Washington, D.C., July 1992.
4. Ravi Ponnusamy, Joel Saltz, and Alok Choudhary. Runtime compilation techniques for data artitioning and communication schedule reuse. In *Procs. Supercomputing '93*, pages 361–370, Portland, Oregon, November 1993.
5. Sanjay Ranka, Jhy-Chun Wang, and Manoj Kumar. Irregular personalized communication on distributed memory machines. *Journal of Parallel and Distributed Computing*, 25(1):58–71, 1995.
6. Gagan Agrawal, Alan Sussman, and Joel Saltz. An integrated runtime and compile-time approach for parallelizing structured and block structured applications. Technical Report CS-TR-3143, University of Maryland, Department of Computer Science, 94.
7. Rajev Thakur, Alok Choudhary, and Geoffrey Fox. Runtime array redistribution in HPF programs. In *Procs. Scalable High Performance Computing Conference SHPCC-94*, pages 309–316, Knoxville, Tennessee, May 1994.
8. S.D. Kaushik, C.-H. Huang, R.W. Johnson, and P.Sadayappan. Multi-phase array redistribution: Modeling and evaluation. In *Procs. 9th International Parallel Processing Symposium*, Santa Barbara, California, April 1995.
9. Shankar Ramaswamy and Prithviraj Banerjee. Automatic generation of efficient array redistribution routines for distributed memory multicomputers. In *Procs. The 5th Symposium on the Frontiers of Massively Parallel Computation*, pages 78–87, McLean, VA, February 1995.
10. Edgar T. Kalns and Lionel M. Ni. DaReL: A portable data redistribution library for distributed-memory machines. In *Procs. Scalable Parallel Libraries Conference*, pages 78–87, Mississippi State University, Mississippi, October 1994.
11. Ken Kennedy Seema Hiranandani and Chau-Wen Tseng. Compiler optimizations for fortran D on MIMD distributed-memory machines. In *Procs. Supercomputing '91*, pages 86–100, Albuquerque, NM, November 1991.
12. Charles H. Koelbel, David B. Loveman, Robert S.Schreiber, Guy L.Steele Jr , and Mary E.Zosel. *The High Performance Fortran Handbook.* The MIT Press, 1994.
13. Samuel P. Midkiff. Local iteration set computation for block-cyclic distributions. In *Procs. the 1995 International Conference on Parallel Processing*, pages II/77–84, Boca Raton, FL, August 1995.
14. Hans Zima and Barbara Chapman. *Super Compilers for Parallel and Vector Computers.* ACM Press, 1990.
15. Kazuaki Ishizaki and Hideaki Komatsu. Loop Parallelization Algorithm for HPF Compiler. In *Eighth Annual Workshop on Language and Compilers for Parallel Computing*, pages 12.1–15, Ohio, August 1995.
16. Toshio Suganuma, Hideaki Komatsu, and Toshio Nakatani. Detection and global optimization of reduction operations for distributed parallel machines. In *Procs. 10th ACM International Conference on Supercomputing*, Philadelphia, Pennsylvania, USA, May 1996.
17. M.Gupta, S.Midkiff, E.Schonberg, P.Sweeney, and K.Y.Wang. PTRAN II: A compiler for high performance fortran. In *Procs. 4th Workshop on Compilers for Parallel Computers*, pages 479–493, Delft, Netherlands, December 1993.

Improving the Performance of Automated Forward Deduction System EnCal

Kazunori Nishi, Jingde Cheng and Kazuo Ushijima

Department of Computer Science and Communication Engineering
Kyushu University, 6-10-1 Hakozaki, Higashi-ku, Fukuoka 812-81, Japan

Abstract

Automated forward deduction system is a very important component of many application systems such as theorem finding systems, active databases, and learning systems where the automated forward deduction system plays a key role as an autonomous reasoning engine. The performance of automated forward deduction system is crucial to its successful applications in practices. This paper presents some techniques which are effective in improving the performance of EnCal, an automated forward deduction system. We show how to keep many logical theorem schemata (LTSs for short) in low memory and how to avoid unnecessary pattern matching in forward deduction by using hash table and cache. We also show that these techniques can be applied to other automated forward deduction systems, and that some of them can be applied to almost all applications which deal with logical formulas.

Keywords

Entailment calculus, Forward deduction, Hashing, Common sub formula sharing

1 Introduction

Reasoning rule generation and automated theorem finding [7][8] are two important fundamental issues in knowledge engineering. To solve the above two problems, it is indispensable to establish a domain-independent fundamental theory that underlies an autonomous reasoning mechanism and then develop automatic reasoning tools working based on the fundamental theory to support the autonomous reasoning mechanism.

Recently, Cheng has proposed some paradox-free relevant logics and shown that an entailment calculus based on the paradox-free relevant logics can underlie reasoning rule generation in knowledge-based systems and automated theorem finding[3][4][5]. We are developing an automated forward deduction system for general purpose entailment calculus, named EnCal, which can support entailment calculus based on the paradox-free relevant logics as well as other logics[6].

Automated forward deduction system is a very important component of many application systems such as theorem finding systems, active databases, and learning systems where the automated forward deduction system plays a key role as an autonomous reasoning engine. The performance of automated forward deduction system is crucial to its successful applications in practices. This paper presents some techniques which are effective in improving the performance of EnCal, an automated forward deduction system. We show how to keep many logical theorem schemata (LTSs for short) in low memory and how to avoid unnecessary pattern matching in forward deduction by using hash table and cache. We also show that these techniques can be applied to other automated forward deduction systems, and that some of them can be applied to almost all applications which deal with logical formulas.

2 Terminologies

In logic, the notion abstracted from various conditionals is called "entailment." In general, an entailment, for instance, "A entails B" or "if A then B," must concern two parts which are connected by connective "... entails ..." and called the antecedent and the consequent of that entailment, respectively. The truth-value and/or validity of an entailment depends not only on the truth-values of its antecedent and consequent but also more essentially on a necessarily relevant and/or conditional relation between its antecedent and consequent [1] [2].

An entailment calculus is a formal logical system where the notion of entailment is represented by a primitive connective and a part of its logical theorems are entailments.

A formal logic system L is a triplet (F(L), $\vdash_L$, Th(L)) where F(L) is the set of all well formed formulas of L, $\vdash_L$ is the logical consequence relation of L such that for P⊆F(L) and C∈F(L), P$\vdash_L$C means that within the framework of L taking P as premises we can obtain C as a valid conclusion, and Th(L) is the set of logical theorems of L such that $\emptyset \vdash_L$ t holds for any t ∈ Th(L).

For a formal logic system where the notion of entailment is represented by primitive connective "⇒", a formula is called a zero degree formula if and only if there is no occurrence of ⇒ in it; a formula of the form A⇒B is called a first degree formula (also called a first degree entailment) if and only if both A and B are zero degree formulas; a formula of the form ¬A is called a first degree formula if and only if A is a first degree formula; a formula of the form A * B, where * is conjunction or disjunction connective, is called a first degree formula if and only if both A and B are first degree formulas, or one of A and B is a first degree formula and the other is a zero degree formula. Let k be a natural number. A formula of the form A⇒B is called a kth degree formula (also called a kth degree entailment) if and only if both A and B are $(k-1)^{th}$ degree formulas, or one of A and B is a $(k-1)^{th}$ degree formula and the other is a jth $(j < k-1)$ degree formula; a formula of the form ¬A is called a kth degree formula if and only if A is a kth degree formula; a formula of the form A * B, where * is conjunction or disjunction connective, is called a kth degree formula if and only if both A and B are k^{th} degree formulas, or one of A and B is a kth degree formula and the other is a jth $(j < k)$ degree formula.

Let (F(L), $\vdash_L$, Th(L)) be a formal logic system and k be a natural number. The kth degree fragment of L, denoted by $Th^k(L)$, is a set of logical theorems of L which is inductively defined as follows (in the terms of Hilbert style formal system): (1) if A is an axiom of L, then A∈ $Th^k(L)$, (2) if A is a jth ($j \leq k$) degree formula which is the result of applying an inference rule of L to some members of $Th^k(L)$, then A∈ $Th^k(L)$, and (3) Nothing else are members of $Th^k(L)$, i.e., only those obtained from repeated applications of (1) and (2) are members of $Th^k(L)$.

3 EnCal and Its Problems

EnCal consists of the following major parts. EnCal-P is a pattern-driven implementation of the inference rule of Modus Ponens (MP for short) for propositional logics. EnCal-Q is an extension of EnCal-P to deal with first order predicate logics. EnCal-E is a tool for reasoning about empirical entailments with logical theorem schema (LTS for short) generated by EnCal-P and EnCal-Q. EnCal-Q2 is an extension of EnCal-Q to deal with second order predicate logics. EnCal-E2 is an extension of EnCal-E to deal with second order theories. EnCal-T is a tool kit for the user to edit input data for EnCal, transform the reasoning results into various forms specified by the user, and provide the user with various set operations on the reasoning results.

This paper focuses on EnCal-P which is the most basic one among components of EnCal. EnCal-P works with a pattern pool and a datum pool. For the given axiom schemata of a specified logic L and the degree k, it puts all axiom schemata of L in the pattern and datum pools at first, and then repeatedly applies MP to every element of the pattern pool and every element of the datum pool such that once a new LTS whose degree is not higher than k is reasoned out, it is added in both the pattern and datum pools. This process continues until no new LTS whose degree is not higher than k can be reasoned out.

The effectiveness and efficiency in practices of a forward deduction system depends on its performance. However, until the work presented in this paper was done, EnCal is just a prototype to verify our idea of automated forward deduction. Therefore, no optimization techniques have been applied on it. From the standpoint of practical use, EnCal-P had two problems. First, although enormous LTSs may be deduced in forward deduction, no attempts are made to store them efficiently. Second, it needs a long time to deduce a fragment of a logic because there are enormous suits of two LTSs to deduce a new LTS, and a deduction itself has a heavy load.

4 Hashing

In order to improve the performance speed of EnCal, we apply hashing technique to the pattern pool and the datum pool. Since we can not decide the total number of the LTSs in k^{th} degree fragment of a logic system, we can not make a complete hash table for the fragment. Then, we fix the number of entries of a table and allow each entry to keep elements as a linear list. In this case, since we must search an element linearly after we get a table entry, it is desirable that there are no entries whose linear list is very long and that all entries are balanced. Figure 1 is an example of this hash table.

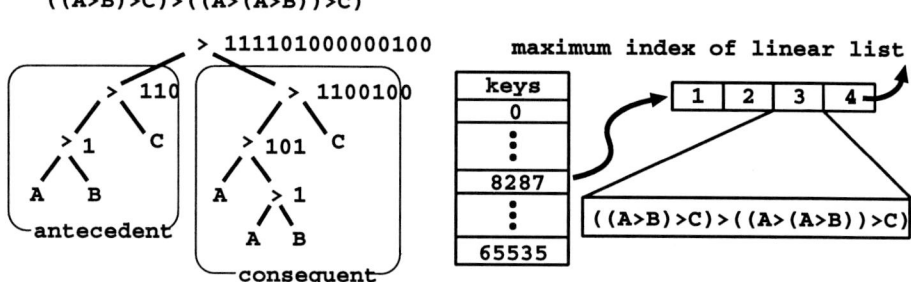

Fig. 1. Structure of hash table

When we consider a LTS as a tree structure, we can group LTSs which have same structure by their connectives such as 'Implication' and 'Negation', and divide them into classes by their pattern variables such as 'A' and 'B'.

Therefore, a hash code is created by parsing a LTS from the head as a character string, and computing according to each character. Moreover, the length of the string is given as initial value so that various initial values are used. We make a hash table by the following Algorithm 1, where we can get a hash code after we give a character string, which terminates at 'NULL', of a LTS as input.

Algorithm 1: Computing a hash code

```
1. give the length of the string as initial value, and point
   the head of the string.
2. if the pointed character is 'NULL', return the current
   value as a hash code, and exit.
3. if the pointed character is 'implication', multiply the
   value by 2.
4. if the pointed character is 'negation', add 3 to the value.
5. if the pointed character is not match above cases, add its
   character code to the value.
6. point the next character, and go back to '2'.
```

Table 1 shows that this kind of hash table is practical enough, since we have to check only under 20 LTSs linearly for the worst case in order to get the one we want from large number of LTSs in such as CMLin(3)' and Te(5), where CMLin(3) denotes the 3rd degree fragment of classical mathematical logic with material implication and negation, and Ren(3) denotes the 3rd degree fragment of relevant logic system R with entailment and negation, and Te(5) denotes the 5th degree fragment of relevant logic system T with entailment.

In linear search, currently we check whether a LTS is the one that we want by comparing as a character string. This means that each of all LTSs in hash table contains own character string and this inefficiency cannot be ignored as a whole.

Table 1. Hash table for logic systems

logic systems (k^{th} degree)	number of LTSs	number of keys used as hash code	maximum index of linear list
Ren(3)	6,520	6,008	4
CMLin(3)'	81,311	38,644	14
Te(5)	300,000+	64,472	18

(containing '+' means EnCal-P is now deducing
CMLin(3)' consists of CMLin(3) and its instances)

5 Saving memory

Since EnCal frequently uses all LTSs in its pool, it is desired that LTSs are in memory for performance, EnCal requires too many memory in an ordinary way. Therefore we must attempt to store them efficiently. Let's consider a "LTS class", Well Formed Form, in C++ for storing LTS in memory.

Example of sub LTS class in C++
```
class Wff {
  char symbol;      // value of this node
  Wff  *left;       // pointer to the left child
  Wff  *right;      // pointer to the right child
}
```

In a general compiler, such as gcc, the size of 'Wff' sums up to 12 bytes. We can describe a LTS as a complex of those objects of the class. For example, '$(A > A)$' consists of three Wff's such as 'A', '$>$' and 'A'.

We have currently deduced 300,000 LTSs in the 5th degree fragment of the relevant logic Te, and 202M bytes memory is needed for the space of 17,700,088 Wff's used in the fragment. Therefore we can not run such an enormous memory needed application on our computer. As a result, since the number of LTSs in the fragment is finally estimated to several million, we must take a step to manage it.

5.1 Reconfiguring a LTS

We improve our system so that it allocates each (sub-)LTS only once in its memory even if the LTS occurs more than once in other LTSs, in other words, they own same pointer jointly. All the sub LTSs are controlled by a hash table to satisfy the unification of sub LTSs on pointers.

Thus we define an operation "reconfiguring a LTS" where all of the sub LTSs are forced to be substituted by the one which belongs to the hash table. Algorithm 2 shows the operation, where give a character string of a LTS as input and get correspond pointer, if any, in a hash table.

Algorithm 2: reconfiguring a LTS

1. convert a sub LTS to a character string.
2. consult a hash table by using the string as key.
3. if the hash table has the same string as key, release the memory used by above sub LTSs and return the pointer in the hash table whose key is same as the string.
4. if not, entry the string as key and its pointer to the hash table, then return own pointer.

When we deduce a new LTS, we must "reconfigure" it before we put it into the LTS pools. Thus we can save many memory by this optimization technique because there are generally many duplicated sub LTSs in a fragment of a logic.

Moreover, we apply a limit of degree to the sub LTSs on a hash table. The $(k-1)^{th}$ degree limit is a nice choice because it is a rare case that same sub LTSs, which are k^{th} degree formulas, exist in k^{th} degree fragment.

5.2 Experiment result

By using pointers which are registered to hash table in programming languages like C, we can store LTSs only 10%-20% memory in comparison with storing them in the ordinary way. The important point here is the reconfiguring overhead is very small as compared with the whole time to deduce a fragment. For example, the time to deduce a fragment of CMLin(3) is 7 minutes and the time to reconfiguring all LTSs in the pool is only 1.0 second, where we experiment on Sun Ultra SPARC(200MHz).

This technique which saves us from wasting many memory without any extra load will contribute not only to an automated forward deduction system such as EnCal but also to all tools which deals with logical formulas.

Table 2. Result of saving memory

logic systems (k^{th} degree)	number of LTSs	memory used in general	limit by $(k-1)^{th}$ degree	limit by $(k-2)^{th}$ degree
CMLin(3)'	81,311	38M	3.7M(9%)	8.6M(26%)
Rc(4)	32,448+	24M	6.3M(26%)	4.7M(20%)
Te(5)	300,000+	202M	29M(14%)	22M(10%)

(containing '+' means EnCal-P is now deducing.
CMLin(3)' consists of CMLin(3) and its instances)

We have referred to the performance that $(k-1)^{th}$ degree limit to sub LTSs in a hash table is a nice choice on k^{th} degree fragment. But Table 2 shows that it is more efficient to limit $(k-2)^{th}$ degree rather than $(k-1)^{th}$ degree on Te(5) and Rc(4), where Rc denotes a sub system of R.

This is caused by that there are many sub LTSs which are referred only once in the hash table. In other words, although those sub LTSs are never used for

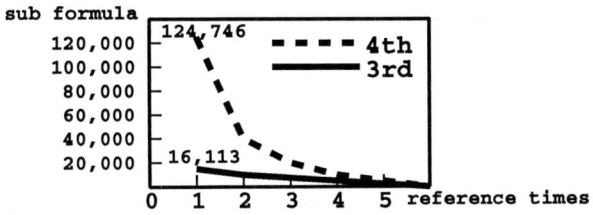

Fig. 2. Sub LTSs and their reference times

reconfiguring, they occupy memory which is unnecessary for efficient reconfiguration. Figure 3 is the graph for reference and the number of sub LTSs by changing limit of degree to sub LTSs from 3rd to 4th on Te(5).

By 4th degree limit at Figure 3, we see that there are 124,746 sub LTSs referred only one time. On the other hand, there are only 16,113 sub LTSs on 3rd degree limit.

Here we point out that the memory used for 5 th degree fragment of Te where sub LTSs are limited to 4th degree is more efficient than that with 3rd degree limit, because we can reduce one time referred sub LTSs by deducing all LTSs. In any case, compressing labels used in a hash table is a good idea.

6 Improvement of performance

The main process of EnCal-P consists of a part to deduce a LTS by applying MP to LTSs on each pool and a part to examine whether a deduced LTS is a discovery of new LTS. Those parts are in a similar situation that there is a heavy load in examining whether MP can be applied or whether it is a discovery of new LTS. Here we explain how to reduce those heavy operations.

6.1 Reducing verbose application of MP

By MP, we can get 'B' from two LTSs, '$A > B$' and 'A'. A LTS deduced by using MP is always the consequent part of a LTS, such as 'B', or an instance of that. Thus we can examine whether or not the consequent part of a LTS is an instance of other LTS in pool before we apply a MP, if yes, there is no need to apply that since a LTS deduced from the consequent part must be instance of one of LTSs.

Table 3 shows number of LTSs which is verbose LTSs for MP on logic systems, where Ten means system of Ticket entailment with entailment and negation, Ren means system of Relevant implication with entailment and negation, and Te means system of Ticket entailment with only entailment.

Table 3. Verbose LTSs for MP

logic systems (k^{th} degree)	sum of LTSs	verbose LTSs	percentage(%)
Ten(3)	263	48	18
Ren(3)	1,397	0	0
Te(4)	10,118	3,168	31

Table 4 is the result, where '%' means the percentage of LTSs which we can avoid by previous check as the real value. This improvement prevents us from deducing verbose LTSs, 35% as the whole on Te(4).

Table 4. Avoiding to deduce verbose LTSs

logic systems (k^{th} degree)	deduced count as a general rule	avoided count by previous check(%)
Ten(3)	26,433	0(0%)
Ren(3)	888,504	0(0%)
Te(4)	55,423,964	19,412,407(35%)

To get complete k^{th} degree fragment for a logic system, we must apply MP to all combinations for already known LTSs in each pool. Although it is possible that the combination sums up to n^2 when we have n LTSs, there is no need for us to apply all MP on that. Because there must be some LTSs whose antecedent parts are duplicated, and these LTSs always make a same binding. The only thing we have to do is that after checking whether or not an antecedent part of a LTS in the pattern pool matches a LTS in the datum pool, if yes, then deduce a new LTS as the binding for all LTSs in the pattern pool.

Table 5 shows the number of LTSs which share the same antecedent with another LTS. And table 6 shows how many verbose application of MP we can avoid in practice, where '%' means the percentage of them.

Table 5. LTSs with same antecedent

logic systems (k^{th} degree)	sum of LTSs	kind of antecedent	the number of LTSs with same antecedent
Ten(3)	263	61	4.31
Ren(3)	1,397	262	5.33
Te(4)	10,118	1,409	7.18
Te(5)	387,503+	45,597	7.8

(containing '+' means EnCal-P is now deducing)

This technique reduces time to deduce k^{th} degree fragment on the ground that the above result comes near 20%, the theoretical value, in spite of our anxiety. Further high degree fragment get much effect since the number of LTS which share duplicated antecedent part with another LTS becomes larger as the degree of fragment becomes higher.

Table 6. Ordering by antecedent part

logic systems (k^{th} degree)	deduced count as a general rule	avoided count
Ten(3)	69,169	22,165(32%)
Ren(3)	2,232,036	701,273(31%)
Te(4)	121,198,081	25,269,423(20%)

6.2 Examining efficiently whether a deduced LTS is new

We apply 'cache' technique to an automated forward deduction system since the cost for checking whether a deduced LTS is an instance of already known LTSs is very high, and the check is not simple such as comparing two character strings. We record history of which LTS is not new by checking with string comparison so that next appearance of that will be rejected by checking as a string before checking actually as a LTS with heavy load.

Fortunately we see a same LTS which is an instance of already known LTSs many times.

Table 7. Using cache

logic systems (k^{th} degree)	target LTSs	avoided by cache	pattern matching count
Ten(3)	2,123	1,841(86%)	2,522(11%)
Ren(3)	43,110	41,186(95%)	93,303(5%)
Te(4)	2,057,086	2,025,743(98%)	32,355,214(8%)

Table 7 shows that this caching technique reduces time to deduce k^{th} degree fragment very much on the ground that 90% deduced LTSs are rejected as duplicated one without pattern matching about duplication, and the total count of actually applying pattern matching sums up only to 10% of general one.

6.3 Performance result

Table 8 shows the execution time to deduce k^{th} degree fragment by those optimization techniques, where we experiment on Sun Ultra SPARC(200MHz). We know that as the number of LTSs increases, our optimizations work well except a logic system Te; this is because Te is a logic system with only entailment, hashing does not work efficiently on the logic system.

Table 8. execution time

logic systems (k^{th} degree)	the number of LTSs	time in ordinary way	time by our optimizations
Ten(3)	263	0:00:04.27	0:00:00.77
Ren(3)	1,472	0:03:30.32	0:00:14.15
CMLin(3)	11,888	10:30:49.39	0:07:43.58
Te(4)	11,156	3:44:23.68	0:34:08.21

7 Concluding remarks

In this paper, we have presented some techniques to improve the performance of EnCal, and some results to show the effectiveness of the techniques.

It follows from this that saving memory where the hash table manages sub LTSs and the performance speed where a duplicated LTS is cached by using hash table can be applied to many other symbolic computing applications which deal with logical formulas, and the improvement to prevent us from applying verbose MP by careful selection can be also applied to many other automated forward deduction systems.

Acknowledgements

We thank Junya Ohori and Hiroshi Mochio for their useful suggestions for this work.

References

1. A. R. Anderson and N.D. Belnup jr, "Entailment: The Logic of Relevance and Necessity," Vol.1, Princeton University Press, 1975.
2. A. R. Anderson and N.D. Belnup jr and J. M. Dunn, "Entailment: The Logic of Relevance and Necessity," Vol.2, Princeton University Press, 1992.
3. J. Cheng, "Entailment Calculus as the Logical Tool for Reasoning Rule Generation and Verification," in J. Liebowitz (Ed.), "Moving Towards Expert Systems Globally in the 21st Century," pp. 386-392, Cognizant Communication Co., 1994.
4. J. Cheng, "Entailment Calculus as the Logical Basis of Automated Theorem Finding in Scientific Discovery," in "Systematic Methods of Scientific Discovery - Papers from the 1995 Spring Symposium," AAAI Technical Report SS-95-03, pp. 105-110, 1995.
5. J. Cheng. "The Fundamental Role of Entailment in Knowledge Representation and Reasoning," in Journal of Computing and Information, Vol.2, No.1, Special Issue: Proc. 8th International Conference on Computing and Information,853-873, 1996.
6. J. Cheng, "EnCal: An Automated Forward Deduction System for General-Purpose Entailment Calculus," in N. Terashima and E. Altman (Eds.), "Advanced IT Tools, IFIP World Conference on Advanced IT Tools, IFIP96 - 14th World Computer Congress," pp. 507-514, Chapman & Hall, 1996.
7. L. Wos, "Automated Reasoning: 33 Basic Research Problems," Prentice-Hall, 1988.
8. L. Wos, "The Problem of Automated Theorem Finding, Journal of Automated Reasoning," Vol.10, No.1, 137-138, 1993.

Efficiency of Parallel Machine for Large-Scale Simulation in Computational Physics

Hiroshi Mizuseki, Keivan Esfarjani, Zhi-Qiang Li, Kaoru Ohno, Yoko Akiyama, Kyoko Ichinoseki and Yoshiyuki Kawazoe

Institute for Materials Research, Tohoku University, Sendai 980-77, Japan

Abstract. In this paper, we report on the efficiency of parallelization for atomistic-level large-scale simulations. Tight-binding and *ab-initio* molecular dynamics simulations are carried out on a supercomputer HITAC S-3800/380 and on a parallel computer HITAC SR2201. We compare the efficiencies of the two different machines based on large scale simulations to investigate advantages and disadvantages of parallel architecture.

1 Introduction

The main purpose of our research is to establish an environment to predict properties of new materials before experimental measurement. To this aim, we are developing computer codes to simulate electronic, magnetic and mechanical properties of new materials in atomistic level. For this purpose, the highest speed machines in the world should be employed. Actually, a variety of large-scale simulations are being carried out by using a vector supercomputer HITAC S-3800/380. However, since few years ago the development of conventional vector machines almost stopped, because of large electricity usage. The trend of computer technology goes toward parallel machines with RISC architecture processor elements. As a result, in the near future, the main platform of scientific simulations is expected to be parallel machines. Recently, the scale of our simulations has gone beyond the ability of conventional vector machines and the memory size of one RISC chip. It is necessary that we prepare programs that are well tuned for parallel machine and investigate the processing speeds of the two different kinds of machines: vector and parallel machine.

The supercomputer we are using for the simulations is HITAC S-3800/380 vector machine with tightly connected 3CPU's; each CPU has 8GFLOPS processing speed and 2GB common main memory. It has further 14GB of extented memory.

Our parallel machine is a HITAC SR2201[1]. This machine has a cutting-edge line of technology that is a pseudo-vector processing function and a 3-dimensional crossbar network. This pseudo-vector processing supports high-speed numerical calculations using multiple processing elements with large memory size. Conventional RISC processors run into difficulty when data for large-scale numerical calculations cannot be located in their caches. On the other hand, SR2201 employs pseudo-vector processing, a pipelining technique which fetches data (operands) directly from the main memory (bypassing the cache) and into floating point resisters, without holding up the execution of subsequent instructions. This approach contributes significantly to enhance the performance of large-scale numerical calculations. However, at the moment this pseudo-vector function is not available for integer calculation in SR2201.

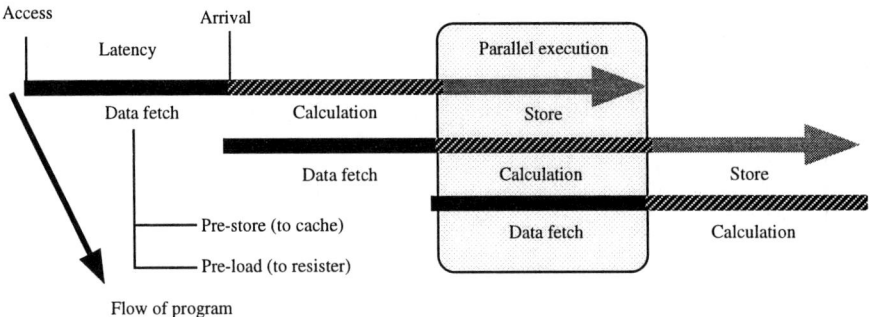

Fig. 1. Pseudo-vector processing function on SR2201 parallel machine.

One of the key techonologies for a parallel machine is the networking between processing elements. The parallel machine SR2201 that we investigated has a 3–dimensional crossbar network. With this switch, only three output lines are necessary from any processor element (PE). This simple layout achieves almost the same performance as the configuration which interconnects all processing elements directly, yet at a much lower cost.

These specifically developed features support high-speed numerical calculations with large scale memory usage distributed on multiple processing elements and achieve an excellent cost/performance ratio.

2   Calculation schemes

We chose two calculation shemes: Tight-binding (TB) method and *ab-initio* molecular dynamics. Unlike the *ab-initio* code, the TB method is easily paralleled. These calculations are performed by 1 processor of the supercomputer and up to 64 processors of the parallel computer.

2.1   Tight-binding method

The force calculation for relaxation is done by using the tight-binding method using the parametrization of Xu *et al.* [2] determined for carbon systems. In general the LDA cohesive energy curves of graphite and diamond are reproduced very accurately with this orthogonal scheme, which has also predicted well the atomic coordinates and bond lengths in C_{60}. For the electronic structure calculation, however, we use the parametrization used by Saito *et al.* [3].

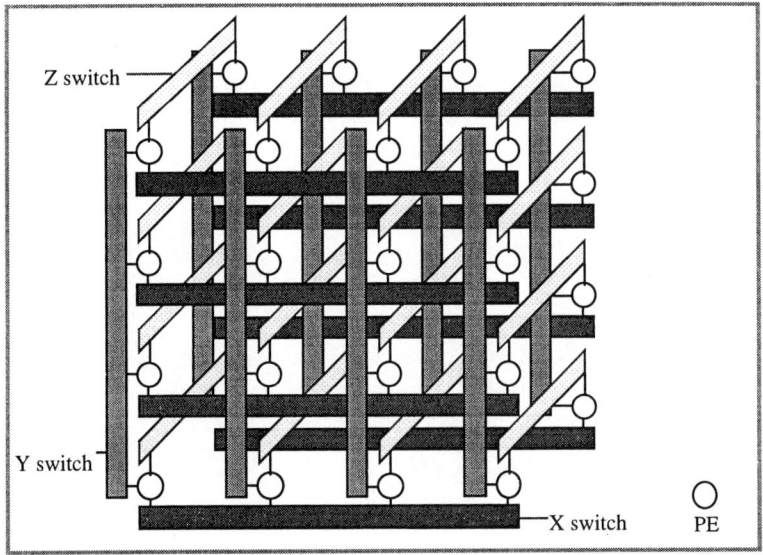

Fig. 2. 3–dimensional crossbar netwrok in SR2201 parallel machine.

This non-orthogonal parametrization is obtained from a fit to the LDA band structure of graphite and can reproduce very well the occupied energy levels in C_{60}. We therefore adopt this formulation for the electronic structure calculations for the cluster and crystalline phases.

2.2 *ab-initio* molecular dynamics

We use a first-principle code based on a plane wave expansion and a norm-conserving pseudopotential, to find an optimal adsorption height of one C_{60} molecule adsorbed on the valley of the GaAs 2×4 surface. In a tetragonal unit cell, we put 40 As atoms, 44 Ga atoms (altogether 6 layers), fixing all the positions to ideal ones at and near the surface.

3 Results

3.1 Tight-binding method

Motivated by previous predictions[4] and recent discoveries of larger fullerenes[5], we have considered 6 isomers of C_{240} for which we calculate the electronic structure in the cluster and crystal phase. These isomers have the simplest nontrivial topological form after the spheroidal C_{60} and the well-known nanotubes. They have toroidal forms, and can be obtained by bending a graphite sheet along both x and y directions, or in other words, by bending and connecting the two ends of an open nanotube. In order to relax the resulting stress, one has to introduce defects such as pentagons in the outer ring and heptagons in the inner ring of the torus. The details of such constructions can be found in ref. [6]. The considered structures have a 5 fold axis and contain 10 pentagons and 10 heptagons. The properties of the considered 6 isomers have been summarized in ref. [7]. Their geometry can be viewed in Fig. 3.

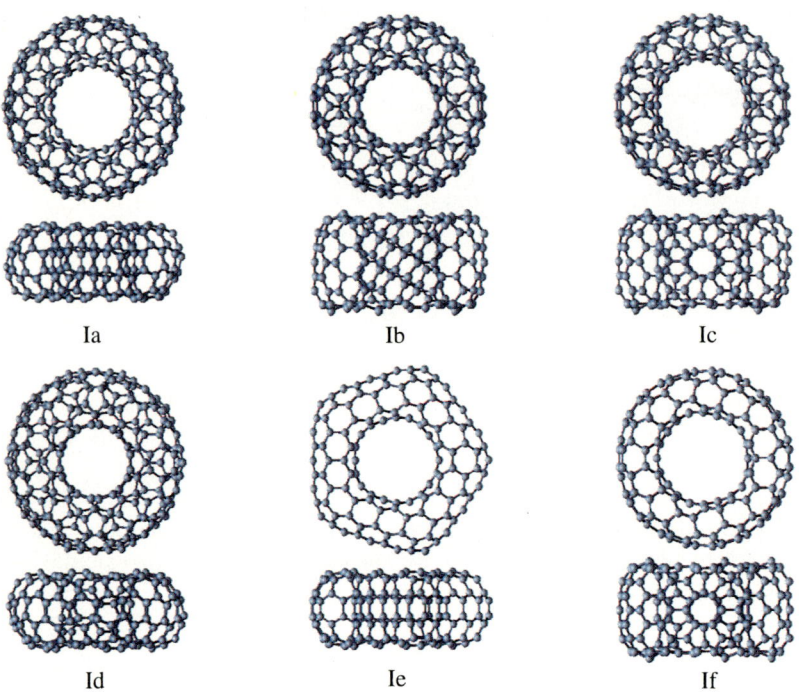

Fig. 3. Orthographic projections of the 6 isomers of C_{240} fullerene estimated by the TB method.

Figure 4 shows the parallel efficiency as a function of number of processors for various number of atoms by using the TB method. Increasing the number of atoms, we observe a linear increment of the efficiency on SR2201, which is normally not easy to be achieved without a pseudo-vector facility. For up to 500 atoms, a system with 16 processors is estimated to be most suitable.

The efficinecy is defined as speed up per number of processor. For large-scale simulation, we can obtain high perfomance than the effect of increasing number of processors, because 1 processor can not locate data in its cache. In other words, *"pseudo"*- vector function can not handle pipelining technique completely.

3.2   *ab-initio* molecular dynamics

By *ab-initio* molecular dynamics, we investigated adsorption of C_{60} over GaAs(001) 2×4 surface [8]. From this calculation, the isodensity surface of the resulting to-

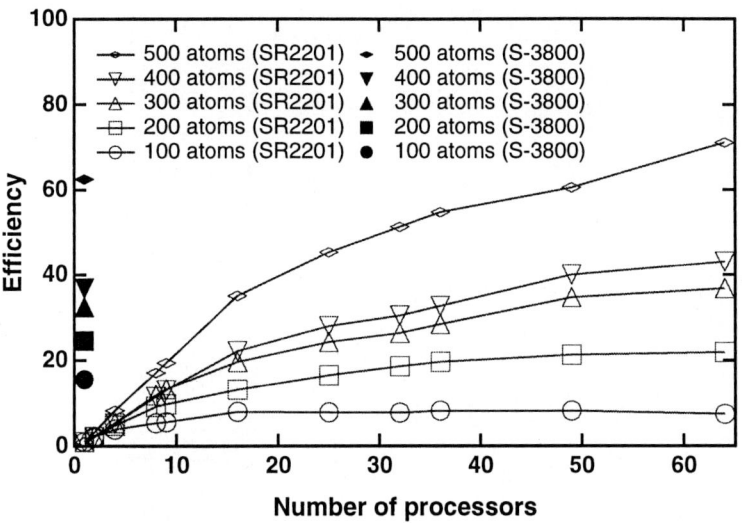

Fig. 4. Parallel efficiency of the TB method as a function of number of processors for various number of atoms.

tal charge distribution is shown in Fig. 5. This result is important to explain the dimerized C_{60} film experimentally grown on GaAs(001) surface up to 10 layers[9], which is very specific only on GaAs surface. Now, from this distribution, we can estimate how many electrons transfered from As atoms located on the surface to the first layer of C_{60}'s. The estimated amount of the charge transfer is about 1.76 electrons per one C_{60}.

Figure 6 shows the performance of parallelized program for the first principle calculation. The achieved performance of 2.4 GFLOPS with 32 processors is almost the same as the most powerful 1CPU of the vector supercomputer with 8 GFLOPS.

4 Discussion

For computer simulations in the field of materials science, supercomputers will never be said to be sufficiently fast since the number of atoms to be treated is the order of 10^{24}. In a pragmatic viewpoint, we should pursuit maximum utilization of existing supercomputers. However, since computer technology is evolving daily, we should construct new methods of simulations which are suitable for those innovative computer architectures. For example, the tight-binding method can be parallelized easily in rather straightforward way. On the other hand, diagonalization of large matrix, which is the most time consuming part of the *ab-initio* quantum method, is known to be difficult to be parallelized. To change the basic formulation of the problem is the most probable way to accomplish parallelization.

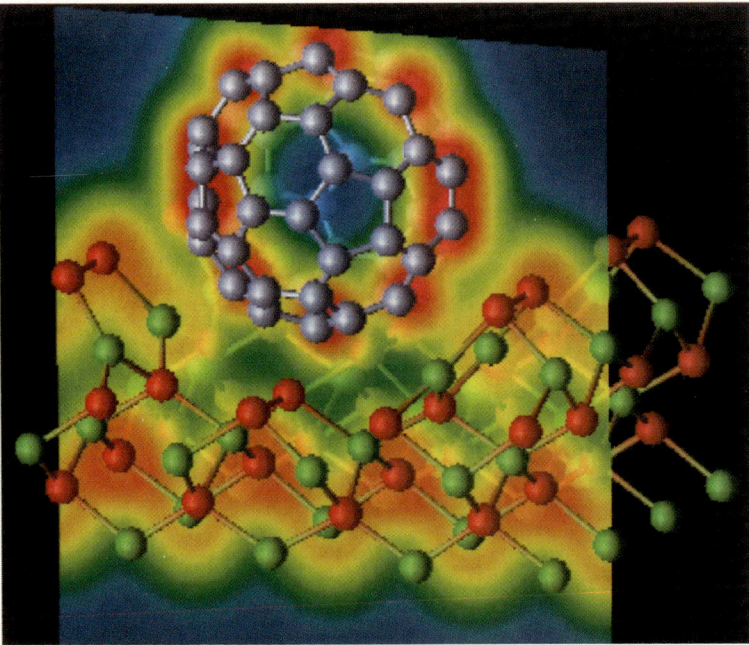

Fig. 5. Electronic charge density in C_{60} adsorbed on GaAs(001) 2×4 surface.

Up to the present, a variety of methods to solve large-scale eigenvalue problem have mostly been designed to work on a vector-architecture, and they have been widely used. Recently, the CG (Conjugate Gradient) method has become popular to overcome the defects of the Hermitin method in parallel architecture. However, CG is not inherently designed for parallel architecture, although it reduces absolute computational complexity. This is a good example to indicate that what is needed mostly is a new method originally designed for parallel computers.

If we discuss performance per processor, some scientific applications can be executed on vector-type supercomputers 10 to 1000 times faster than on a RISC processor. This fact poses severe doubt on common-sense that a RISC-based parallel supercomputer is more cost-effective than conventional vector supercomputers. In other words, distributed processing corresponds to mere human-wave tactics. That is, it can perform simple works with high efficiency, nevertheless it is completely useless for brain works. In some cases, what brings us a breakthrough is the fast speed of a single processor, not the performance of the entire system. The fundamental situation is not changed from those in the history of science: Only special talent of a genius can bring remarkable progress on genuine difficult problems. The reason why performance of single vector processor in practical applications is far greater than that of the single processor of parallel computer is their wide band-width between CPU and main memory.

Theoretical maximum performance of RISC architecture can be obtained

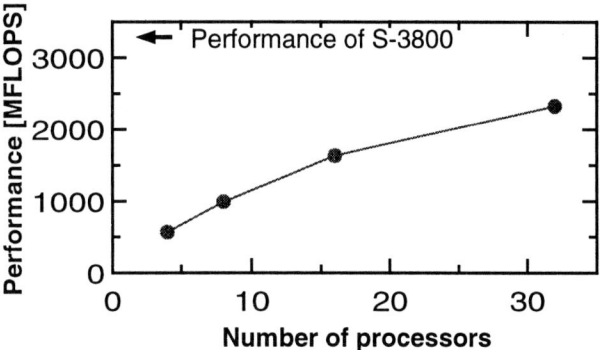

Fig. 6. Parallel efficiency of first principle calculation as a function of number of PE.

only when entire operand array is stored in their cache. On the other hand, vector processors can transfer all operands needed to and from CPUs without disadvantage of cache-miss. Noting these issues, the band-width between registers and local main memory as well as the band-width of interprocessor communications should fundamentally be enlarged. The pseudo-vector processing function employed in SR2201 provides significant reduction of cache-miss of large arrays and improves ability of data transfer.

When parallel computers are free from the bottlenecks mentioned above, and novel parallel computing systems which exhibit high efficiency appear, our ultimate goal of *"Designing materials within computers before experiments on physical system."* will be realized. Of course, several break-throughs are necessary for that purpose. A possibility is to install additional arithmetic pipelines specially designed for scientific calculations as firmwares. Since limited flexibility or room for customization will be available for the pipelines, their usage is similar to that of today's DSP (Digital Signal Processor). For example, if one of such additional pipelines is dedicated to pseudo-RNG (Random Number Generator), the workload of the CPU core is significantly reduced and simplified, hence the performance of Monte Carlo simulations will be largely improved.

5 Conclusions

Large scale computer simulations on fullerene-based materials have been performed on a vector supercomputer and on a parallel computer. Measured efficiencies are satisfactory up to 64 processors. By TB calculation, because of the present limitation of the memory size of 256MB/processor which we used, 500 atoms are the maximum to be treated, and good performance is obtained with up to 32 processors. For the first principle calculation, because the main part the heavy routines is matrix diagonalization, it is not easy to parallelize the existing code. We are developing new formulation suitable for parallel machine.

Acknowledgements

The authors would like to express their sincere thanks to the Materials Information Science Group of the Institute for Materials Research, Tohoku University, for their continuous support of the HITAC S-3800/380 and SR2201 supercomputing facilities. They are also grateful to Hitachi Ltd. for supporting parallelization of our FORTRAN source codes.

References

1. K. Nakazawa, H. Nakamura, H. Imori, and S. Kawabe, Proc. Supercomputing '92 (IEEE/ACM), pp.642–651 (Nov. 1992); H. Nakamura, H. Imori, K. Nakazawa, T. Boku, I. Nakata, Y. Yamashita, H. Wada and Y. Inagami, Proc. Int. Conf. on Supercomputing '93 (ACM), pp.298–307 (1993); H. Nakamura, T. Wakabayashi, K. Nakazawa, T. Boku, H. Wada, and Y. Inagami, Proc. of IEEE TENCON '94, Aug. 1994; K. Shimamura, S. Tanaka, T. Shimomura, T. Hotta, E. Kamada, H. Sawamoto, T. Shimizu, and K. Nakazawa, Proc. Int. Conf. on Computer Design (ICCD '95), pp.102–109, 1995; K. Takeda, Proc. Supercomputer '95, University of Mannheim, June 1995 (FOKUS: Praxis Information und Kommunikation, Band 13, K. G. Saur)
2. C. H. Xu, C. Z. Wang, C. T. Chan and K. M. Ho, J. Phys. Condens. Mat. 4 (1992) 6047–6054.
3. S. Saito, S. Okada, S. Sawada, and N. Hamada, Phys. Rev. Lett. **75** (1995) 685–688.
4. A. L. Mackay and H. Terrones, Nature (London) **352** (1991) 762; T. Lenosky, X. Gonze, M. P. Teter, and V. Elser, ibid. **355** (1992) 333–335; D. Vanderbilt and J. Tersoff, Phys. Rev. Lett. **68** (1992) 511–513; S. J. Townsend, T. J. Lenosky, D. A. Muller, C. S. Nichols, and V. Elser, ibid. **69** (1992) 921–924; R. Phillips, D. A. Drabold, T. Lenosky, G. B. Adams, and O. F. Sankey, Phys. Rev. **B46** (1992) 1941–1943; W. Y. Ching, Ming-Zhu Huang, and Young-Nian-Xu, ibid. **46** (1992) 9910–9912; Ming-Zhu Huang, W. Y. Ching, and T. Lenosky, ibid. **47** (1992) 1593–1606; L. A. Chernozatonskii, Phys. Lett. **A172** (1992) 173–176.
5. S. Iijima, Nature (London) **354** (1991) 56–58; T. W. Ebbesen and P. M. Ajayan, ibid. **358** (1992) 220–222; S. Iijima and T. Ichihashi, ibid. **363** (1993) 603–605; D. S. Bethune, C. H. Kiang, M. S. deVris, G. Gorman, R. Savoy, J. Vazquez, and R. Beyers, ibid. **363** (1993) 605–607.
6. S. Itoh and S. Ihara, Phys. Rev. **B49** (1994) 13970–13974; S. Itoh, S. Ihara and J. Kitakami, Phys. Rev. **B47** (1993) 1703–1704; S. Ihara, S. Itoh and J. Kitakami, Phys. Rev. **B47** (1993) 12908–12911; S. Ihara, S. Itoh and J. Kitakami, Phys. Rev. **B48** (1993) 5643–5647.
7. Y. Hashi, K. Esfarjani, S. Itoh, S. Ihara and Y. Kawazoe, Transactions of the Materials Research Society of Japan **20** (1996) 486–489; K. Esfarjani ,Y. Hashi, S. Itoh, S. Ihara and Y. Kawazoe, Z. Phys. D, to appear
8. K. Ohno, Z. Q. Li, H. Kamiyama, Y. Kawazoe, Q. Xue, T. Hasegawa, H. Shinohara, and T. Sakurai, Sci. Rep. RITU **A43** (1997) 61–65.
9. Q. Xue, T. Hashizume, Y. Hasegawa, H.Kamiyama, Z.-Q. Li, K. Ohno, Y. Kawazoe, H. Shinohara and T. Sakurai, to be submitted.

Parallel PDB Data Retriever "PDB Diving Booster"

Kentaro ONIZUKA, Tamotsu NOGUCHI, Minoru SAITO,
and Yutaka AKIYAMA

Parallel Application TRC Lab., Real World Computing Partnership
Tsukuba Mitsui Building 17F, 1-6-1 Takezono Tsukuba-shi Ibaraki 305, Japan
Phone: +81-298-53-1662, Fax: +81-298-53-1680, Email: onizuka@rwcp.or.jp

Abstract. A powerful PDB-data-retrieval system "PDB Diving Booster," which is a collection of functions/methods for data-retrieval and data-transmission, will enhances high-performance parallel-distributed programming in the field of structural biology. This system provides with 1) functions for reading whatever irregular data in PDB file, 2) methods to access and calculate elementary data in the structured object representing a protein structure, and 3) powerful data-transmission functions/methods for a parallel distributed environment. High performance parallel analysis of protein structures is realized only by writing a simple and short program in C++ and linking over libraries.
Keywords: *PDB, retrieval, library, parallel distributed machines, C++*

1 Introduction and System Design Policy

1.1 Background

PDB (Brookhaven Protein Data Bank) is a widely-used database of protein structure in the field of structural biology. This is actually a collection of files each of which represents a protein 3D structure determined by physical/chemical experimental methods. Each file contains a set of atomic coordinates, amino-acid sequences, and various information on the experimental methods and references. The entries are ever increasing rapidly and acceleratingly, and also increasing are the needs to read and process the contents of PDB by machines.

The PDB format (*i.e.* the data format for the representation of atom coordinates and others in PDB) is quite popular in the field. Most computer software accepts the format as a *de facto* standard. Regarding the actual data files in PDB, however, approximately one third of PDB files violate the standard format, which are, in many cases, difficult to fix without careful examination and deep knowledge of structural biology. As the result, most laboratories have to employ a special technician who sticks to PDB, maintains, corrects, and reformats the files for the researchers. The PDB format is actually out of date. That was originally designed for ancient computers on which old-style FORTRAN was the only language for the programming purposes. The format has, thus, been so far refusing such practical software tools on UNIX/PC, like "awk," "perl."

To meet the ever increasing demands from the new users (*i.e.* UNIX/PC users), several new schemes of PDB standardization and many new formats for conventional software have been proposed so far (for example MMDB [Hogue et al 96]). Needed is, however, neither a new format nor a standardization. We do not want to write another program to accept the new standardized format. Also, the large amount of software which accepts the conventional PDB format will never be deserted.

Truly needed is, thus, a powerful PDB-file reader, which is capable of retrieving whatever PDB file however irregular and catches whatever subtle information hidden behind such descriptions that violate the standard. Chances are that those irregular data often imply some hidden policy of the experimentalists who determined that structure. The hidden information of this kind, often, gives inspirations or suggestions to the researchers of dry labs.

A half decade ago, when PDB did not have more than one thousand entries, researchers had a time to read almost all entries without through the instrumentality of computers. At least, they were able to read all the entries that were rejected by their local PDB file-reader. Now that the number of PDB entries is more than five thousand, and presumably will soon exceed ten thousand. It is, thus, the crucial time for us to develop a standard and powerful PDB reader, which analyze PDB data and load the data onto silicon memory in structure. Once the PDB data is parsed and loaded onto the memory as structured data, the information in need is retrieved rapidly by calling the retrieval function or method in terms of Object-Oriented-Database (OODB) system. The retrieved data are processed by the user's program or transmitted via digital networks.

1.2 Design Policy

The targets of our research are 1) to develop a powerful PDB reader, 2) to provides with a set of functions/methods as the fundamental framework and building blocks of a researcher's program for a particular research. We started, as the beginning, the development of the method-library in C++ for reading variety of irregular PDB-files. Then we gradually added variety of convenient functions to the library for 3D-structural and statistical analysis.

Considering the number of entries and presumable amount of computation required for such as the all-possible-pairwise comparison of protein structures, most kinds of computation carried out with the library must be executed in parallel to rapidly meet the various requirement from the on-going research activities. The system is, thus, the collection of library-functions/methods for 1) PDB data reading, 2) accessing the elementary data of PDB, 3) the inter-processor-data-transmission for parallel processing, 4) the 3D structural calculation, and also 5) statistical analysis.

To avoid complex parallel programming, we restricted the programming style as those based on SPMD (single-program multiple-data), where the library users write a small program for elementary processing or analysis, and also a host program for process management and data collection. The elementary program executed at each processor element has the interface for sending/receiving protein structure data and simple integer/floating-point array for the computation result. The interface, however, does not support the remote-procedure-call except for those limited number of built-in data-request messages.

2   Tendency of irregular PDB files

This section exemplify the data and lines in irregular format frequently found in PDB files.

1. **EXPDTA**
   The experimental method adopted in the structure determination is described in EXPDTA line. The experimental method is, however, not categorized in terms of description format.
2. **RESOLUTION in REMARK section**
   The resolution of the experimentally determined structure data is described in RESOLUTION line within REMARK section. Some PDB data has a peculiar format, such as column-shift.
3. **MODEL and ENDMDL lines**
   Most of the structure data determined by NMR has plural structure models. Each model should be blaketted by "MODEL <number>" line and "ENDMDL" line. However, there are some PDB files in which MODEL line or ENDMDL line is omitted.
4. **HETATM and ATOM section**
   When an atom in a crystal structure does not belong to a protein or a nucleotide chain, it depends on the author of the PDB file whether the atom is labeled HETATM (*i.e.* hetero atom) or ATOM. The difference depends on the experimentalist's values.
5. **Chain ID in SEQRES and ATOM lines**
   The order of the chain ID occurring in SEQRES section is not always consistent with that in ATOM section.
6. **Sequence in SEQRES section and that in ATOM section**
   The sequence of residues in SEQRES lines is often different from that in ATOM section. Considering that the serial ID of residues in ATOM section is not reliable enough, there is almost no way to guess the residue-correspondence between the two sequence. SEQRES lines have the columns which represent how many residues are in the chain, though the PDB reader cannot rely on the number in the columns.
7. **Serial ID of Residues**
   The serial ID of residues is not the serial number of residues. They are IDs and IDs should not be changed even when a new residue is discovered and inserted. To indicate which residue is inserted, official PDB format requires an insertion code at the column immediately after the serial ID columns. The insertion code is, however, very frequently omitted.
8. **Order of atoms within a residue**
   The official documents of PDB [PDB Guide] define the order of atoms in each residue in ATOM section, though it is also unreliable in actual entries.

Atoms are usually omitted according to the experimental result of structure determination.

9. **Atom names**

   Atom names are often different from the standard. For example, when the side chain has a branch at "CG" atom and there are two "CD" atom, the first one should be labeled "CD1," and the other should be "CD2." In many cases, CD1 is just labeled as "CD" and only "CD2" is correctly named.

10. **ALTLOC**

    When an atom occurs at more than two different locations in a conformation, ALTLOC code is used to indicate which is the major conformation in the PDB file. It is, however, not always the case.

11. **Residue names**

    There are many residue names in PDB. PDB official documents deal with part of them. All the residue names found in PDB80 are below.

Standard residues	ALA,ARG,ASN,ASP,CYS,GLN,GLU,GLY,HIS,ILE, LEU,LYS,MET,PHE,PRO,SER,THR,TRP,TYR,VAL
Ambivalent Residues	ASX, GLX, ALI, ARO, ACD, BAS
Rare Residues	ABU, ALB, HYP, INI, PCA, SAR
Nucleic Acids	A,C,G,T,U,I,+A,+C,+G,+T,+U,+I
Rare Nucleic Acids	1MA,2MA,6IA,4AC,5MC,OMC,1MG,2MG, M2G,7MG,OMG, OMU,YG,H2U,4SU,5MU,PSU
Non-residues	ACE, FOR, HOH
Unknown residue	UNK

3 System Design Overview

In this section, the schematic design of our system is overviewed.

Our PDB retriever system "PDB Diving Booster" consists of four library modules, 1) PDB reader, 2) structural and statistical analysis module, and 3) data-transmission module for parallel distributed programming ??. In the following subsections, the basic design and the specification overview of each module is described.

3.1 PDB file-Reader

The PDB file-reader analyzes PDB files one by one. The contents of PDB files are structured into C++ objects and loaded onto the memory. To improve the system performance, the file-reader outputs each structured PDB data from memory to a binary file respectively. The binary file here is generated by packing the contents of PDB file into a binary data-string. Thus, the loading process of a binary file is exactly an unpacking process from a binary string to the structured object.

The framework of data structure is shown in Fig. 2.

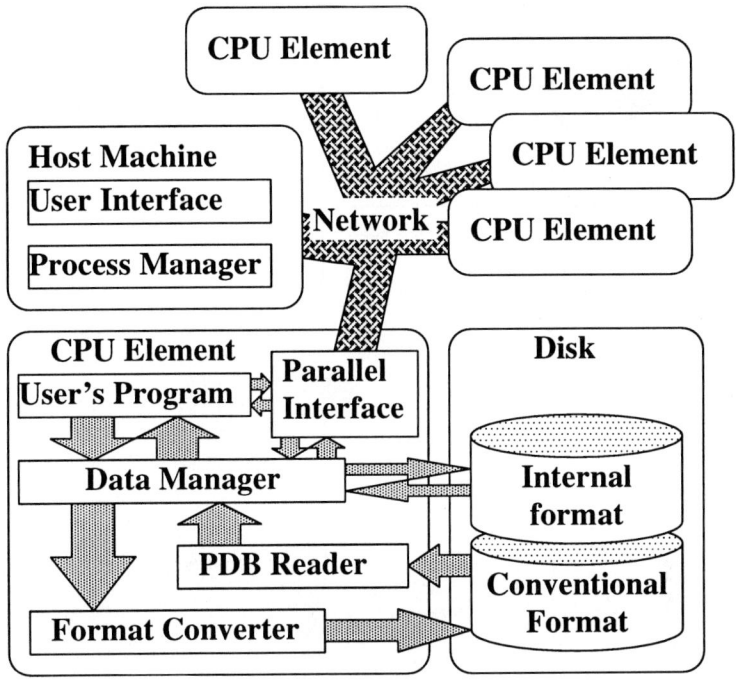

Fig. 1. System Overview

3.2 C++ Library for structural analysis

The system provides with a various functions to support users' particular data analysis. Fundamental computational functions are available.

1. **PDB Diving Booster Library package**
   This package is used to call functions of PDB Diving Booster modules.
2. **3D Vector and Matrix (3D Rotation) Library Package**
   For such purposes like dihedral/torsion angle and distance calculation, 3D vector and 3D matrix library package provides with the class and the member functions for such calculation. The class "rotation" defines the 3D rotation matrix for the given Euler's angle.
3. **Vector and Matrix Library Package**
   This library package provides with, 1) class library for vector and matrix (the dimension is assumed more than three). Addition to matrix or vector addition, production, eigenvalue and vector calculation is also available. These library functions are useful for statistical or multi-variate analysis.
4. **Conformation Library package**
   The class conformation is defined as a set of 3D vectors (of class "vector3D") which are about to represent the atom coordinates. The main usage of this class is to define its member function for least-root-mean-square-deviation

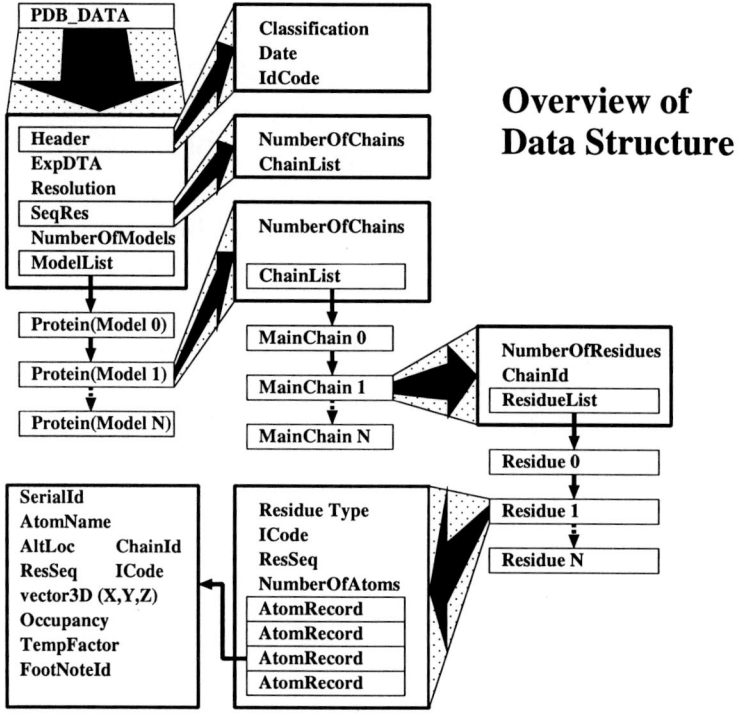

Fig. 2. Internal Data Structure

matching. Also the calculation of centroid (mass-center) and distance geometry map are provided with as member functions.

5. **Pairwise Sequence Alignment package**
   To find homologous sequence within PDB data, pairwise sequence alignment package is indispensable. This package make it possible for the users to produce their own representative PDB data set such as [Noguchi et al 97] or to analyze 3D-identification in terms of sequence homology.

3.3 C++ Library for parallel computing

To simplify the programming task and to avoid the complex debugging for parallel execution, the system provides users with a basic functions or object-methods for parallel programming.

Firstly, the library has a method for packing structured PDB data into datapackets for easy data-transmission among processing elements. Secondly, the library provides with the basic functions for structured-data request among processing elements.

As was mentioned above, the library provides with most of the methods (functions) to access structured objects/data. In the case an object exists in that memory space with which the program is running, the process has means to access whatever element in need of the data by calling that method of the object. Whereas in the case an object about to be accessed is on the memory of different processing elements, there are two alternatives to implement the means to access the data elements of the *far* object. One is to provide with elementary data request messages (or remote method call), and the other is to provide with only the means for the complex object transmission. The first one is interesting in terms of computer science. It is, however, difficult to design the remote-message/function-call protocol if OS or programming language does not support such remote-message-call like that of MPC++[Ishikawa 96]. And then, the users of the library also would have to master how to utilize the protocol before they start the programming. Our system does not support remote function/method call. The users have to, therefore, explicitly write the statements for far-object-accessing.

4 Performance

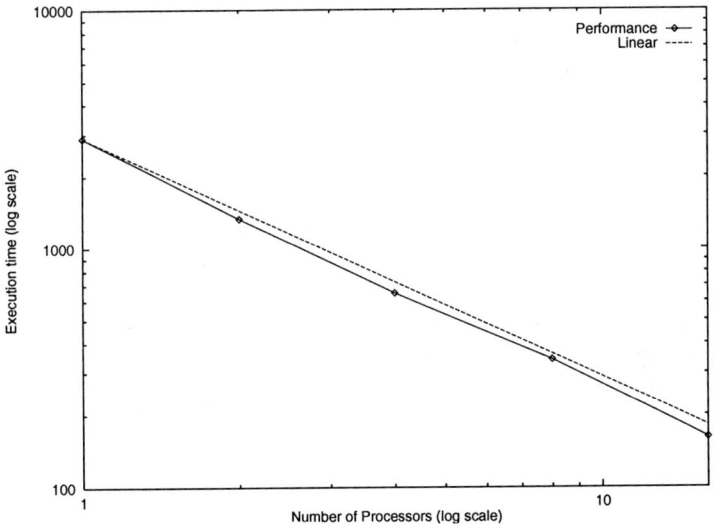

Fig. 3. Number of processors vs. Reading time

At the moment, all of the C++ library packages for PDB reading, structure and statistical computation have been implemented. Regarding the reading performance (accuracy and coverage) of PDB files, almost all lines in whole PDB data are accepted without problems, though there are actually 213 lines yet to read in PDB release 78 with 4873 entries. All of them are in ATOM section which represent the alternative location of atom without the label of ALTLOC ID.

When the system reads 5478 raw PDB files of PDB Release 80, the performance of parallel execution is slightly super linear with respect to the number of processors as is shown in Fig. 3. This is due to the intense memory allocation when executed with small number of processors.

5 Discussion

There are many commercial software programs for protein structure analysis, which are equipped with a brilliant 3D-graphic viewer and a sophisticated graphical user interface. Most of those biologists who normally stick to a particular protein get a profit from such commercial software, analyze and investigate the protein structure of their interest simply with the dancing mouse cursor and agreeable sound of clicking. These software programs are, however, not suitable for a particular analysis throughout all PDB data. Suppose there be a need to statistically analyze the distance between "CA" and "CB" atom of residues throughout all PDB entries. Do you think there is anybody who wants to point "CA" and "CB" atoms by mouse and drag menu to calculate that distance for more than hundred thousand times?

"PDB Diving Booster" is the solution. Things the researcher has to do is to write a short program in C++ language and then link the program with "PDB Diving Booster" library. The program is actually executed in parallel. The tasks are processed in data-parallel against each PDB entry. If the result is not that they expect, waiting is only a *simple* debugging.

This system will enhance the study of protein structure and might lead to discover some precious law or truth of protein structure formation.

References

[Hogue et al 96] Hogue, C. W.V.; H. Ohkawa; and S. H. Bryant 1996. "A dynamic look at structures: WWW-Entrez and the Molecular Modeling Database". *Trends in Biochemical Sciences 21*: 226-229

[Bryant 89] Bryant, S. H. 1989. "PKB: a program system and data base for analysis of protein structure." *Proteins 5*: 233-47 (1989)

[PDB Guide] PDB release 80. 1997. "Protein Data Bank Contents Guide" included in *BrookHaven Protein Data Bank* release 80., http://www.pdb.bnl.gov

[Noguchi et al 97] Noguchi, T ; K. Onizuka; Y. Akiyama; and M. Saito 1997. "PDB-REPRDB, A Database of Representative Protein Chains in PDB (Protein Data Bank)" *Proc. of ISMB'97*: 214-217, The AAAI Press.

[Ishikawa 96] Ishikawa, Y. 1997."Multi Thread Template Library – MPC++ Ver.2.0 Level 0 Document –" *RWC Technical Report TR-96012*,
http://www.rwcp.or.jp/lab/mpslab/mpc++/mpc++.html

A Parallelization Method for Neural Networks with Weak Connection Design

Alexandra I. Cristea* and Toshio Okamoto*
Graduated School of Information Systems
University of Electro-Communication

Abstract. Hereby we present the construction and usage of "Weak Connection"(WeCo) on Neural Networks(NN). We will show how these parallelization hypothesis increases the final system flexibility. The net design is based on standard procedures, but changed accordingly to *WeCo* parallelization principles. *WeCo* means parallelization with less weight on communication systems, as in: fine, medium and coarse grain parallelism, or between the parts of the implementation program. *WeCo* lays in-between parallel computers and sequential machines, building the bridge between them.

Keywords: Neural Networks, Weak Connections, Parallelism, Access Points

1 Introduction

The design problem of a NN can be defined as: *"Given a specific problem, a net has to be constructed that, with a minimal number of neurons and a minimal number of layers, has to fulfill the learning support requirements."* Theoretical approaches were made in the past (see [1], [5], [7], [8], [10, 11]), but there exists no general NN construction method.

The present study aims at a different approach, and introduces the concept of *WeCo* and other additional parallel techniques, that improve and simplify the constructional aspect of a NN problem. WeCo is a step towards parallelization, that can be taken on sequential machines.

The studied problem is of Time Series(**TS**) prediction[1]. The data are some real-world data from [2], some user-designed test series, and are values (daily prices on SE market) over any desired period of time. The data are scaled over the interval [0, 1]. The considered time-period is usually a month(rounded to 30 days), of which about 20 values for 20 days are known, and the remaining 10 days are to be predicted.

The designed net is based on **Lyapunov Gradient Descent NN**. The simplest net construction would be an 1 layer NN (1 input layer, 1 output layer) which isn't enough for the complexity of the analysed data. The next step is a 2 layer NN. Previous results (see for e.g. [2]) showed that a 2 layer net is enough. Therefore, if the Lyapunov Gradient Descent NN idea shows itself to be successful, it should at least be able to work on a net of a similar dimensional complexity. But different problems offer different complexities, so that a growing architecture can lead to serious problems, which we try to solve by parallelization. We compare our WeCo result with a classical BP (Backpropagation) net. Chapter 2 presents the parallelization methods, while chapter 3 considers the communication requirements implied by the parallelization. In chapter 4 we show the parallelization methods implemented until now, and in chapter 5 some intermediary results. Chapter 6 states some conclusions.

2 Parallelization

In this section, we introduce parallelization methods for our WeCo NN, based on a combination use of fine, medium and coarse grain parallelism. Such an integrated parallelization, namely our final goal, is known to be a current trend [9]. First, we give a target parallel system on which parallelism can be exploited the most for our WeCo NN. Then, we present the method for each grain parallelism. Furthermore, we explain a simpler but still effective parallelization scheme on a PVM environment by workstation clusters, which is our current study.

---

* Chofu, Chofugaoka 1-5-1, Tokyo 182, Japan
  e-mail: alex@ai.is.uec.ac.jp; phone: 0424-40-7956 : fax: 0424(89)6070
[1] more specifically, Stock Exchange(**SE**) forecasting

2.1 Parallel System Environment

Parallelizing our WeCo NN efficiently, the newest parallel computer architecture should be assumed. A scalable shared memory multiprocessing(SSMM) system shown in fig.1 can be a good example for the best parallelization of our WeCo NN.

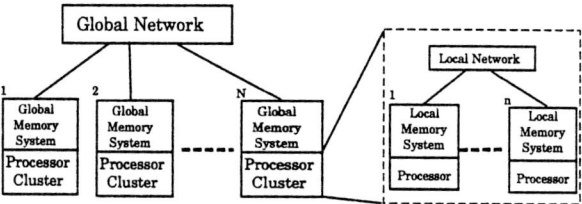

Fig. 1. Example of Scalable Shared Memory Parallel Computer

As the name says, SSMM is a shared memory parallel computer with a hierarchical configuration to be scaled. The SSMM architecture is found in [4] or [6], for example.

In our example architecture shown in fig.1, the system consists of N processor clusters and a global memory system connected by a global network. Each processor cluster is constructed by n processors and local memory system connected by a local network. The global network should be an Interconnection Network while the local network can be a simple bus network.

Each processor of the system can access any part of the global memory while local memory can be accessed by the processors which are in the same processor cluster. We also assume each processor is a SuperScalar RISC processor with m function units.

2.2 Multi-Grain Parallelization

We find the three levels of parallelism in our WeCo NN from the viewpoint of granularity. The coarse one is given by the concept of our WeCo. In the other words, it can be viewed as a layer oriented parallelism in a pipelining fashion. The middle one is neuron parallelism which takes advantage of the fact that each neuron in the same layer can be calculated independently. The fine one, *synapse level parallelism*, is based on the computational characteristics of NNs: dot product.

The coarse grain parallelism can be processed at processor clusters and global memory level on the global network of the example architecture: Each processor cluster corresponds to one layer of our WeCo NN. The middle one is suitable for processors and local memory: Each processor does the independent calculation of each neuron. The fine one can be represented as Instruction Level Parallelism (ILP) which makes an efficient use of function units on the SuperScalar Processor.

2.3 Layer Level (coarse) Parallelism

Layer level parallelism implies a dimension N for the exploited parallelism, where N is the number of layers in a NN. We will hereby show the construction of such a parallel structure.

The WeCo Net: Instead of rewriting for each new layer all connections with all other layers, and redesigning the architecture, we just define an elementary NN, consisting of 1 input and 1 output layer. In the case that a new layer is needed, we will simply add a fully independent NN, that will hardly influence the previous design. Using this idea, the result is a new net, similar to a 2-layer BP net, with the significant difference, of having actually 2 NNs[2], which are connected only by *"Weak Connection"*. The information exchange between layers is WeCo, therefore reduced as much as possible, and occurs only at some well-defined access points (**AP**). The net can be seen as consisting of 2 separate networks, that are "weakly connected" to each-other (fig. 2, a)), that is, all computed data are exchanged at the AP through data communication channels, through shared resources or message passing[3]. That implies that the algorithms for the two sets of weights of the

---

[2] The Network consists of a **Lyapunov Gradient Descent NN** and a **BP NN**, that share a hidden layer.

[3] as in workstation clusters

2-layer design can be separated from each-other, whereas in standard BP that is not possible. This fact allows a separate processing of the two weight-sets, as well as a *flexible design* of the net. The deletion or addition of a new layer doesn't influence the computations done by the preceding or following layers, so that the calculus is *transparent*.

This idea, we think, is valuable from the constructional point of view, for generating, even with the simplest trial-and-error technique, networks for different types of problems. The same algorithm can be applied on various nets during the net construction procedure.

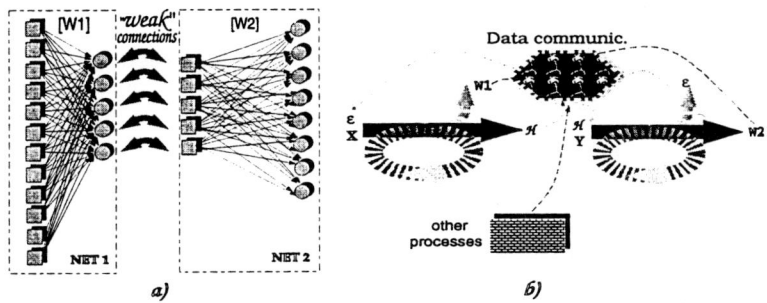

Fig. 2. a) The NN: A "weak" connection and b) Data flow in WeCo

Further optimisation for pipelined processor clusters:

Let's consider the following parallelization optimisation, for the NN: Input, NN 1, NN 2, ... , NN N, Output (a N-layered NN). The parallel computer is: N processors connected by a network (shared/distributed memory type). Let communication latency between Processor i and j be Cm(i,j). Let computational time of NN i be Cp(i). The following technique is valid if $Cp(i) \geq Cm(i,j)$. In processor i, the basic procedure order is: Cm(i-1,i), Cp(i), Cm(i,i+1). Here, using a "double buffering" and asynchronous communications, the optimisation results:

	s1	s2	s3	s4	s5	s6
Buffer1:	Cm(i-1,i) (1)	Cp(i) (1)	Cm(i,i+1) (1)			
Buffer2:		Cm(i-1,i) (2)	Cp(i) (2)	Cm(i,i+1) (2)		
Buffer1:			Cm(i-1,i) (3)	Cp(i) (3)	Cm(i,i+1) (3)	
Buffer2:				Cm(i-1,i) (4)	Cp(i) (4)	Cm(i,i+1) (4)

In the initial state (s1), there is no overlapping: just reading data 1 from processor i-1. In s2, the computation of data 1 and reading data 2 from processor i-1 are overlapped. In s3, writing data 1 to processor i+1, the computation of data 2, and reading data 3 from processor i-1 are overlapped. Basically, there is a N (pipelined) parallelism (N-layer). Plus here there is a 3 (read/compute/write overlap) parallelism. Therefore, the total parallelism is 3*N. The requirements for such a parallelization to work is that the number of communication ports must be more than 1. Namely, this method is not valid for, for example, a bus network.

2.4 Neuron Level (medium) Parallelism

The exploited parallelism here is of dimension n, where $n_i = k_i * n$, n_i - the number of neurons in layer i and k_i - is a coefficient. By defining a neuron class, a neuron becomes an independent item. In order to really function without any influences from other neurons in the same layer, some hardware support is required. In a layer, each neuron calculates from its own weights of connections and the same input which is the output from the lower layer. In short, weights are local but the inputs can be shared by all neurons in the same layer. Comparing the amount of time for communication of the input with the overall calculation time of each neuron, the sharing overhead should be reasonably low. Therefore, the neuron level parallelism requires a shared memory mechanism of low memory latency.

In the example SSMM architecture shown in fig.1, the shared memory mechanism can be achieved at individual processor clusters: Each processor in the same processor cluster take care of k_i neurons, and the input to the layer i is shared by all the processors in the same processor cluster through shared local memory.

2.5 Instruction Level (synapse, fine) Parallelism

Here, the dimension of exploited parallelism is m, considering m as the number of existing scalar units.

Usage of parallelism on Super Scalar RISC processors: We assume the processor element of the parallel computer is a Super Scalar RISC processor. Here, we can exploit instruction level parallelism (ILP) as well as layer oriented parallelism and communication-computation overlap. Since the calculation in each layer is dominated by dot products, it is very easy to apply software pipelining technique to make the full use of several scalar units of the Super Scalar RISC processors.

This combination of coarse (by layer), medium (communication/computation overlap) and fine (ILP) grain parallelism achieves the maximum performance. Therefore, the overall parallelism is: $3 * N * n * m$.

3 Communication

1) The data flow at layer level: Fig.2,b) shows the data flow structure of the newly constructed system . The first network (left side in fig.2, b)[4]) has as inputs the general input of the system, X, and the backpropagated error of the second NN, ε. It outputs the hidden layer values, H and, as secondary outputs, its own weights, $W1$. The second NN (right side in fig.2, b), corresponding to NET2 in fig.2, a)) inputs the hidden layer values, shared with NET1, as well as the general output of the system, Y. This NN outputs primarily its own weights, $W2$, and secondarily the computational error, ε.

With *WeCo* we defined some NN access points, through which *and only through which* the information interchange is possible. As a result, other, even unknown[5] systems, can have access at the system's partial outputs, read and/or change them, without interfering with the actual internal processing in either of the two NNs. For instance, all the test programs have parallel access at the data computed by the system. What we created is a flexible shared system, based on building blocks.

2) The data flow at neuron level: For input and output between these independent neurons we selected the *shared memory*[6] (fig.3), that can be reached by different processors (here, neurons). In order to have a correct traffic of data, we also have to set some rules: for instance, a neuron can read from several locations (from all its inputs), but can write only in one (its own output). In

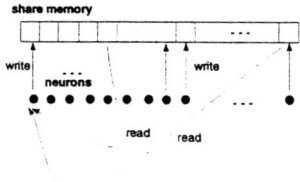

Fig. 3. The share memory role in the neuron to neuron communication: a neuron writes in only one location, but can read from several others.

---

[4] corresponding to NET1 in fig.2, a)
[5] Unknown to the two computing NNs, NET1 and NET2, in fig.2.
[6] as stated in paragraph 2.4

this way, a neuron is completely independent, unaware of the processing of other neurons, working parallel to other neurons. Also, the input data doesn't have to be copied locally for each neuron, thus saving memory space.

3) The data flow at synaptic level: The multi-grain parallelization has, as could be seen, different communication characteristics for each level. Synapse level requires intra-processor, namely inter-function units of the superscalar processor communication. This is achieved by using register files in the processor. Neuron Level requires inter-processor communication in the same processor cluster. In this case, the local shared memory must be used. The layer level requires inter-processor cluster communication. It is possible by using global shared memory in the example architecture. It can also implemented by sending messages in loosely coupled parallel computers such as workstation clusters. In general, the amount of communication and required computation must be carefully considered to gain the maximum performance.

4 Preliminary and Practical Implementation

As a prototype parallel system of our WeCo NN, a limited version of the multi-grain parallelization is under implementation. The target parallel system is a workstation cluster with PVM. In this prototype, we employ just two level parallelism: Layer Level and Synapse Level. Since the layer level parallelism is the main concept of our WeCo NN, its implementation could be done easily. As for the synapse level, it is basically the parallelism among dot products which calculates the behaviour of the synapse. Dot products can be optimized on spurscalar processors by using software pipelining[3].

1) the Global Program: is based on NN traditional features. An "Epoch" that computes through

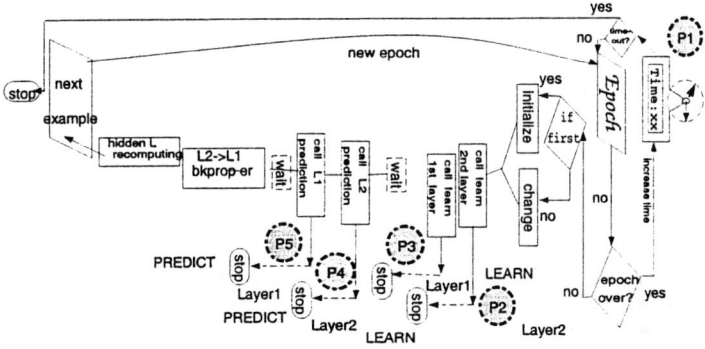

Fig. 4. The levels of program interaction during learning

the examples, on a pseudo-random basis, can be seen in picture 4. After the finishing of each "Epoch", the time is increased. The *WeCo* concept is hereby applied through the fact that the learning and prediction processes are, as can be seen in fig.4, independent/parallel processes/programs (P1-P5).

A standard cyclus is as follows: at the start, there are some initializations, replaced by some variable changes in the further steps. Then, learning processes (stage1, 2 from fig. 5) are called: process P2 for NET1 (fig. 2, a)), and in parallel P3 for NET2 (same figures). After waiting for the two processes to finish, stage3 (graphical display) is called for both NET1 and NET2 in parallel. After a new waiting time, some error backpropagation is done (through the file system) from NET1 to NET2, the hidden layer values are recomputed and the system goes on with the next example, till the end of example supply. Only then the global time is increased. After a timeout interval, the global process, P1, finishes, too.

2) WeCo for the Training and Testing: The system can perform: **training** (learning) through weight computation, **forecasting** with a 2-layer NN, **user interfacing**. These processes are implemented as independent, parallel programs. The communication between these processes is established by shared memory or message passing. Although current implementation of our system uses files for the communication, we intend to replace the common resources with message passing of PVM(parallel virtual machine) on workstation clusters, for example. Of course, our final target is the massively parallel processing environment described at the beginning of section 2. The **AP** are in each of the cases similar with the AP described in 2.3, in what the internal features of the net are concerned. The 3 independent processes build a 3 level data hierarchy, as can be seen in fig.5. The interaction of the user with these levels is made possible through a similar button input from interface windows, that correspond exactly to the computation stages in fig. 5.

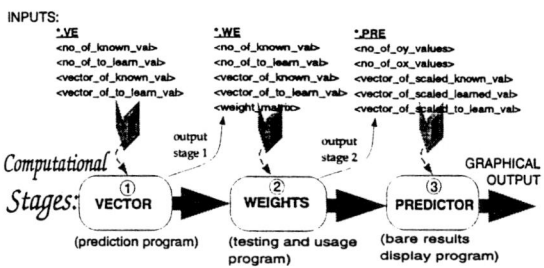

Fig. 5. The 3 input levels

These processes work with some standard format files, of the form shown in picture 5: *.VE, *.WE, *.PRE. The contents of these files as well as order of contents are also depicted in the same figure: For the prediction (VECTOR) stage, past SE values (vector_of_known_val) and the desired future values (vector_of_to_learn_val), together with their dimension, are inputted through the *.VE file. For the testing (WEIGHTS) stage, the inputs are similar, but the weight matrix computed in the previous step is added to the *.WE file. The PREDICTOR stage deals only with scaled values, from the same two vectors of past SE values and desired SE values, to which the predictor vector (vector_of_scaled_learned_val) is added. Also, the *.PRE file contains some dimension indicators for the display: the density of elements displayed on the Ox, respectively Oy axis.

For the training, the system/user enters at stage1 of the structure. For testing, the same enter at stage2. Stage3 is only the connection to the user interface. A full training implies the stage1 to stage3 suite. Testing, however, requires only stage2 and stage3. Then again stage3 can be skipped in order to gain speed during learning. Therefore, the graphical output can be reduced to a minimum, comprising only the start and stop states. As any entry level in the computational stages is therefore possible, and the particular stages are *WeCo* to each-other, a high flexibility of the system is achieved.

The **learning process** is supervised gradient-descent based Lyapunov. The computed forecasted data are compared to the true future values. Triggered visualization of the learning process during training is possible. A full help support is provided at each step of the program usage, both for current state explanations as well as for advice about possible steps to take by the system user.

5 Results

We will introduce the elements of result display and analysis.

a) System Learning Display: We present here a learning example of an equal share of 10 input data and 10 outputs (fig.6 a)). That means that out of 10 input daily SE values, 10 output daily values were to be learned. The weight matrix dimension had therefore a 10x10=100 member dimension. The continuous, zig-zag line on the left side of the graphical chart are the past (or input) data, while on the right side, the desired outputs[7] and the prediction can be seen. The prediction is normally displayed by a dotted line, while the desired outputs are displayed by a continuous line[8] but here, with "0" error, the two outputs overlap, and therefore, a single line is visible. The program displays[9] here, the learning of the correspondence of 10 past prices with 10

---
[7] as in *Supervised Learning*
[8] just like the inputs
[9] with the stage 3 PREDICTOR program, as in fig. 5

future prices. The outputs are scaled from 0 to 1 (Oy axis), and the time (days) is represented on 0x.

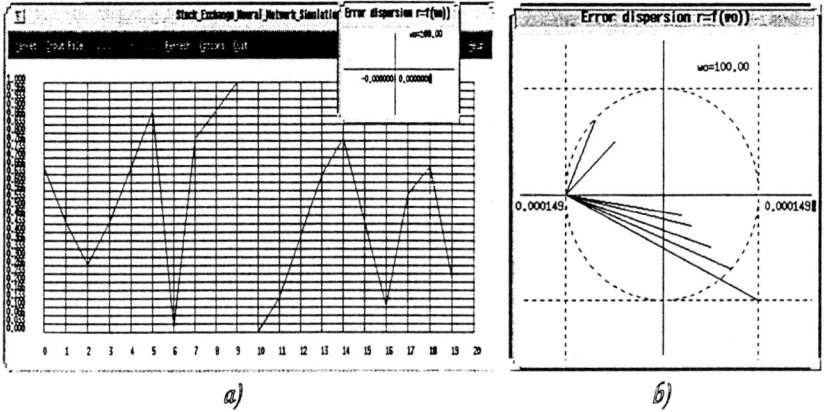

Fig. 6. a) A "perfectly" trained net, b) Error display window

b) Error display: The error display is shown, for a better understanding, in a separate, two-dimensional error-display window (fig.6 a)) upper right corner, fig.6 b))[10] The display of error was designed so that a qualitative reading should be done easily. For this reason, on both vertical and horizontal axis of the error-window, individual errors at different times are represented. Errors can occur if the desired values and the predicted output values don't overlap. The dimension of the error vector is the same as the dimension of the output vector (here, 10). Of interest are: first, the display of the maximal error of the whole interval - represented in the window by the exterior square - and then, the other errors, with lower values - represented by the lines starting in the left corner of the picture and ending at the intersection with the second diagonal, on which all the errors (including the maximal) are represented. In this way, out of a single glance, the error structure can be understood.

c) WeCo Display: There are *three patterns* to be observed (fig.7): one is the approximation of the system's general output, compared to the general input and general desired pattern. This pattern is a complex and derivated pattern, out of the two secondary ones: the general system output, the general system desired output versus the hidden layer values (output of NET2 in fig. 2,a)), and the internal output of the first weights computation, versus hidden layer values and general system input (output of NET1 in fig. 2, a)). Due to the independence of the user-graphical-interface to the learning algorithm and procedures, as a result of the *WeCo* design, these three patterns can be viewed by a simple triple call of the graphical display program, each time with the corresponding [input data, output data] pair.

Observation patterns:

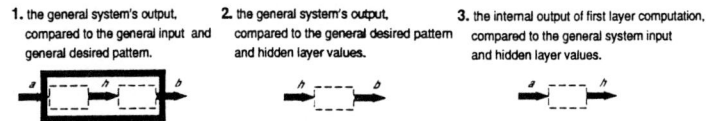

Fig. 7. The required observation values(two parallel subnets forming a global NN)

These were the elements of the result display. We used them to view first the behaviour of a regular BP net, then the *WeCo* net training. There were some little differences regarding the convergence. One of the similarities that appear is that both methods tend to increase the weights

---

[10] The **prediction error** is displayed in percentage to the maximal value(price) that occured in the given time-period in a "Error dispersion" window.

during learning (if no momentum term is added). As a convergence difference we mention that the growth-rate is larger in the BP case, while our method seem to show a smoother pattern. The convergence time of our system though, is still rather long (in the range of hours), even for a small number of examples. The differences were though significant regarding the programming effort, first during the initial net construction step, but even more stressed when changes on the NN had to be performed. In that case, the *WeCo* net showed itself to be clearly superior.

6 Conclusions

A SE forecasting tool was constructed, based on the TS behaviour of SE and on Lyapunov theory. The NN design was done based on a *"WeCo"* concept and other parallel techniques. We applied *WeCo* in various ways, to produce fine, medium and coarse parallelism. *WeCo* allowed constructional flexibility. Even with only a few of the presented parallel features implemented, our WeCo net proved to be superior to the classical BP net.
For further work we intend to explore optimizations by implementing the other *WeCo* parallelization techniques and to find an optimal share key between them.

References

1. D.Dasgupta and D.R.McGregor,1992. *Designing Application-Specific Neural Networks using the Structured Genetic Algorithm.* Proceedings of COGANN-92 (Internat. Workshop on Comb. of Genetic Alg. and NN, June6,1992, Ed.Whitley and Schaffer, Publ.: IEEE Computer Society Press).
2. D.Komo, C.I.Chang and H.Ko,1994. *Neural Network Technology for Stock Market Index Prediction.* International Symposium on Speech, Image Processing and Neural Networks, 13-16 April 1994, Hong-Kong, IEEE, pages 543-546.
3. M.Lam. Software Pipelining: An Effective Scheduling Technique for VLIW Machines. In *Proc. ACM SIGPLAN*, pp.318-328, 1988.
4. Daniel E. Lenoski *The Design and Analysis of Dash: A Scalable Directory-based Multiprocessor.* PhD Thesis, Stanford University, CSL, 1992.
5. A.U.Levin, T.K.Leen,J.E.Moody,1994. *Fast Pruning Using Principal Components.* Advances in Neural Information Processing 6, J.Cowan, G.Tesauro, J.Alspector, eds., Morgan Kaufmann, San Mateo, CA, 1994.
6. S. Mori and et al. *A Distributed Shared Memory Multiprocessor: ASURA - Memory and Cache Architectures-.* In *Proc. Supercomputing 93*, pp.740-749, 1993
7. N. Murata, S. Yoshizawa and S. Amari,1995. *Network Information Criterion - Determining the Number of Hidden Units for an Artificial NN Model.* Dept. of Mathematical Eng. and Info. Physics, Fac. of Engineering, Univ. of Tokyo,ftp-source.
8. M.E.Nelson, W.Furmanski, J.M.Bower,1989. *Simulating Neurons and Networks on Parallel Computers.* Methods in Neuronal Modeling, From Synapses to Networks, ed. C.Koch and I.Segev, MIT, 1989,chapt. 12.
9. C. Brownhill, A. Nicolau, S. Novack and C. Polychronopoulos *The Promis Compiler.* In *Proc. PACT97,* will appear, 1997
10. G.Thimm, E.Fiesler,1996. *A Neural Network Construction Method based on Boolean Logic.* accepted for IEEE International Conference on Tools with Artificial Intelligence proceedings, Toulose, France, 1996.
11. G.Thimm, E.Fiesler,1996. *Neural Network Pruning and Pruning Parameters*, 1st OWSC. http://www/bioele.nuee.nagoya-u.ac.jp/wsc1, 1st Online Workshop on Soft Computing, Nagoya, Japan, 1996.

Exploiting Parallel Computers to Reduce Neural Network Training Time of Real Applications

Jim Torresen[1/2] Shin-ichiro Mori[1], Hiroshi Nakashima[1], Shinji Tomita[1], Olav Landsverk[2]

[1]Department of Information Science
Faculty of Engineering, Kyoto University
Yoshida-hon-machi, Sakyo-ku,
Kyoto 606-01, Japan

[2]Department of Computer and Information Science
Norwegian University of Science and Technology,
N-7034 Trondheim, Norway
Email: jim@nera.no

Abstract. Neural networks have been proposed to solve difficult problems like speech and character recognition. However, there has so far not come up any revolutionary system. This paper gives the results of a survey of the ongoing research on neural network applications. Moreover, we point out the demands for the mapping of neural applications onto parallel computer hardware. We propose a flexible mapping of back propagation trained neural networks onto a highly parallel computer.

The experiments undertaken show the need for application specific mapping of the given neural network and training set.

1 Introduction

Lots of neural network research has taken place for the last ten years. Some ask if neural networks really can do more than well-known statistical methods [1]. This paper surveys some of the present neural network applications trained by the backpropagation (BP) algorithm [2].

Speeding up neural network computation by using parallel processing machines has been deeply studied. Several neurocomputers and neurochips have been designed. However, there seems to be a gap between the research on computer architecture and neural applications. The performance of parallel implementations are often measured by non-real large network sizes [3]. Whereas, real applications often use small networks. One reason for this may be that application researchers often implement their programs on a single processor machine like a PC or a UNIX workstation. Therefore, they may have to scale down the network size. From the applications studied in this paper it is not possible to draw a single conclusion that recognition rate can be improved be making the networks larger. However, some applications will require parallel hardware to perform real-time recognition.

In this paper we propose a parallel BP implementation that runs real neural network applications efficiently. It is evaluated on Fujitsu AP1000, a message passing MIMD computer with two dimensional torus topology network [4]. AP1000 has distributed memory and each cell consists of a Sparc CPU, a FPU, a message passing chip, 128 KB cache and 16 MB main memory. The system used in this research consists of 512 cells.

2 Neural network applications

The main contributors to the neural network research, Widrow and Rumelhart, recently published a paper: "Neural networks: Applications in Industry, Business and Science" [5]. They list a wide variety of commercial applications of neural networks. The increase in products in the last years is partly explained by the availability of an increasingly wide array of dedicated hardware. However, many of the applications (e.g. control application) ought to be cheap to be of commercial interest. In [6] a survey of neural network applications, trained by backpropagation, is given. Mainly large applications — i.e. where parallel processing is of interest, was included.

Application	Network size (I x H x O)	No of tr. pat.
Cancer cell classification	3600 x 20 x 1	467
Coin recognition	259 x 5 x 6	
Handwritten digit recognition (P)	32 x 15 x 10	2000
Image compression	64 x 8,6,4 (x 64)	
Locating plant cells in image	100 x 15 x 10	52;67;69
NETtalk text-to-phonemes	203 x 60-120 x 26	5438
Object detector	961 x 50 x 1	50
Object inspection	4096 x 64 x 2	
Optical Character Recognition	3000 x 20 x 94	94
Optical Character Recognition	2500 x 100 x 94	1,128
Papanicoloau Smear Cell Classif.	80 x 6 x 1	702
Paper currency recognition	128 x 64 x 12	
Road classifications in satellite images	56 x 20 x1	552
Speech recognition	234 x 1000 x 61	1,300,000
Speech recognition	351 x 4000 x 61	6,000,000
Welding Defect Indentific.	15 x 10 x 9	1024

Table 1. Some neural network applications based on feed-forward neural networks. (P) indicate that a pre-processing network is used in front of the feed-forward meural network inputs. I, H, O indicate the number of input, hidden and output neurons, respectively.

Table 1 lists a summary of the survey. As far as possible, variations in the network structure are indicated. See [6] for references to each work.

3 Parallel processing for neural networks

From the survey in the previous section we see at least two common characteristics of many applications. First, some of the networks are small. The survey was concentrating on large networks, thus many other applications using small networks were omitted. Second, the output layer often consists of a small number of neurons. The latter indicates that it may be in vain to allocate many processors for computing the output layer.

It is obvious that some of the neural networks, like those recognition speech and vision applications, need parallel processing. This is needed both for training and when performing recognition. However, quite many applications can do without expensive parallel hardware. This may also be necessary for a product to be marketable, e.g. control applications in consumer products. Still, parallel processing can be used to reduce the training time of an application.

Below, first the parallel nature of the BP algorithm is described. Then, it is proposed how to train both small networks and networks with few output units on a parallel computer.

3.1 Parallel implementations of BP

The BP algorithm reveals several different kinds of parallelism, as described in [8, 9]:

- *Training session parallelism*, Starts training sessions with different initial starting values on different processors.

- *Training set parallelism*, Splits the training set across the processors. Each processor has a local copy of the complete weight matrix and accumulate weight change values for the given training patterns.

- *Pipelining*, Pipelines the training patterns in the layers, i.e. compute hidden and output layer on *different* processors. While the output layer processor calculates output and error values, the hidden layer processor concurrently processes the next training pattern.

- *Node parallelism*, The neurons within a layer run in parallel (named neuron parallelism). Further, the computation within each neuron may also run in parallel (Nordström [9] names this weight or synapse parallelism).

The weight update interval, μ, denotes the number of training patterns that is presented between weight updates.

Several BP implementations have been conducted on general purpose computers. Node parallelism is used to implement BP on Intel iPSC/860 hypercube in [10] and on MasPar MP-1 SIMD computer in [11]. A combination of training set and node parallelism for MasPar MP-1216 is given in [12]. Common to all these implementations is the lack of possible adaption to the large range of neural network applications. E.g., in the case of a node parallel implementation, it scales badly for *small* feed-forward networks.

Kumar et al. [3] propose a hybrid scheme using node and training set parallelism for hypercube architectures. The method makes it possible to vary the number of processors assigned to each degree of parallelism and utilizes the hypercube communication network. However, the effect of μ is not considered. As shown in [13], less frequent weight updates makes the convergence rate slower (i.e. the decrease in error per iteration is smaller) during training. Since the main interest is to minimize the *total* training time, it is impossible to omit the selection of a proper μ, when training set parallelism is used.

3.2 Small networks

To obtain high speedup on a highly parallel computer, we showed [14, 15] that *multiple* inherent degrees of parallelism in the BP algorithm should be combined. This is highly needed if the neural networks are of small size. The partitioning of the network is given in Figure 1, where pipelining and neuron parallelism are combined. Moreover, we included training set parallelism. In this paper, we propose a flexible and scalable mapping based on these results. While the previous work was on fixed combinations of the degrees of parallelism, the new scheme allows arbitrary combinations of training set parallelism and neuron parallelism. The proposed the mapping is shown in Figure 2. It combines neuron parallelism and training set parallelism. We allow the number of processors assigned for each degree of parallelism to be changeable. Within each dotted rectangle, neuron parallelism is used. Pipelining may also be included [16].

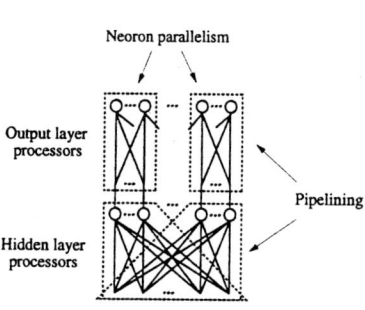

Fig. 1. Partitioning of the network, when pipelining and neuron parallelism are combined. The dotted lines show which *neurons* are mapped to each processor.

Below each processor mapping, the number of training set partitions (N_{TSP}) and the number of processors for neuron parallelism (N_{NP}) are given. The relation between them can be expressed by:

$$N_{TSP} = \frac{C_x C_y}{N_{NP}} \quad (1)$$

where C_x and C_y is the number of processors in the horizontal and vertical directions, respectively. For a given weight update interval, a larger number

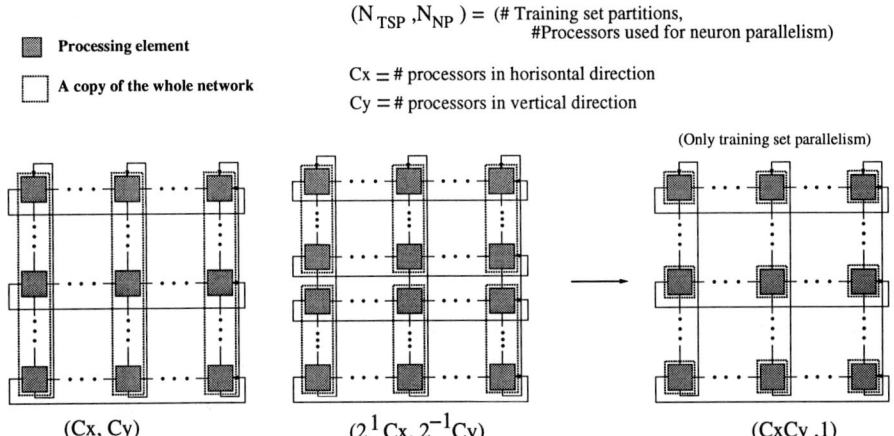

Fig. 2. Combination of training set and neuron parallelism.

of training set partitions imply fewer training pattern computations on each processor between weight updates.

Before running a neural application, we have to select the mapping giving shortest possible total training time. Thus, for each mapping, we have to predict the total training time for a set of weight update intervals. The total training time is given by:

$$T_{total} = T_{1it}(\mu)N(\mu) \qquad (2)$$

where T_{1it} is the time for one iteration and $N(\mu)$ is the number of iterations needed. T_{1it} can be found either by estimation or by running each of the possible mappings for one iteration. $N(\mu)$ can be estimated based on the error after a few iterations [17]. We should select the mapping giving smallest possible T_{total} and use the corresponding μ-value.

3.3 Few output units

To train a network efficiently, when few output units is present, the number of processors assigned to the output layer computation should be reduced. Thus, we have to pipeline the computation so that hidden and output layer outputs are computed on different processors.

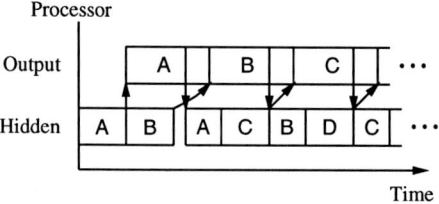

Fig. 3. Pipelining of the training patterns.

Figure 3 shows a pipelining example. First, the hidden layer processor computes output values of training pattern A. The output processor reads the values and computes output and error values of A. The hidden processor concurrently process the next training pattern (B). Then, it reads the hidden error for A and both processors accumulate the weight change values for A.

Figure 4 shows a possible pipelined mapping. The dotted rectangle represent one dotted rectangle in Figure 2. The processors marked **H** compute the hidden layer part, while the output layer part is computed on the processors marked **O**. For most cases, the number of processors assigned to the hidden layer, C_h, will be larger than the number of processors assigned to the output layer, C_o. The best configuration can be found by measuring the time for one pass through the training set for different processor assignments. Network contention occurs in this implementation, i.e. several processors may send messages concurrently to the same link. However, the communication time on AP1000 is reasonable, when network contention exists. It is approximately doubled if 5 messages are on the same channel, indicating that much time is spent within each processing element for communication overhead and less on the physical transfer of data [18].

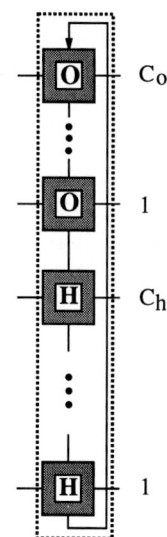

Fig. 4. Pipelined mapping of BP.

4 Results and discussion

To be able to evaluate the benefit of a parallel back propagation algorithm that can vary the amount of each degree of parallelism, we implemented the mapping shown in Figure 2 on AP1000.

NETtalk

First, we use the NETtalk [19] neural network application in our experiments. The number of hidden units is 120. The training set consists of the 1000 most common English words (total 5438 characters). We have undertaken experiments with $\mu = 63, 259, 494, 906, 1360, 2719, 5438$.

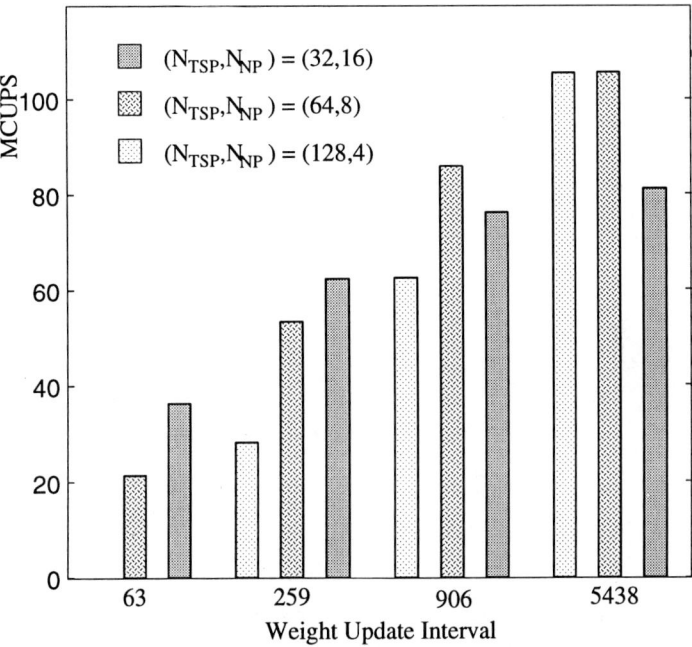

Fig. 5. MCUPS performance for different combinations of neuron and training set parallelism on a 16 x 32 processor configuration running the NETtalk application. The number of hidden neurons is 120 and the performance for $\mu = 63, 259, 906, 5438$ are given.

Figure 5 shows the performance of NETtalk on a large system, for some μ-values. Performance is measured in MCUPS (Million Connections Updated Per Second). The configuration for each bar is indicated by (N_{TSP}, N_{NP}), as explained in the theory section. The performance improves with less frequent weight updates. The most efficient system is $(N_{TSP}, N_{NP}) = (32, 16)$ for frequent updates and $(N_{TSP}, N_{NP}) = (64, 8)$ for rare updates. This means that using 64 training set groups is better than using 32, even though network contention occurs.

Minimizing the total training time

In this section we show the relation between weight update interval and the number of iterations needed to obtain convergence for the NETtalk training set.

Moreover, we find the best weight update interval when running on AP1000.

The convergence was measured in percentage of the characters that have obtained an error of less than 0.1. Due to representation limitations of the network, the number of patterns that is not trained stabilized slightly below 5%. Figure 6 shows the number of iterations needed $(N(\mu))$ to reduce the error to 5% for the investigated μ-values.

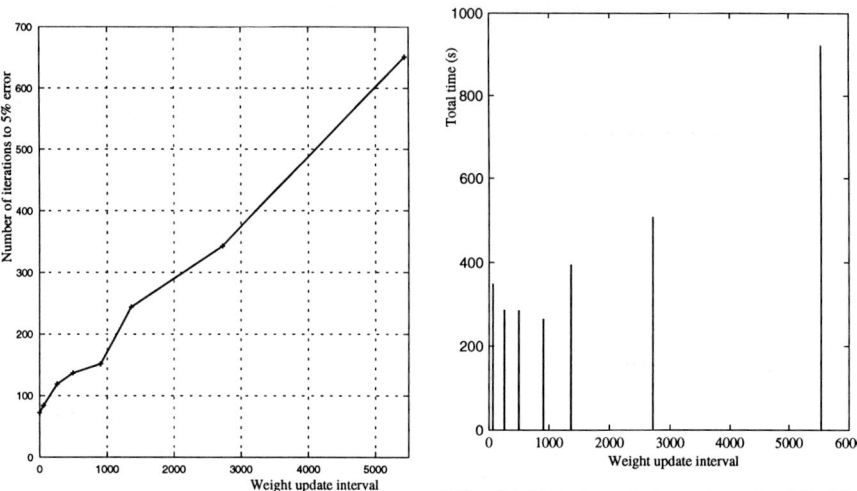

Fig. 6. Number of iterations needed to obtain convergence.

Fig. 7. Total training time for NETtalk running on 512 cells, using 120 hidden units.

The time for one iteration, T_{1it}, can be computed from Figure 5 (not all μ-values are shown). Hence, we can find the best weight update interval by employing (2) for $\mu = 63, 259, 494, 906, 1360, 2719, 5438$. We get the total times as shown in Figure 7. We omit to accumulate weight changes for a pattern when the error is sufficiently small to avoid over-training, thus T_{1it} is decreasing with the convergence. Since we use a constant T_{1it}, the T_{total} will be larger than the real execution time. The best weight update interval is for every 906 pattern when $(N_{PSP}, N_{NP}) = (64, 8)$. Convergence is obtained after 265 s.

Running the same program (with parallel parts removed) on a Sparc10 workstation, using *learning by pattern*, converged after 73 iterations and 169 minutes. AP1000 speedup based on total execution time $(T_{1it}N)$ is then 38 times. This is a conservative measure, since the decreasing T_{1it} for AP1000 is not accounted for. Thus, training time can be reduced from hours to minutes by using parallel processing.

5 Conclusion

The results of a survey of neural network applications trained by back propagation have been given. A mapping of the BP neural network training algorithm onto a highly parallel computer has been proposed. The mapping is flexible and designed to efficiently train both small networks and networks with different numbers of units in each layer. The results indicate the need for making a parallel mapping flexible to run the different real applications efficiently. Moreover, we have shown how the weight update interval influences on the total training time.

References

1. Tom Kavli. Nevrale nett: Hvor vil vi de neste årene ? In *Proc. of the Norwegian Neural Network Seminar*. SINTEF Instrumentation, November 1994.
2. D.E. Rumelhart, G.E. Hinton, and R.J. Williams. Learning internal representation by error propagation. In *Parallel Distributed Processing*, volume 1, pages 318–362. The MIT Press, 1986.
3. Vipin Kumar et al. A scalable parallel formulation of the back propagation algorithm for hypercubes and related architectures. *IEEE Trans. on Parallel and Distributed Systems*, 5(10):1073–1090, October 1994.
4. Hiroaki Ishihata et al. Third generation message passing computer AP1000. In *Proc. of the International Symposium on Supercomputing*, pages 46–55, Nov. 1991.
5. Bernard Widrow et al. Neural networks: Applications in industry, business and science. *Communication of ACM*, 37(3):93–105, March 1994.
6. Jim Tørresen. *Parallelization of Backpropagation Training for Feed-Forward Neural Networks*. PhD thesis, Norwegian University of Science and Technology, 1996. ISBN 82-7119-906-4.
7. Terrence J. Sejnowski. NETtalk corpus, obtainable from ftp.idiap.ch in pub/benchmarks/neural/nettalk.tar.z.
8. Alexander Singer. Implementation of artificial neural networks on the Connection Machine. *Parallel Computing*, 14:305–315, Summer 1990.
9. Tomas Nordstrom and Bertil Svensson. Using and designing massively parallel computers for artificial neural networks. *Journal of Parallel and Distributed Computing*, 14(3):260–285, March 1992.
10. Darin Jackson and Dan Hammerstrom. Distributing back propagation networks over the Intel iPSC/860 hypercube. In *Proc. of Int. Joint Conference on Neural Networks*, volume I, pages 569–574, 1991.
11. G. Chinn et al. Systolic array implementations of neural nets on the MasPar MP-1 massively parallel processor. In *Proc. of Int. Joint Conference on Neural Networks*, volume II, pages 169–173, 1990.
12. Andreas Zell et al. Problems of massive parallelism in neural network simulation. In *Proc. of IEEE Int. Conference on Neural Networks*, pages 1890–1895, 1993.
13. Helene Paugam-Moisy. Parallel neural computing based on neural network duplicating. In Ioannis Pitas, editor, *Parallel algorithms for digital image processing, computer vision and neural networks*, chapter 10, pages 305–340. John Wiley & Sons, 1993.

14. Jim Torresen, Shin-ichiro Mori, Hiroshi Nakashima, Shinji Tomita, and Olav Landsverk. Parallel back propagation training algorithm for MIMD computer with 2D-torus network. In *Proceedings of International Conference On Neural Information Processing (ICONIP'94), Seoul, Korea*, volume 1, pages 140–145, October 1994.
15. Jim Torresen, Shin-ichiro Mori, Hiroshi Nakashima, Shinji Tomita, and Olav Landsverk. Exploiting multiple degrees of BP parallelism on the highly parallel computer AP1000. In *Fourth International Conference on Artificial Neural Networks (ANN'95)*, pages 483–488, Cambridge, UK, June 1995. IEE.
16. Jim Torresen, Hiroshi Nakashima, Shinji Tomita, and Olav Landsverk. General mapping of feed-forward neural networks onto an MIMD computer. In *Proc. of IEEE Int. Conference on Neural Networks (ICNN'95)*, Perth, Western Australia, 27 November – 1 December 1995. IEEE.
17. Jim Torresen, Shinji Tomita, and Olav Landsverk. The relation of weight update frequency to convergence of BP. In *Proc. of World Congress on Neural Networks (WCNN'95)*, volume 1, pages 679–682, Washington, D.C., July 1995. INNS Press.
18. Hiroaki Ishihata. Performance evaluation of the AP1000. In *Proc. of CAP workshop*, pages N–1–8, 1991.
19. Terrence J. Sejnowski and Charles R. Rosenberg. Parallel networks that learn to pronounce English text. *Complex Systems*, 1:145–168, 1987.
20. Kwang Bo Cho et al. Image compression using multi-layer perceptron with block classification and SOFM coding. In *Proc. of World Congress on Neural Networks*, volume 3, pages 26–31, 1994.

Author Index

Akiyama, Y., 281, 389
Akiyama, Yk., 381
Amamiya, M., 91, 243
Amano, H., 171
Ando, M., 281
Arabnia, H.R., 72
Araki, K., 231
Baba, T., 305
Bik, A.J.C., 1
Brownhill, C.J., 183
Cheng, J., 371
Cristea, A.I., 397
Cybenko, G., 71
Eickemeyer, R.J., 75
Esfarjani, K., 381
Evripidou, S., 107
Fukuda, A., 351
Funahashi, A., 171
Gallivan, K.A., 1
Gannon, D., 42
Gao, G.R., 30
Goshima, M., 195
Hanawa, T., 171
Hiraki, K., 255
Hiyama, M., 305
Horiguchi, S., 327
Huang, R., 159
Ichikawa, A., 143
Ichinoseki, K., 381
Imlig, N., 131
Inenaga, K., 243
Ishizaki, K., 217
Joe, K., 231, 351
Johnson, R.E., 75
Kacsuk, P., 91
Kaneda, Y., 143
Kawata, H., 267
Kawazoe, Y., 381
Komatsu, H., 217, 361
Kubota, A., 195
Kudoh, T., 171
Kunieda, Y., 205
Kunkel, S.R., 75
Kusakabe, S., 243

Landsverk, O., 405
Li, Z.-Q., 381
Lim, B.H., 75
Loher, D., 293
Ma, J., 159
Marsolf, B.A., 1
Matsumoto, H., 281
Matsumoto, T., 255
Misoo, Y., 281
Mizuseki, H., 381
Mori, S., 195, 405
Morimoto, T., 243
Mutenda, L., 305
Nakajo, H., 143
Nakashima, H., 195, 405
Nakatani, T., 217
Nanri, T., 335, 343
Nicolau, A., 183
Nishi, K., 371
Nobukuni, Y., 255
Noguchi, T., 281, 389
Novack, S., 183
Ogasawara, T., 361
Ohno, K., 381
Okamoto, T., 397
Omori, Y., 351
Omura, Y., 281
Onizuka, K., 281, 389
Polychronopoulos, C.D., 183
Saito, M., 281, 389
Sarkar, V., 30
Sasakura, M., 231
Sato, H., 335, 343
Sato, T., 119
Shimasaki, M., 335, 343
Shimizu, N., 317
Squillante, M.S., 75
de Sturle, E., 293
Tamura, F., 267
Tanaka, T., 195
Tatsumi, S., 195
Tomita, S., 195, 405
Tooyama, T., 327
Torresen, J., 405

Tsuboi, E., 159
Tsuda, T., 205
Tsutui, A., 131
Uchihira, N., 267
Uehara, T., 205
Ushijima, K., 371
Veidenbaum, A.V., 51
Watanabe, T., 73
Watanabe, Tk., 317
Wijshoff, H.A.G., 1
Wu, C.E., 75
Yoshinaga, T., 305

Springer and the environment

At Springer we firmly believe that an international science publisher has a special obligation to the environment, and our corporate policies consistently reflect this conviction.

We also expect our business partners – paper mills, printers, packaging manufacturers, etc. – to commit themselves to using materials and production processes that do not harm the environment. The paper in this book is made from low- or no-chlorine pulp and is acid free, in conformance with international standards for paper permanency.

Lecture Notes in Computer Science

For information about Vols. 1–1265

please contact your bookseller or Springer-Verlag

Vol. 1266: D.B. Leake, E. Plaza (Eds.), Case-Based Reasoning Research and Development. Proceedings, 1997. XIII, 648 pages. 1997. (Subseries LNAI).

Vol. 1267: E. Biham (Ed.), Fast Software Encryption. Proceedings, 1997. VIII, 289 pages. 1997.

Vol. 1268: W. Kluge (Ed.), Implementation of Functional Languages. Proceedings, 1996. XI, 284 pages. 1997.

Vol. 1269: J. Rolim (Ed.), Randomization and Approximation Techniques in Computer Science. Proceedings, 1997. VIII, 227 pages. 1997.

Vol. 1270: V. Varadharajan, J. Pieprzyk, Y. Mu (Eds.), Information Security and Privacy. Proceedings, 1997. XI, 337 pages. 1997.

Vol. 1271: C. Small, P. Douglas, R. Johnson, P. King, N. Martin (Eds.), Advances in Databases. Proceedings, 1997. XI, 233 pages. 1997.

Vol. 1272: F. Dehne, A. Rau-Chaplin, J.-R. Sack, R. Tamassia (Eds.), Algorithms and Data Structures. Proceedings, 1997. X, 476 pages. 1997.

Vol. 1273: P. Antsaklis, W. Kohn, A. Nerode, S. Sastry (Eds.), Hybrid Systems IV. X, 405 pages. 1997.

Vol. 1274: T. Masuda, Y. Masunaga, M. Tsukamoto (Eds.), Worldwide Computing and Its Applications. Proceedings, 1997. XVI, 443 pages. 1997.

Vol. 1275: E.L. Gunter, A. Felty (Eds.), Theorem Proving in Higher Order Logics. Proceedings, 1997. VIII, 339 pages. 1997.

Vol. 1276: T. Jiang, D.T. Lee (Eds.), Computing and Combinatorics. Proceedings, 1997. XI, 522 pages. 1997.

Vol. 1277: V. Malyshkin (Ed.), Parallel Computing Technologies. Proceedings, 1997. XII, 455 pages. 1997.

Vol. 1278: R. Hofestädt, T. Lengauer, M. Löffler, D. Schomburg (Eds.), Bioinformatics. Proceedings, 1996. XI, 222 pages. 1997.

Vol. 1279: B. S. Chlebus, L. Czaja (Eds.), Fundamentals of Computation Theory. Proceedings, 1997. XI, 475 pages. 1997.

Vol. 1280: X. Liu, P. Cohen, M. Berthold (Eds.), Advances in Intelligent Data Analysis. Proceedings, 1997. XII, 621 pages. 1997.

Vol. 1281: M. Abadi, T. Ito (Eds.), Theoretical Aspects of Computer Software. Proceedings, 1997. XI, 639 pages. 1997.

Vol. 1282: D. Garlan, D. Le Métayer (Eds.), Coordination Languages and Models. Proceedings, 1997. X, 435 pages. 1997.

Vol. 1283: M. Müller-Olm, Modular Compiler Verification. XV, 250 pages. 1997.

Vol. 1284: R. Burkard, G. Woeginger (Eds.), Algorithms — ESA '97. Proceedings, 1997. XI, 515 pages. 1997.

Vol. 1285: X. Jao, J.-H. Kim, T. Furuhashi (Eds.), Simulated Evolution and Learning. Proceedings, 1996. VIII, 231 pages. 1997. (Subseries LNAI).

Vol. 1286: C. Zhang, D. Lukose (Eds.), Multi-Agent Systems. Proceedings, 1996. VII, 195 pages. 1997. (Subseries LNAI).

Vol. 1287: T. Kropf (Ed.), Formal Hardware Verification. XII, 367 pages. 1997.

Vol. 1288: M. Schneider, Spatial Data Types for Database Systems. XIII, 275 pages. 1997.

Vol. 1289: G. Gottlob, A. Leitsch, D. Mundici (Eds.), Computational Logic and Proof Theory. Proceedings, 1997. VIII, 348 pages. 1997.

Vol. 1290: E. Moggi, G. Rosolini (Eds.), Category Theory and Computer Science. Proceedings, 1997. VII, 313 pages. 1997.

Vol. 1291: D.G. Feitelson, L. Rudolph (Eds.), Job Scheduling Strategies for Parallel Processing. Proceedings, 1997. VII, 299 pages. 1997.

Vol. 1292: H. Glaser, P. Hartel, H. Kuchen (Eds.), Programming Languages: Implementations, Logigs, and Programs. Proceedings, 1997. XI, 425 pages. 1997.

Vol. 1293: C. Nicholas, D. Wood (Eds.), Principles of Document Processing. Proceedings, 1996. XI, 195 pages. 1997.

Vol. 1294: B.S. Kaliski Jr. (Ed.), Advances in Cryptology — CRYPTO '97. Proceedings, 1997. XII, 539 pages. 1997.

Vol. 1295: I. Prívara, P. Ružička (Eds.), Mathematical Foundations of Computer Science 1997. Proceedings, 1997. X, 519 pages. 1997.

Vol. 1296: G. Sommer, K. Daniilidis, J. Pauli (Eds.), Computer Analysis of Images and Patterns. Proceedings, 1997. XIII, 737 pages. 1997.

Vol. 1297: N. Lavrač, S. Džeroski (Eds.), Inductive Logic Programming. Proceedings, 1997. VIII, 309 pages. 1997. (Subseries LNAI).

Vol. 1298: M. Hanus, J. Heering, K. Meinke (Eds.), Algebraic and Logic Programming. Proceedings, 1997. X, 286 pages. 1997.

Vol. 1299: M.T. Pazienza (Ed.), Information Extraction. Proceedings, 1997. IX, 213 pages. 1997. (Subseries LNAI).

Vol. 1300: C. Lengauer, M. Griebl, S. Gorlatch (Eds.), Euro-Par'97 Parallel Processing. Proceedings, 1997. XXX, 1379 pages. 1997.

Vol. 1301: M. Jazayeri, H. Schauer (Eds.), Software Engineering - ESEC/FSE'97. Proceedings, 1997. XIII, 532 pages. 1997.

Vol. 1302: P. Van Hentenryck (Ed.), Static Analysis. Proceedings, 1997. X, 413 pages. 1997.

Vol. 1303: G. Brewka, C. Habel, B. Nebel (Eds.), KI-97: Advances in Artificial Intelligence. Proceedings, 1997. XI, 413 pages. 1997. (Subseries LNAI).

Vol. 1304: W. Luk, P.Y.K. Cheung, M. Glesner (Eds.), Field-Programmable Logic and Applications. Proceedings, 1997. XI, 503 pages. 1997.

Vol. 1305: D. Corne, J.L. Shapiro (Eds.), Evolutionary Computing. Proceedings, 1997. X, 307 pages. 1997.

Vol. 1306: C. Leung (Ed.), Visual Information Systems. X, 274 pages. 1997.

Vol. 1307: R. Kompe, Prosody in Speech Understanding Systems. XIX, 357 pages. 1997. (Subseries LNAI).

Vol. 1308: A. Hameurlain, A M. Tjoa (Eds.), Database and Expert Systems Applications. Proceedings, 1997. XVII, 688 pages. 1997.

Vol. 1309: R. Steinmetz, L.C. Wolf (Eds.), Interactive Distributed Multimedia Systems and Telecommunication Services. Proceedings, 1997. XIII, 466 pages. 1997.

Vol. 1310: A. Del Bimbo (Ed.), Image Analysis and Processing. Proceedings, 1997. Volume I. XXII, 722 pages. 1997.

Vol. 1311: A. Del Bimbo (Ed.), Image Analysis and Processing. Proceedings, 1997. Volume II. XXII, 794 pages. 1997.

Vol. 1312: A. Geppert, M. Berndtsson (Eds.), Rules in Database Systems. Proceedings, 1997. VII, 214 pages. 1997.

Vol. 1313: J. Fitzgerald, C.B. Jones, P. Lucas (Eds.), FME '97: Industrial Applications and Strengthened Foundations of Formal Methods. Proceedings, 1997. XIII, 685 pages. 1997.

Vol. 1314: S. Muggleton (Ed.), Inductive Logic Programming. Proceedings, 1996. VIII, 397 pages. 1997. (Subseries LNAI).

Vol. 1315: G. Sommer, J.J. Koenderink (Eds.), Algebraic Frames for the Perception-Action Cycle. Proceedings, 1997. VIII, 395 pages. 1997.

Vol. 1316: M. Li, A. Maruoka (Eds.), Algorithmic Learning Theory. Proceedings, 1997. XI, 461 pages. 1997. (Subseries LNAI).

Vol. 1317: M. Leman (Ed.), Music, Gestalt, and Computing. IX, 524 pages. 1997. (Subseries LNAI).

Vol. 1318: R. Hirschfeld (Ed.), Financial Cryptography. Proceedings, 1997. XI, 409 pages. 1997.

Vol. 1319: E. Plaza, R. Benjamins (Eds.), Knowledge Acquisition, Modeling and Management. Proceedings, 1997. XI, 389 pages. 1997. (Subseries LNAI).

Vol. 1320: M. Mavronicolas, P. Tsigas (Eds.), Distributed Algorithms. Proceedings, 1997. X, 333 pages. 1997.

Vol. 1321: M. Lenzerini (Ed.), AI*IA 97: Advances in Artificial Intelligence. Proceedings, 1997. XII, 459 pages. 1997. (Subseries LNAI).

Vol. 1322: H. Hußmann, Formal Foundations for Software Engineering Methods. X, 286 pages. 1997.

Vol. 1323: E. Costa, A. Cardoso (Eds.), Progress in Artificial Intelligence. Proceedings, 1997. XIV, 393 pages. 1997. (Subseries LNAI).

Vol. 1324: C. Peters, C. Thanos (Eds.), Research and Advanced Technology for Digital Libraries. Proceedings, 1997. X, 423 pages. 1997.

Vol. 1325: Z.W. Raś, A. Skowron (Eds.), Foundations of Intelligent Systems. Proceedings, 1997. XI, 630 pages. 1997. (Subseries LNAI).

Vol. 1326: C. Nicholas, J. Mayfield (Eds.), Intelligent Hypertext. XIV, 182 pages. 1997.

Vol. 1327: W. Gerstner, A. Germond, M. Hasler, J.-D. Nicoud (Eds.), Artificial Neural Networks – ICANN '97. Proceedings, 1997. XIX, 1274 pages. 1997.

Vol. 1328: C. Retoré (Ed.), Logical Aspects of Computational Linguistics. Proceedings, 1996. VIII, 435 pages. 1997. (Subseries LNAI).

Vol. 1329: S.C. Hirtle, A.U. Frank (Eds.), Spatial Information Theory. Proceedings, 1997. XIV, 511 pages. 1997.

Vol. 1330: G. Smolka (Ed.), Principles and Practice of Constraint Programming – CP 97. Proceedings, 1997. XII, 563 pages. 1997.

Vol. 1331: D. W. Embley, R. C. Goldstein (Eds.), Conceptual Modeling – ER '97. Proceedings, 1997. XV, 479 pages. 1997.

Vol. 1332: M. Bubak, J. Dongarra, J. Waśniewski (Eds.), Recent Advances in Parallel Virtual Machine and Message Passing Interface. Proceedings, 1997. XV, 518 pages. 1997.

Vol. 1333: F. Pichler. R.M. Díaz (Eds.), Computer Aided Systems Theory – EUROCAST'97. Proceedings, 1997. XI, 626 pages. 1997.

Vol. 1334: Y. Han, T. Okamoto, S. Qing (Eds.), Information and Communications Security. Proceedings, 1997. X, 484 pages. 1997.

Vol. 1335: R.H. Möhring (Ed.), Graph-Theoretic Concepts in Computer Science. Proceedings, 1997. X, 376 pages. 1997.

Vol. 1336: C. Polychronopoulos, K. Joe, K. Araki, M. Amamiya (Eds.), High Performance Computing. Proceedings, 1997. XII, 416 pages. 1997.

Vol. 1337: C. Freksa, M. Jantzen, R. Valk (Eds.), Foundations of Computer Science. XII, 515 pages. 1997.

Vol. 1338: F. Plášil, K.G. Jeffery (Eds.), SOFSEM'97: Theory and Practice of Informatics. Proceedings, 1997. XIV, 571 pages. 1997.

Vol. 1339: N.A. Murshed, F. Bortolozzi (Eds.), Advances in Document Image Analysis. Proceedings, 1997. IX, 345 pages. 1997.

Vol. 1340: M. van Kreveld, J. Nievergelt, T. Roos, P. Widmayer (Eds.), Algorithmic Foundations of Geographic Information Systems. XIV, 287 pages. 1997.

Vol. 1341: F. Bry, R. Ramakrishnan, K. Ramamohanarao (Eds.), Deductive and Object-Oriented Databases. Proceedings, 1997. XIV, 430 pages. 1997.

Vol. 1342: A. Sattar (Ed.), Advanced Topics in Artificial Intelligence. Proceedings, 1997. XVIII, 516 pages. 1997.

Vol. 1344: C. Ausnit-Hood, K.A. Johnson, R.G. Pettit, IV, S.B. Opdahl (Eds.), Ada 95 – Quality and Style. XV, 292 pages. 1997.